Artificial Intelligence for Precision Agriculture

Precision agriculture is a next-generation farming management concept that optimizes resource use, productivity, quality, profitability, and sustainability by observing and responding to crop variability. Precision agriculture employs digital technologies such as the Internet of Things (IoT), artificial intelligence (AI), 5G communication, cybersecurity, edge computing, cloud-native principles, and blockchain to ensure crops and soil receive exactly what they need for optimal health and productivity.

Artificial Intelligence for Precision Agriculture explores the latest developments in precision agriculture, detailing how AI contributes to its goals. The book discusses how precision agriculture solutions use IoT devices, data storage, AI analytics, connectivity, and cloud infrastructures to analyze factors such as soil type, terrain, weather, plant growth, and yield data. It also examines edge technologies—sensors, microchips, beacons, RFID tags, robots, drones, and actuators—that collect field data and transmit it to cloud-based AI platforms for analysis. The book shows how AI-driven insights guide actions in the field, such as crop rotation, optimal planting and harvesting times, and soil management, and help farmers apply the right amounts of water, fertilizers and pesticides, reducing waste and environmental impact. Applications covered in the book include:

- Drone-based high-resolution field mapping
- Tracking crops
- Crop yield assessments
- Data collection for irrigation, fertilization, and crop management
- Advanced weather monitoring
- Equipment management

With chapters on AI model development, plant disease detection and remediation, sustainable farming techniques, data integration, AI-enabled data analytics, and knowledge visualization, this book is a comprehensive guide to technologies and applications in precision agriculture.

Artificial Intelligence for Precision Agriculture

Edited by
Pethuru Raj, N. Gayathri, and
G. Jaspher Willsie Kathrine

CRC Press
Taylor & Francis Group
Boca Raton London New York

CRC Press is an imprint of the
Taylor & Francis Group, an **informa** business
AN AUERBACH BOOK

Cover: Web Large Image (Public)

First edition published 2025
2385 NW Executive Center Drive, Suite 320, Boca Raton FL 33431
and by CRC Press

4 Park Square, Milton Park, Abingdon, Oxon, OX14 4RN

CRC Press is an imprint of Taylor & Francis Group, LLC

ISBN: 9781032462349 (hbk)
ISBN: 9781032825106 (pbk)
ISBN: 9781003504900 (ebk)

DOI: 10.1201/9781003504900
Typeset in Garamond
by KnowledgeWorks Global Ltd.

Contents

Preface

Precision agriculture (PA) is a new-generation farming management concept based on observing, measuring, and responding to inter- and intra-field crop variability. PA is emerging as a potent domain focusing on resource utilization efficiency, heightened productivity, quality, profitability, and sustainability of agricultural production.

PA uses digital technologies such as the Internet of Things (IoT), artificial intelligence, 5G communication, cybersecurity, edge computing, cloud-native computing principles, and blockchain to ensure that crops and soil receive exactly what they need for optimum health and productivity. When managing crops, PA diligently considers different aspects such as soil type, terrain, weather, plant growth, and yield data. PA relies on scores of IoT edge devices, data storage appliances, AI-powered data analytics platform solutions, connectivity, and cloud-native infrastructures. Edge technologies such as infinitesimal sensors, microchips and microcontrollers, beacons, LED lights, RFID tags, stickers, robots, drones, and other actuators are meticulously leveraged to collect fields' various structural and operational aspects. Any suitable IoT middleware solution can be used to transmit aggregated data to cloud-based AI platform solutions to do deeper investigations on the collected and cleansed data to facilitate knowledge discovery quickly. The actionable insights gained after that are fed back to different actuators positioned in the field to activate and accomplish the desired activities with all the clarity and alacrity.

The knowledge discovered provides farmers with helpful advice and guidance about crop rotation, optimal planting times, harvesting times, and soil management. Also, farmers can be empowered to identify fields that require treatment and determine the optimum amount of water, fertilizers, and pesticides to apply. This helps farmers avoid wasting resources. This ensures the soil receives the correct amount of additives for optimum health while reducing costs and controlling the farm's environmental impact.

There are potent applications with the PA paradigm. Cameras-attached drones can create high-resolution maps of fields. This can help identify problem areas, track crops, and assess yields. Smartphones can collect valuable data through a host of embedded applications to embark on multiple beneficial things (irrigation, fertilization, crop management, etc.). Weather monitoring, prediction, and labor and equipment management will become hugely advanced.

This book is destined to cover all the latest happenings in this PA field. We have incorporated chapters explaining how AI immeasurably contributes to achieving the ideals and goals of PA. AI model development processes are incorporated in this book. Plant disease detection and remediation steps are delineated through a couple of chapters. Sustainable farming techniques are described in one well-written chapter. A chapter articulates and accentuates field and farming data integration and ingestion mechanisms. Several chapters detail how AI-enabled data analytics is being accomplished to inform our esteemed readers. Finally, knowledge visualization through a host of 360-degree dashboards is expounded in the book's last chapter.

Editors

Pethuru Raj is the chief architect at Edge AI Division, Reliance Jio Platforms Ltd. (JPL), Bangalore, India.

N. Gayathri is an assistant professor in the Department of Computer Science and Engineering at GITAM University, Bengaluru, India.

G. Jaspher Willsie Kathrine is an associate professor and programme coordinator – CSE at the Department of Computing Science and Engineering, Karunya Institute of Technology and Sciences, Coimbatore, India.

Contributors

Bulbul Ahmed
Galgotias University
Greater Noida, Uttar Pradesh, India

E. John Alex
Department of ECE
CMR Institute of Technology
Hyderabad, Telangana, India

Saiganesh B.
Department of Electronics and Communication Engineering
Rajalakshmi Engineering College
Chennai, Tamil Nadu, India

R. Babu
Department of Computational Intelligence
Faculty of College of Engineering and Technology
SRM Institute of Science and Technology
Chennai, Tamil Nadu, India

Rahul Banerjee
ICAR – Indian Agricultural Statistics Research Institute
New Delhi, India

Bharti
ICAR – Indian Agricultural Statistics Research Institute
New Delhi, India.

Pankaj Das
ICAR – Indian Agricultural Statistics Research Institute
New Delhi, India

N. Deepa
Saveetha School of Engineering
Saveetha Institute of Medical and Technical Sciences
Saveetha University
Chennai, Tamil Nadu, India.

M.G. Dinesh
Department of CSE
EASA College of Engineering and Technology
Chavadi, Tamil Nadu, India

V. Ebenezer
Division of Computer Science and Engineering
Karunya Institute of Technology and Science
Coimbatore, Tamil Nadu, India

E. Bijolin Edwin
Division of Computer Science and Engineering
Karunya Institute of Technology and Sciences
Coimbatore, Tamil Nadu, India

Eugene Berna I.
Department of Artificial Intelligence and Machine Learning
Bannari Amman Institute of Technology
Erode, Tamil Nadu, India

Deva Priya Isravel
Department of Computer Science and Engineering
Karunya Institute of Technology and Sciences
Coimbatore, Tamil Nadu, India

Titus Issac
Department of Computer Science and Engineering
Karunya Institute of Technology and Sciences
Coimbatore, Tamil Nadu, India

G. Gandhi Jabakumar
Department of Computer Science and Engineering
Prince Shri Venkateshwara Padmavathy Engineering College
Chennai, Tamil Nadu, India

Ayush Jain
Department of Computer Science and Engineering
Noida Institute of Engineering and Technology
Greater Noida, Uttar Pradesh, India

Raashi Samir Jatakia
Vidyalankar Institute of Technology
Mumbai, Maharashtra, India

J. Jeyalakshmi
Department of Computer Science and Engineering
Amrita School of Computing
Amrita Vishwa Vidhyapeetham
Chennai, Tamil Nadu, India

K. Arul Jeyaraj
Department of Electronics and Communication Engineering
PSNA College of Engineering & Technology
Dindigul, Tamil Nadu, India

M. Robinson Joel
Department of Information Technology
Kings Engineering College
Chennai, Tamil Nadu, India

Vijay K.
Department of Computer Science and Engineering
Amrita School of Computing
Amrita Vishwa Vidhyapeetham
Chennai, Tamil Nadu, India

N. Karthik
Department of Computer Science and Engineering
National Institute of Technology
Puducherry, India

G. Jaspher Willsie Kathrine
Department of Computer Science and Engineering
Karunya Institute of Technology and Sciences
Coimbatore, Tamil Nadu, India

S. Stewart Kirubakaran
Division of Computer Science and Engineering
Karunya Institute of Technology and Sciences
Coimbatore, Tamil Nadu, India

Eashaan Manohar
Rajalakshmi Engineering College
Mevalurkuppam, Tamil Nadu, India

Attiuttama Mishra
Department of Applied Computational Science and Engineering
G. L. Bajaj Institute of Technology and Management
Greater Noida, Uttar Pradesh, India

Gajanan Dattatraya Nagare
Vidyalankar Institute of Technology
Mumbai, Maharashtra, India

Neethu Narayanan
Department of Computer Science
Rajagiri College of Social Sciences (Autonomous)
Kochi, Kerala, India

Arjun K.P.
Department of Computer Science and Engineering
GITAM University
Bengaluru, Karnataka, India

Girija P.
Department of CSE
Vel Tech Rangarajan Dr. Sagunthala R&D Institute of Science and Technology
Avadi, Tamil Nadu, India

D. Prabha
Department of M.Tech. CSE
Sri Krishna College of Engineering and Technology
Coimbatore, Tamil Nadu, India

S. Prithi
Department of Computer Science and Engineering
Rajalakshmi Engineering College
Chennai, Tamil Nadu, India

Siva Subramanian R.
Department of Computer Science and Engineering
RMK College of Engineering and Technology
Puduvoyal, Tamil Nadu, India

Sowmia K.R.
Department of Computational Intelligence
School of Computing, SRM Institute of Science and Technology
Chennai, Tamil Nadu, India

Udendhran R.
Karunya Institute of Technology and Sciences
Coimbatore, Tamil Nadu, India

Elijah Blessing Rajsingh
Division of Computer Science and Engineering
Karunya Institute of Technology and Sciences
Coimbatore, Tamil Nadu, India

A. K. Reshmy
Department of Computational Intelligence
School of Computing, SRM Institute of Science and Technology
Chennai, Tamil Nadu, India

Ravikumar S.
Department of Computer Science & Engineering
Vel Tech Rangarajan Dr. Sagunthala R&D Institute of Science and Technology
Chennai, Tamil Nadu, India

Prithi Samuel
Department of Computational Intelligence
School of Computing, SRM Institute of Science and Technology
Chennai, Tamil Nadu, India

V. Saranya
Saveetha School of Engineering
Saveetha Institute of Medical and Technical Sciences
Saveetha University
Chennai, Tamil Nadu, India

Salaja Silas
Karunya Institute of Technology and Sciences
Coimbatore, Tamil Nadu, India

Sahadeva Singh
Galgotias University
Greater Noida, Uttar Pradesh, India

Devi T.
Saveetha School of Engineering
Saveetha Institute of Medical and Technical Sciences
Saveetha University
Chennai, Tamil Nadu, India

Ayushi Bhavesh Tanna
Vidyalankar Institute of Technology
Mumbai, Maharashtra, India

M. Roshni Thanka
Division of Computer Science and Engineering
Karunya Institute of Technology and Sciences
Coimbatore, Tamil Nadu, India

M. Varghese
Department of Computer Science and Engineering
PSN College of Engineering and Technology
Tirunelveli, Tamil Nadu, India

Chapter 1

Machine Learning Algorithms and Applications

Aradhna Saini, Gaurav Dhuriya, Ayush Jain, and Attiuttama Mishra

1.1 Introduction to Machine Learning

1.1.1 Definition and History

Machine learning is a technique that uses mathematics and statistical learning to create a model that helps in predicting the values which are not known prior. In machine learning, we try to solve problems with computers without explicitly programming them for a specific task. It means learning from past data. It makes use of historical information/data for making predictions. In Traditional Programming, we give data and program to the computer to give output, but in Machine learning, we give data and output to the computer to get the program as output. In simple language, whenever a machine tends to learn from its experience, we say that machine learning is playing its role. Based on the experience, it improves its performance on every iteration [1].

The more information we feed into the machine, the greater performance is seen. Over the past years, there have been several use cases of machine learning in day-to-day applications. It helps in various tasks such as spam filtering, recommendation systems, image, and speech detection. For example, spam filtering uses machine learning to filter out flag emails and primary emails. Machine learning has helped a lot in simplifying our daily routine. It helps to simplify complex tasks.

DOI: 10.1201/9781003504900-1

Data is inputted as training data, which the system learns from the data entered when applying various algorithms. It builds a model on top of it, and upon getting new data, it predicts the output.

Machine learning helps in decreasing the amount of time required by a person to do the same set of tasks; some tasks might not be possible for humans to perform, but the machine does that. It saves time and money being invested. Several industries and companies and like Amazon, YouTube, and Netflix use machine learning models that use large amounts of data to analyze and help in product recommendation. This helps businesses to become profitable and customize their user need for more engagement with the user. With more data being generated, more uses of machine learning are seen in recent trends. It helps in locating the patterns in data and generating insight from it. Over the past years, almost in each domain some or the other way we have been using machine learning applications. For example, in the agriculture sector, machine learning plays a very important role in helping the farmer to protect their crop and prevent it from various damage. Several devices are installed on farmland and areas that warn of the unwanted intrusion of animals in field that might harm the crop and their growth. With the help of the machine, the production of quality crops has increased compared with previous human-centric method. The number of pesticides and insecticides that must be used based on the current crop condition is maintained with the help of a machine learning algorithm working behind in the system that helps farmers in their production.

The use of machine learning in agriculture is providing immense growth in several areas with exceptional results. It helps in predicting quality of soil, detecting disease, collecting data, providing insights and analytical reports, and performing several other tasks. It also helps farmers to maximize their production and yield. It is interesting to study how this technology evolved.

1.1.2 History of Machine Learning

Machine learning seems to have evolved in recent years, but it is not. The idea is old, and it has been in use for long time [2].

Early History

- 1936 – A scientist namely Alan Turing gave the theory that a set of instructions is executed by a machine itself and is determined by a machine.
- 1940 – "ENIAC", the first manually operated computer, was invented.
- 1943 – A human neural network was modelled with an electric circuit. After that in 1950, scientists started studying how a human neuron works.
- 1952–1959 – Real invention happened in this era, and the term "ML" was coined by Arthur Samuel. In this era, Samuel created a program that led IBM computers to play checkers. Also, it is used in most of the machine learning based game.

AI Winter

- 1974–1980 – In this era, various failures of machine occurred, and audience lost interest in ML and its field which led to various downsides of ML.
- 1985 – A neural network NET talk was invented which was trained to teach itself the correct pronunciations of 20,000 words in a week.
- 1997 – IBM Deep Blue computer won the game of chess against a chess expert. This led to the boom of AI.

ML in Recent Decade

- 2006 – New name to neural network "deep learning" was given.
- 2011 – In this year, a deep neural network model was created by Google that led to the recognition of the image of a cat and a human.
- 2014 – Facebook created a Deep Face neural network that helped machines to recognize a person the same as a human does.

In recent years, machine learning and Artificial Intelligence are almost in each sector contributing their advancement such as in Alexa, robotics, automatic cars, and chatbots. In the agriculture field, machine learning plays a vital role in areas such as prediction of crop disease, weather forecasting, pesticide concentration, and the impact of insecticides on farmland and crop fertility. Machine learning helps in the management of soil health and warns farmers about the spread of disease and pests. It helps in optimal irrigation schedule and planting. It helps farmers in warning system for different weather changes.

Three classifications of machine learning are as follows:

1. Supervised machine learning
2. Unsupervised machine learning
3. Reinforcement learning

1.2 Supervised Machine Learning

In supervised machine learning, we train our model on the labelled dataset. The user provides labelled data to the machine, which creates a model on top of it and predicts the output. Each data point hence "supervises" the learner as to the best output to assign to a given input. A sample labelled test data is provided to check the correctness of predicted output. The name *supervised* means under supervision; i.e., the model is trained with labelled data. It maps the input data to the output data [3].

1.2.1 Classification

Supervised machine learning classifies the data into a new category based on the input data. It divides the dataset into a number of groups of different classes. For example, to get whether the crop is infected by any disease or not (Yes/No classification). Classification is done through various machine learning applications that include disease prediction, and image recognition. Depending upon the problem, we choose the appropriate algorithm for classification. Based on the pattern it learned from the past data, it will classify and predict the output into different classes or categories.

There are some classifications of supervised machine learning.

1.2.1.1 Logistic Regression

In this regression algorithm, we solve both classification as well as regression problems. It classifies the data into different categories having binary format output. It is a probabilistic model used to predict the dependent variable given features in a dataset. It uses the sigmoid function to calculate the data in this logistic regression.

$$f(x) = \frac{1}{1+e^{x}}$$

In this algorithm, the probability value lies between 0 and 1 (Figure 1.1). It gives the "S" shaped graph on plotting. This algorithm helps to classify new and unseen data into discrete and continuous datasets. It uses the sigmoid function whose value ranges between 0 and 1 which illustrates the probability of occurrence

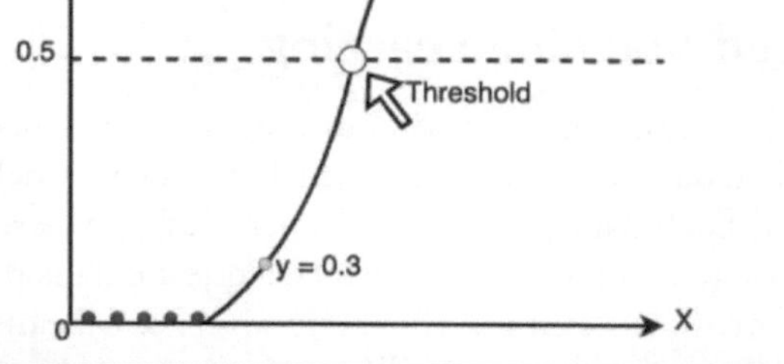

Figure 1.1 Logistic regression.

of output. Between the given range, it is assumed that the independent feature must not exhibit any collinearity among its features.

Following are the types of logistic regression:

1. Binomial regression
2. Multinomial regression
3. Ordinal regression

For example, in crop disease prediction, logistic regression is used to predict the probability of a crop being infected based on various features such as soil concentration, disease data, etc. In soil quality, logistic regression is used to predict the quality of soil.

Create a model – Lr=LogisticRegression ()

Train the model – Lr.fit (X_train, y_ train)

Make Prediction on test Data – Y_pred= Lr. Predict(X_test)

1.2.1.2 K-Nearest Neighbor

It is a supervised machine learning algorithm technique that works on the most similar categories of new data that matches with the already existing data (Figure 1.2). In simple words, it works on the most similar feature and the already existing data feature. Whenever a new data comes, it classifies it into a category that is like the already present category. This algorithm is mostly used in classification problems. This algorithm does not learn at the time of training the model, but it stores the data entities, and while classifying new data, it performs the highest

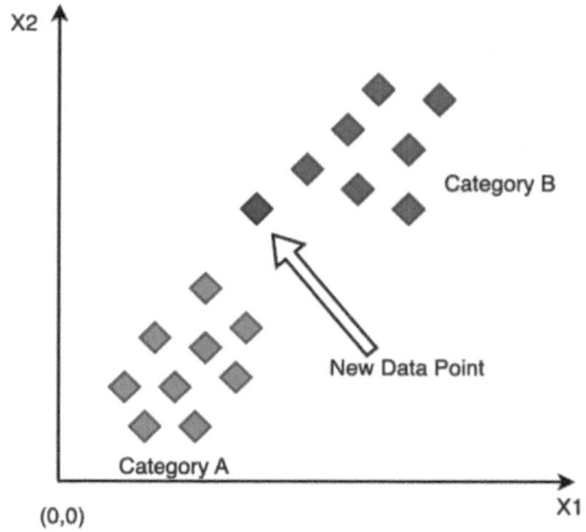

Figure 1.2 KNN.

similar feature to categories. For example, new data of a cat is input into the model, the model based on the already classified categories of dog and cat matches with the feature and then new data is put into most similar categories.

How does KNN work

1. We need to select the value of K initially.
2. Calculate the Euclidean distance.
3. With the K value obtained, we need to count the entities in each category.
4. For new coming data points, it categorizes the point to the most similar category present.

It works well for large datasets also.

1.2.1.3 Support Vector Machine (SVM)

It is an algorithm used for both classification and regression problems. In this, a hyperplane is drawn between data points to classify the new incoming data points. Figure 1.3, in this algorithm, tries to locate the most similar feature between two classes called support vector. It chooses the extreme points on the dimensional plane to draw the hyperplane. Whenever a new data entity is coming, it tries to classify the data that stands closest to the categories within the hyperplane.

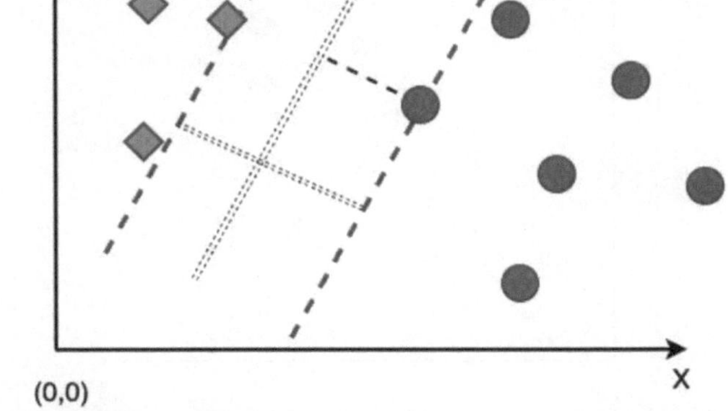

Figure 1.3 SVM.

Hyperplane: It is a decision boundary that separates different classes into the dimensional plane. Here, hyperplane is the best plane which is at the maximum distance from the support vector. Dimension means a number of features present in the dataset. Two-dimension data means we have three features present in the dataset.

1. **Linear SVM**: In this, if we draw a straight line and dataset is separated into two classes, then we say it is a linearly separable dataset and the term is called linear SVM.
2. **Nonlinear SVM:** If on drawing a straight line, we are unable to separate the dataset into two classes, we call it as nonlinear SVM.
3. **Neural Network:** It is a network that has similar characteristics to biological neural networks. It could be learned with an example. Networks are fast in their computation as compared to their traditional system. Neural network tasks include optimization of end result, classification, pattern recognition, etc.

1.2.1.4 Decision Tree

It is a supervised machine-learning technique used for mainly classification problem. The decision is taken on the features present in the dataset. There are two entities namely decision nodes and leaf nodes. Leaf nodes are the output to the decision node, whereas decision nodes make multiple decisions based on the dataset. Basically, a graphical structural pattern of data is represented using this algorithm. The values are compared between the root and the record entity. On that basis, a decisive branch is selected and the next node is taken.

1.2.1.4.1 Types of Decision Tree

If we have a categorical target variable, it is a categorical variable decision tree. If we have a continuous target variable, we say it is a continuous variable decision tree.

How decision tree works

1. Its criterion for decision is different for regression and classification problems.
2. It uses a different algorithm to decide to split a node into several sub-nodes.
3. In this, the purity of the node increases with respect to the target variable.
4. It splits the nodes on all variables and then selects the split which results in the most homogeneous sub-nodes.

There are various algorithms in decision tree such as ID3, C4.5, CART, CHAID, and MARS etc.

1.2.2 Regression

In Regression, a relationship is established between dependent(target) and independent(feature) variable. It helps to study the changes in the values of the dependent variable whenever we change the independent term, and it helps to find the continuous values such as predicting marks, land fertility, etc. For example, in agriculture, we consider various factors such as area under curve (AUC), annual rainfall, and food price that contribute to the fertility of crops. In this situation, regression helps to establish the relationship among features and to identify their influence on crop yield. In simple words, it establishes the relationship between the explanatory and response.

There are some regression supervised machine learning:

1.2.2.1 Linear Regression

Regression is the method of statistics in which a relationship between the dependent and independent features is studied. Figure 1.4 helps in analyzing the change in dependent i.e., target value when there is a change in the independent feature. It is used to predict continuous values. For example, it can predict the likelihood of admission of the student based on several features such as marks, performance, and research paper. It is a supervised learning technique that helps in predicting the continuous output variable based on several independent features. The graph is plotted between variables according to the dataset between the dependent and independent entities. The line plotted is called the regression line. Linear regression

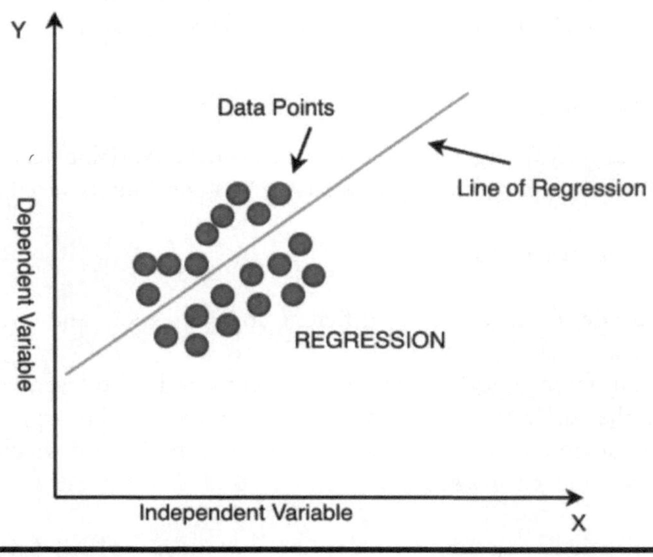

Figure 1.4 Linear regression.

is used for predictive analysis and establishes a linear relationship between continuous variables, hence called linear regression. Regression can be multiple when there is more than one independent feature in a dataset.

Dependent variable: Response variable whose values need to be predicted.
Independent variable: The predictor variable used to predict the response variable.

Several applications of linear regression are as follows:

- Analyzing sales trend
- Price prediction
- Crop disease prediction
- Weather forecasting

1.2.2.2 Random Forest

This algorithm works for both classification and regression problems. It is based on using multiple algorithms on top of the dataset to find the best desirable outcome, a method known as the ensemble technique; this improves the model performance. It uses a number of decision tree algorithms on top of different subsets of datasets to get the most accurate result. Basically, in a random forest, we use multiple decision tree outputs and choose the most accurate one. In this, overfitting can be prevented by increasing the number of decision tree to get higher accuracy. There are various applications of random forests in the domain of agriculture, banking, etc.

Advantage

- It maintains the accuracy, even though some data is missing.
- It is used to solve complex problems.
- It usually gives higher accuracy.
- It takes less model training time to generate the output.

1.3 Unsupervised Machine Learning

In supervised learning, we entered the labeled data as input data and training is done on top of predefined labelled data. In an unsupervised machine learning algorithm, data entered is not labelled and is completely unknown to the model for finding the pattern from the data. In this, no supervision of data is required; it learns from the unlabeled data which is not predefined. Unsupervised machine learning algorithm groups the dataset based on the similarities and features into separate groups [4].

There are some important key points in unsupervised machine learning:

1. It works on an unlabeled data set i.e., input data is not tagged and predicts the output based on model training and learning from the data.
2. It is based on real-life human-centric approach where similar to humans, machine is trained to learn by its own experience and learnings.
3. It is trained on unlabeled data; it is more complex compared with supervised machine learning which is fully trained on labelled data.

Unsupervised machine learning is further divided into two categories:

1.3.1 Clustering

In clustering, data possessing similar features are grouped together and form clusters of the same kind, while dissimilar data forms another cluster possessing similarities among each other. In simple, grouping of data is done based on similar features present in a dataset. It categorizes the data based on similar features represented by surrounding data. Clustering helps in simplifying the data pattern and recognizing the data based on common features.

1.3.1.1 K-Means Clustering

Cluster is a group of data points that are similar to other data points within the cluster and dissimilar to the objects in another cluster. It is an unsupervised machine learning technique that is mainly used for clustering problems in which grouping of unlabeled datasets into cluster happens wherein K is the predefined cluster (Figure 1.5).

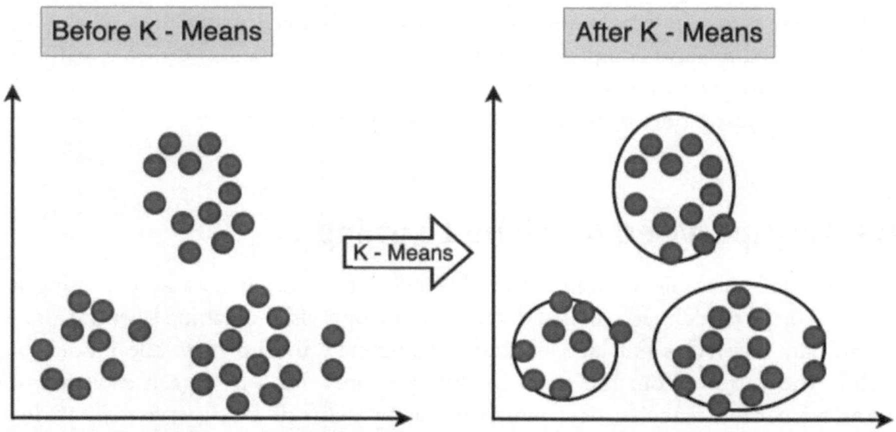

Figure 1.5 K-means clustering.

Suppose, we have K=2, there are two clusters of the whole dataset possessing similar characteristics in the respective cluster. K-means clustering is a centroid-based algorithm in which a particular cluster is aligned with a centroid. It minimizes the sum of the distance between the cluster and data points.

It works on unlabeled datasets. Hence, when we input the dataset, it segregates the data into the given K clusters and finds the best cluster by adjusting the centroid. It iteratively determines the best K centroids and determines each coming data to the nearest K-center thereby assembling similar features among surrounding data points. The K value in K-Means clustering is determined using the Elbow method, where a graph between WCSS and K value is plotted, wherever we see a sharp bend in the graph, that value of k is taken into consideration.

1.3.2 Association

In this method, the relationship is established among data in a dataset. It helps in associating the group of objects. The association technique is usually seen in the market basket analysis method that describes how variable tends to occur in a large dataset. For example, in the crop rotation method in agriculture, we use an association technique such that the production of crops increases and the risk factor of disease decreases.

1. Preparation of data is done according to the association rule mining technique. The dataset contains a row that represents the planting season and columns that represent the weather conditions and types of crops.
2. Association rule mining – In this we apply certain conditions for positive result generation. If crop A is giving better results on X amount of pesticide use, then the crop might be suitable for growth using the same pesticide in the near future.
3. Result – In this, based on the above factors, results are a mode for making correct decision and optimizing growth.
4. Checking – The result generated must be re-checked why it generated a better result using the support variable, such as confidence score or some other metrics.

The Limitation in Unsupervised Learning is that accuracy might be less as compared with supervised machine learning due to an unlabeled dataset used as input, and here we do not know the outputs of the model prior.

1.3.2.1 Apriori Algorithm

These algorithms work mainly on transaction database. Here an association rule is established, and with the help of that how weekly or strongly data is connected is studied. They use the hash map and BFS (Breadth–First Search) technique to find the item association efficiently. It finds the most occurring item from the data given i.e., an entity whose support is greater than the threshold value.

Basic steps:

1. Figure out the support of data items in the database and select the minimum confidence and support.
2. Pick all the higher support values.
3. Figure out all the rules of these sub-problems having higher confidence value than the threshold.
4. Sort in decreasing order of fit.

This is easy to understand and can be easily implemented.

1.3.2.2 Association Rule Learning

It is an unsupervised machine learning technique that matches the dependency of one data item on another data item and maps accordingly to gain maximum profit.

It is used in various tasks such as market basket analysis.

1.4 Reinforcement Learning

It is a learning which is based on feedback generation. In this learning, each time the machine processes a step or move, feedback is generated for each positive feedback; a reward point is generated else a penalty is imposed on each wrong step. In this learning, agent is not given a fixed data set; instead, data is collected as the agent interacts with the environment over time. It works on the learning from its own previous steps or experiences. It also works on unlabeled datasets. The main objective of reinforcement learning is to improve each time it is iterated and to maximize the reward. For example, in a game of chess, if reinforcement learning is applied, for each move, a computer will learn and, based on its experiences, try to optimize its current move and generate results accordingly.

There are some important terms in reinforcement learning which are:

- Reward: A feedback is generated to the agent for evaluation of its move.
- Agent: The variable that acts on the environment.
- Environment: The surroundings in which the agent is playing.
- Action: The step taken by the agent in environment.

There are three ways in which reinforcement learning is implemented:

Value Function: Value-based method tries to maximize a value function where an agent expects long-term return.

Policy Based: In this method, we try to maximize the reward by using a policy that generates the same.

Model Based: In this method, agent tries to learn from the virtual model created in the surrounding environment. Positive reinforcement learning occurs in the event where model takes positive step towards its goal hence reward is generated, whereas in negative reinforcement learning negative reward is generated due to bad learning. Hence, further it tries to optimize its performance based on its past learning.

1.5 Machine Learning Workflow

Till now, we have studied how machine learning algorithm is used in various fields and how it helps in various domains. But how it works, and process implementation is not known. Now we will study how machine learning models and built, and how training happens on a given dataset [5].

1.5.1. **Data Cleaning**: In this step, all the related data is gathered. Data can be from primary as well as secondary sources based on the user requirement and problem specific. This is the initial step to begin the workflow. Without data, nothing can be done. There can be several sources through which data can be gathered such as the internet, companies database, feedback, etc.

1.5.2. **Data Preparation**: After data gathering is done, we select the part of data that is required for model formation or our specific use case. In this step, we explore the feature of data and their dependencies on each other, find correlations and understand each aspect of data.

1.5.3. **Data Description**: In this step, we describe the data. The story of what the idea is all about and its features is studied from the data.

1.5.4. **Data Wrangling**: In this step, we interpret the data according to our use case. In this step, cleaning of data is done which is the most important step before proceeding further. Unwanted features and parameters are removed which least add their significance in the output and that might alter our result in future prediction. In this, missing values are filled, duplicity in data is removed and unwanted data is also removed.

1.5.5. **Analysis of data**: This is the most important step where we analyze the final dataset we have prepared till now. Visualizations of data are important to generate insight beforehand. Various patterns and studies of data are done at this step. We plot graphs and charts to study the data. After completing data analysis, model building is done which is where we experiment with multiple algorithms and choose the one that gives the best accuracy out of many.

1.5.6. **Model Building**: In this step, we apply various machine learning algorithms such as regression, classification, and clustering based on the dataset. Here, we train our model using the final dataset prepared and predict the output. Once training is done, testing of desired model becomes important.

In this phase, unknown or unseen data is given to the model to test its performance and accuracy. If the accuracy on unseen data is not coming as we wanted, we apply different algorithms on the dataset or change certain parameters to get better satisfactory results.

1.5.7. **Deployment:** After successful result generation, deployment of the model is done for the public use case. Deployment is the last step where our best model is taken. It is mainly done on cloud platforms such as AWS, Azure, Google Cloud and Heroku.

1.6 Current Applications of ML in Agriculture

We live in an intriguing time when there are stories about new machine learning (ML) applications that could benefit consumers and companies virtually every week. The orchard industry, horticulture, floriculture, and farming are not outliers. Agriculture adopts ML technologies even if it does not do so as quickly as some other industries. Here are a few examples of artificial intelligence applications in upstream agriculture that is not all-inclusive (Figure 1.6):

1.6.1. **Smart Sprayers:** If you could only target the weeds, why would you spray pesticides on plants as well? If it is not necessary to use herbicides, computer vision might assist save money by preventing chemical contamination of plant goods.

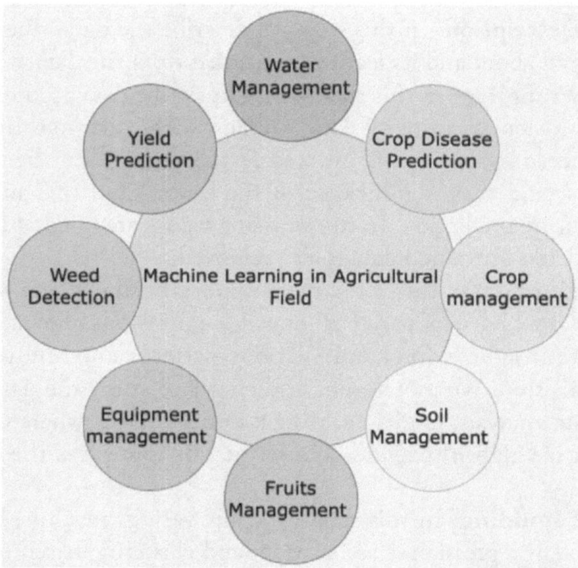

Figure 1.6 ML applications in agriculture.

1.6.2. **Yield Prediction:** A level of accuracy and scale that is not possible without ML is brought to the agricultural sector. This could encourage farmers to adopt preventative practices rather than reactive ones.

1.6.3. **Smart Greenhouses:** The market for smart greenhouses is anticipated to grow from USD 1.4 billion in 2020 to USD 2.1 billion by 2025. Growers employ sensors and gather data more often than farmers because of the regulated and semi-controlled environment in greenhouses. If data from numerous sensors and actuators are combined while streaming, ML algorithms open up new possibilities for giving farmers insights.

1.6.4. **Integrated Pest Management Using Computer Vision:** Today more than ever, a difficulty for growers is the lack of skilled scouting people. What if farmers might benefit from a computer program that was linked to stationary or moving cameras? This entails the ability to view a greenhouse (or farm) in real time even when no one is there to inspect each plant individually.

1.6.5. **Plant Disease Prediction:** Based on past experience and the forecasted weather, illnesses can be expected to arise. Farmers will benefit financially and be able to use pesticides more efficiently as a result.

1.7 Issues Slowing the Adoption of Machine Learning in Agricultural Production

Any genuine effort to implement machine learning in agriculture must look at the practical challenges that have prevented this sector from evolving as quickly as other sectors. Even though the landscape of technology use in agriculture is always evolving, we must first understand where it came from and where it is today in order to forecast where it will go in the future [6]. There were (and to some extent still "are") four major roadblocks in the way of the upstream agriculture sector implementing the right ML-based technology.

1.7.1. **Data Collection:** Despite the growth of precision agriculture over the past three decades, there have been two major obstacles that have prevented machine learning predictions from being widely used in the sector. First, many farmers and producers have not used precision agriculture tools (although there is a lot of evidence and justification for this). Second, using precision-Ag technology does not necessitate obtaining all necessary data via internet of things (IoT) or other media in order to make predictions. Technically speaking, the latter component could be the tipping point for how beneficial machine learning is. Due to external factors that are outside of our control, the majority of agricultural activities have some inherent uncertainty (or irreducible error, in machine learning language) [7]. This suggests that farmers, growers, agronomists, and consultants on the one side, and data scientists with domain experience in agriculture on the other, should work together continuously to coordinate data collection operations. This widespread awareness and comprehension clear the way for developing models with trustworthy utilities for farmers and producers.

1.7.2. **Data Ownership and Privacy:** In the agriculture industry, whose data are they? The answer to this query is not as straightforward as you may think. Yes, producers – farmers, greenhouse managers, breeders, agronomists, etc. – hold and have access to data, but it does not tell the whole picture. Although owners have access to certain essential data, they do not have quick access to many other sources of pertinent data, such as satellite imaging or historical environmental data for the area, unless they pay for access and receive training on how to use it. Data scientists and data engineers, who are trained to collect and synthesize data from many sources, may be able to quickly close this gap (particularly the farmers' lack of access to environmental and satellite data). Another issue that makes the exchange of agricultural data more difficult is data privacy. When using ML in agriculture on a large scale, there are a number of ways to address issues with intellectual property, including anonymizing, encrypting, and aggregating. When the granularity of data matters or data contains a geographical aspect, there are acceptable non-disclosure agreements (NDAs) and machine learning algorithms to help analyze the data and extract insightful knowledge without dealing with issues surrounding agricultural data's intellectual property.

1.7.3. **Agricultural Data Has Been Scattered:** Even while all relevant data from a farm may be accessible to a farmer (or an adviser), the collection of data from all farms is not centralized. That suggests that the common knowledge and insight gained by ML-based techniques may not fully flourish. Let's contrast agriculture with banking to illustrate the stark differences. For free or for a fee, you may access a variety of financial data for all publicly traded companies with a history spanning many decades through services like Bloomberg. Because of this, the sheer volume of academic, for-profit, and nonprofit research that uses ML in financial data is astounding. The agricultural industry does not have access to a system of unified, high-resolution data [8]. According to FAO estimates, there are 570 million farms worldwide. How might a data scientist utilize so much data to produce more? Utilizing satellite image data is one method of overcoming this challenge without going against privacy laws. For almost the whole planet, historical satellite data with resolutions as good as 30 cm is freely available to download. This suggests that experts in computer vision and data scientists may extract pertinent information (such as NDVI and NDWI) from historical satellite data to build models for the forecasting of production, disease, or stresses in farms.

1.7.4. **Cost of Change Toward ML in Agriculture, Cost of Mistakes:** In the case of the farming business, the high cost of change is particularly true [9]. In comparison to several other businesses that welcomed artificial intelligence with open arms, this cost structure is very different. If Uber launches an email campaign based on recommendations from their data science team and discovers after a few days that it is not as successful as expected, they may promptly stop/modify it. In an agricultural system, the risk factor is of a different magnitude; following bad advice might result in reduced crop yield for an entire farming season. Accepting advances at smaller amounts, even though the farmer farms on a massive scale is one way to get over this obstacle, which is currently practiced in farming. The tiny scale denotes a small region of a farm or subtle changes made over several agricultural seasons while taking input into consideration. Probabilistic techniques are a different method that naturally appears in data science and ML solutions. Although predictions in ML are seldom deterministic, they can be presented as such. Because forecast accuracy is always a probability, a statistical confidence range is usually used next to a prediction. A skilled data scientist will seldom make recommendations based on a model unless they express their level of confidence in the projection. This approach is frequently used to evaluate and reduce the risk connected to changes suggested by ML solutions.

1.8 Application of Machine Learning

1.8.1 Image Recognition

The amalgamation of ML and image recognition within the agricultural sector has yielded a plethora of applications that substantially augment farming methodologies [10]. Crop disease detection uses machine learning algorithms to analyze photos of crops, swiftly recognizing indications of illnesses, pests, or deficits in nutrients. This technology provides farmers with immediate notifications and specific suggestions for treatment, therefore decreasing the need for the use of wide-ranging chemical substances. Weed detection algorithms are capable of differentiating between cultivated crops and undesirable weeds, so enabling accurate and targeted herbicide treatment, ultimately leading to the reduction of environmental consequences. Crop growth monitoring, which is enhanced by machine learning models, allows for the continuous monitoring of crop development, therefore empowering farmers to enhance the efficiency of irrigation, fertilization, and harvesting schedules. The process of yield prediction involves the analysis of both historical and real-time field photographs in order to provide accurate forecasts about crop yields. This information is then used to assist in making informed choices regarding resource allocation and marketing strategies.

The process of Soil Health Assessment employs machine learning techniques to analyze photographs and assess the health of the soil, therefore generating suggestions for improving soil conditions and optimizing crop choices. Crop counting is a precise method used to identify the number of plants in a picture, hence aiding in the assessment of crop density and the optimization of planting and harvesting procedures.

The field of Irrigation Management employs image analysis techniques to assess soil moisture levels and crop water stress, therefore enabling accurate and targeted modifications to irrigation practices.

The use of image recognition by Harvesting Assistance facilitates the identification of mature products for the purpose of automating the harvesting process. This implementation enhances operational efficiency and concurrently diminishes expenditures associated with manpower.

The use of machine learning techniques in the process of quality grading enables the assessment of agricultural goods by analyzing their visual attributes, therefore simplifying the tasks of sorting and packing. The practice of Livestock Monitoring involves the use of image recognition techniques to conduct health assessments and detect indications of sickness or stress in animals [11].

The use of machine learning techniques is employed in Pest and Predator Identification to identify and monitor the presence of pests. On the other hand, animal monitoring is utilized to avoid crop damage by monitoring the activities of animals. The use of machine learning techniques in the field of crop mapping and zoning enables the generation of comprehensive crop maps derived from satellite or drone data, hence facilitating informed decision-making in land-use planning processes.

Finally, ML is employed in the field of Weather and Climate Analysis to conduct an analysis of satellite pictures. This analysis serves the purpose of monitoring the influence of weather patterns on agricultural crops, therefore facilitating the implementation of adaptive farming techniques (Figure 1.7). These applications exemplify the transformative impact of machine learning and image identification

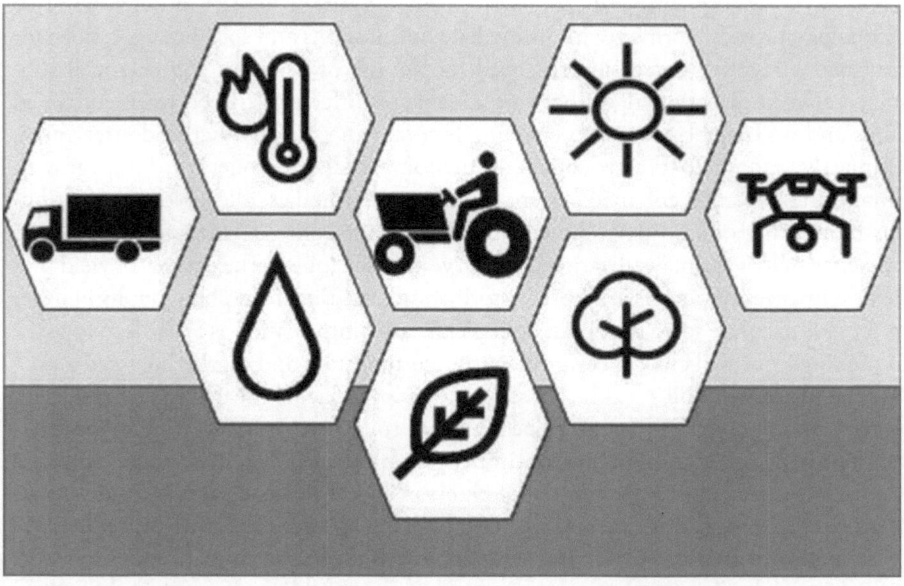

Figure 1.7 ML in agriculture.

on the agricultural sector, as they enhance accuracy, sustainability, and production while mitigating resource inefficiencies.

1.8.2 Natural Language Processing (NLP)

ML with natural language processing (NLP) has several applications in agriculture, helping stakeholders make informed decisions, improve communication, and enhance overall efficiency. Here are some key applications of ML and NLP in agriculture:

1. **Crop and Pest Management:**
   - Pest Identification and Management: ML-based chatbots or mobile apps can identify and recommend treatment options for crop pests and diseases based on text or image descriptions provided by farmers.
   - Crop Advisories: NLP can analyze weather and soil data to provide personalized crop advisories to farmers, helping them make informed decisions about planting, irrigation, and pest control.
2. **Market Analysis and Price Prediction:**
   - Market Sentiment Analysis: NLP can analyze news articles, social media posts, and other text data to gauge market sentiment and predict price fluctuations.
   - Price Prediction: ML models can use historical price data and NLP to forecast crop prices, assisting farmers and traders in decision-making.
3. **Precision Agriculture:**
   - Weather Forecasting: NLP models can interpret weather forecasts and provide real-time weather updates to farmers, helping them plan field operations and irrigation.
   - Soil Analysis: NLP can analyze soil reports and provide recommendations for soil amendments and nutrient management.
4. **Supply Chain Optimization:**
   - Demand Forecasting: NLP can process market data and customer communications to forecast demand for agricultural products, aiding in supply chain planning.
   - Logistics and Inventory Management: NLP can assist in managing logistics, including inventory tracking, routing, and distribution, optimizing the supply chain.
5. **Customer Support and Extension Services:**
   - Chatbots for Farmer Queries: ML-powered chatbots can answer farmer queries, offer guidance, and provide information on farming practices and government schemes.
   - Extension Services: NLP can be used to create digital extension services that deliver advice, recommendations, and educational content to farmers via text and voice interfaces.

6. **Crop Insurance:**
   - Claim Processing: NLP can automate the processing of crop insurance claims by extracting relevant information from textual documents and assessing damage reports.
7. **Regulatory Compliance:**
   - Compliance Monitoring: NLP can help farmers and agribusinesses stay compliant with regulations by analyzing and interpreting legal documents and regulations related to agriculture.
8. **Research and Knowledge Discovery:**
   - Scientific Literature Analysis: NLP can parse and summarize agricultural research papers, helping researchers and policymakers stay up-to-date with the latest advancements.
   - Data Extraction: NLP can extract valuable information from unstructured data sources, such as agricultural reports, to facilitate data-driven decision-making.
9. **Multilingual Communication:**
   - NLP can assist in overcoming language barriers by providing translation services for farmers and agribusinesses, enabling global collaboration and market access.
10. **Food Safety and Traceability:**
    - NLP can help track and trace the origin of agricultural products and their supply chain information, enhancing food safety and transparency.
11. **Climate and Environmental Monitoring:**
    - NLP can analyze climate-related text data and reports to provide insights into climate change impacts on agriculture and recommend adaptive strategies.

These applications demonstrate the versatility of ML and NLP in addressing various challenges in agriculture, from crop management and market analysis to regulatory compliance and environmental monitoring. By harnessing the power of language understanding and data analysis, NLP-based solutions are helping to advance the agricultural sector and promote sustainability [Table 1.1].

1.8.3 Speech Recognition

In agriculture, the integration of machine learning with speech recognition offers a multitude of applications. Farmers can operate farm equipment, log data, and manage labor hands-free through voice commands, enhancing efficiency. Voice-activated systems provide real-time weather updates, assist in crop quality assessment, and offer market information, aiding decision-making. Additionally, speech recognition facilitates the reporting of pests, diseases, and emergencies, enabling swift responses. It's also valuable for livestock management, equipment maintenance scheduling, and language translation, promoting global collaboration and knowledge exchange. Overall, these applications streamline agricultural processes, reduce manual labor, and improve communication, making farming more efficient and accessible [12].

Table 1.1 Concise Overview of the Various Applications of ML and NLP in Agriculture

Application	*Description*
Crop and Pest Management	• Pest Identification and Management: ML-based chatbots or apps for identifying and treating crop pests and diseases. • Crop Advisories: Providing personalized crop advisories based on weather and soil data.
Market Analysis and Price Prediction	• Market Sentiment Analysis: Analyzing text data to gauge market sentiment. • Price Prediction: Using ML and NLP to forecast crop prices.
Precision Agriculture	• Weather Forecasting: Providing real-time weather updates. • Soil Analysis: Recommending soil amendments and nutrient management.
Supply Chain Optimization	• Demand Forecasting: Forecasting demand for agricultural products. • Logistics and Inventory Management: Optimizing supply chain logistics.
Customer Support and Extension Services	• Chatbots for Farmer Queries: Answering farmer queries and providing guidance. • Extension Services: Delivering advice and educational content via text and voice interfaces.
Crop Insurance	• Claim Processing: Automating the processing of crop insurance claims.
Regulatory Compliance	• Compliance Monitoring: Analyzing regulations and legal documents.
Research and Knowledge Discovery	• Scientific Literature Analysis: Parsing and summarizing research papers. • Data Extraction: Extracting information from unstructured data sources.
Multilingual Communication	• Providing translation services for farmers and agribusinesses.
Food Safety and Traceability	• Tracking and tracing the origin of agricultural products and supply chain information.
Climate and Environmental Monitoring	• Analyzing climate-related text data to provide insights into climate change impacts on agriculture.

1.8.4 Recommendation Systems

Machine learning recommendation systems are finding valuable applications in agriculture, offering data-driven insights and personalized recommendations that optimize various aspects of farming. These systems assist in crop selection, advising farmers on suitable crops based on soil quality, weather conditions, and historical performance [11]. They also play a crucial role in fertilizer and nutrient management by providing tailored recommendations and reducing resource wastage. Pest and disease management benefit from these systems, as they suggest precise treatment strategies, minimizing pesticide use while maintaining crop health. Moreover, recommendation systems optimize irrigation schedules, conserve water resources, and promote sustainable practices like crop rotation and reduced tillage. They also contribute to market success by predicting crop prices and recommending strategic selling times. Equipment maintenance, livestock management, weed control, and supply chain optimization are further enhanced by these systems, making agriculture more efficient, sustainable, and profitable [Table 1.2].

1.8.5 Fraud Detection

ML with fraud detection techniques can be applied to various aspects of agriculture to identify and mitigate fraudulent activities and protect the interests of farmers and agribusinesses. Here are some key applications of ML-based fraud detection in agriculture [13]:

1. **Crop Insurance Fraud Detection:**
   - ML algorithms can analyze claims data to detect fraudulent insurance claims, such as false reports of crop damage.
   - This helps insurance companies identify and prevent fraudulent payouts, ensuring the sustainability of crop insurance programs.
2. **Input Supply Fraud:**
   - Fraud detection systems can monitor and verify the authenticity of agricultural inputs such as seeds, fertilizers, and pesticides.
   - This reduces the risk of counterfeit or substandard inputs entering the supply chain.
3. **Market Price Manipulation:**
   - ML can analyze market data and trading activities to detect price manipulation and insider trading in agricultural commodity markets.
   - This promotes fair market practices and protects farmers from price manipulation schemes.
4. **Counterfeit Agricultural Products:**
   - ML-powered fraud detection can identify counterfeit agricultural products and ensure that farmers receive genuine seeds, chemicals, and equipment.
   - This safeguards crop quality and yield.

Table 1.2 Various Applications of ML and Speech Recognition in Agriculture, Offering Improved Efficiency, Communication, and Farm Management

Application	*Description*
Voice-Activated Farm Equipment	ML-driven speech recognition enables the operation of farm machinery using voice commands, enhancing equipment control.
Hands-Free Data Entry	Speech recognition allows for hands-free logging of critical agricultural data, reducing the need for manual record-keeping.
Labor Management	Voice recognition systems assist in labor coordination by assigning tasks and conveying instructions to farm workers through voice commands.
Inventory Management	Farmers can update and manage inventory hands-free, enabling real-time monitoring of stock levels and supplies.
Weather Updates and Alerts	Voice-activated systems provide weather forecasts and alerts, aiding in informed decisions regarding planting, harvesting, and irrigation.
Crop Quality Assessment	ML analyzes spoken descriptions of crop quality and grade, facilitating sorting and packaging decisions.
Safety and Emergency Response	Speech recognition supports hands-free communication during emergencies and safety reporting on the farm.
Market Information	Farmers can inquire about market prices, trends, and demand using voice-activated systems, aiding marketing decisions.
Pest and Disease Reporting	Field workers report pest infestations and crop diseases through speech recognition, expediting response, and treatment.
Irrigation Control	Farmers can control irrigation systems with voice commands, adjusting water flow based on real-time conditions and moisture levels.
Livestock Management	ML-powered voice recognition assists in livestock management, including feeding and health checks, through voice commands.
Equipment Maintenance	Farmers schedule equipment maintenance and receive service reminders using voice-activated systems to ensure machinery remains in good condition.
Language Translation and Communication	Speech recognition facilitates communication between farmers who speak different languages, fostering global collaboration.
Training and Education	Voice-activated educational resources provide training on best practices, pest identification, and crop management, enhancing agricultural knowledge and skills.

5. **Contract and Payment Fraud:**
   - ML can analyze contracts and payment records to detect fraudulent transactions, including cases of double invoicing or misappropriation of funds.
   - This helps prevent financial losses for farmers and agribusinesses.
6. **Supply Chain Transparency:**
   - ML-based tracking systems can enhance supply chain transparency by detecting irregularities in the movement of agricultural products from farm to market.
   - This reduces the risk of theft and fraud along the supply chain.
7. **Contract Farming Compliance:**
   - ML can assess contract farming compliance by analyzing data related to agreements between farmers and buyers.
   - This helps identify deviations from contract terms and ensures fair treatment of farmers.
8. **Grant and Subsidy Fraud:**
   - ML-based fraud detection can identify fraudulent applications for agricultural grants and subsidies, reducing financial losses for governments and organizations.
9. **Labor and Wage Fraud:**
   - ML can analyze labor and wage data to detect fraudulent practices such as underreporting labor hours or wage theft.
   - This protects the rights and incomes of agricultural workers.
10. **Land and Property Fraud:**
    - Fraud detection technology can monitor land and property transactions to detect fraudulent land grabs or property disputes.
    - This safeguards land rights and ownership for farmers.

These applications demonstrate how ML-based fraud detection can play a crucial role in ensuring transparency, fairness, and security within the agricultural sector, ultimately benefiting farmers, agribusinesses, and consumers [Table 1.3].

1.8.6 Autonomous Vehicle

The combination of autonomous vehicles with ML has the potential to greatly impact the agricultural sector. These vehicles have the capability to be outfitted with a diverse range of sensors and use ML algorithms, therefore potentially transforming several crucial facets of agriculture [14]. First and foremost, ML facilitates accurate crop monitoring and management via the analysis of sensor data.

This analysis allows the detection of various concerns such as diseases, pests, and nutrient deficits, hence enabling the implementation of tailored therapies. The use of machine learning models contributes to the improvement of weed identification and management practices, since these models possess the capability to accurately identify and selectively eliminate weeds. Consequently, this targeted approach results in a reduction in the overall consumption of herbicides.

Table 1.3 Various Applications of ML in the Recommendation System

Application	*Description*
Crop Recommendation	ML recommends suitable crops based on soil data, weather conditions, and historical performance, aiding informed planting decisions.
Fertilizer and Nutrient Recommendations	Recommendation systems offer tailored advice on fertilizers and nutrients, optimizing resource use and reducing environmental impact.
Pest and Disease Management	ML-powered systems suggest pest control strategies using real-time and historical data, promoting targeted treatments, and reducing pesticide usage.
Irrigation Optimization	Systems analyze soil moisture, weather forecasts, and crop needs for precise irrigation scheduling, and conserving water and resources.
Crop Rotation Planning	ML recommends crop rotation plans to enhance soil health, reduce disease pressure, and ensure long-term sustainability in agriculture.
Market Price Prediction and Marketing	Systems predict crop prices and recommend optimal selling times, assisting farmers in maximizing profits through strategic sales.
Equipment Maintenance	ML predicts equipment maintenance needs based on usage patterns and sensor data, minimizing downtime and extending machinery life.
Livestock Management	Recommendation systems provide insights into livestock breeding, feeding, and health practices, enhancing animal welfare and productivity.
Crop Variety Selection	ML suggests crop varieties based on performance data and local climate conditions, helping farmers choose disease-resistant options.
Weed Control	Systems recommend weed control strategies based on crop type, weed species, and growth stage, reducing herbicide use.
Agrochemical Usage Optimization	ML analyzes data to optimize agrochemical application timing and quantity, reducing costs and environmental impact.
Supply Chain and Inventory Management	Recommendation systems optimize inventory levels, ensuring timely access to resources and reducing waste in farming.
Sustainable Agriculture Practices	ML recommends sustainable farming practices like cover cropping and reduced tillage, improving soil health and sustainability.

The use of computer vision technology in autonomous vehicles has the potential to optimize the process of harvesting and fruit picking. By using computer vision algorithms, these vehicles are capable of accurately identifying ripe produce, hence reducing labor expenses and avoiding damage during the harvesting process. ML also offers advantages in soil analysis, as it enables the creation of comprehensive maps that may be used for the accurate application of fertilizers and irrigation [15].

In addition, the use of machine learning enables the effective operation of autonomous tractors and ploughs, therefore enhancing the processes of field preparation and navigation.

The use of predictive maintenance strategies assures the continuous operating status of these vehicles via the anticipation and prediction of maintenance requirements. ML algorithms have shown to be valuable tools in assisting farmers in making well-informed choices by providing production projections. The examination of weather and climatic data enables the adaptation of operational strategies in accordance with forecasted conditions. Moreover, the integration of data and the use of decision-support technologies combine many sources of data in order to provide complete insights. Autonomous irrigation systems use machine learning techniques to enhance the efficiency of water distribution.

The field of pest and disease management is enhanced by the advantages of early detection, as it allows for timely intervention and control measures (Figure 1.8).

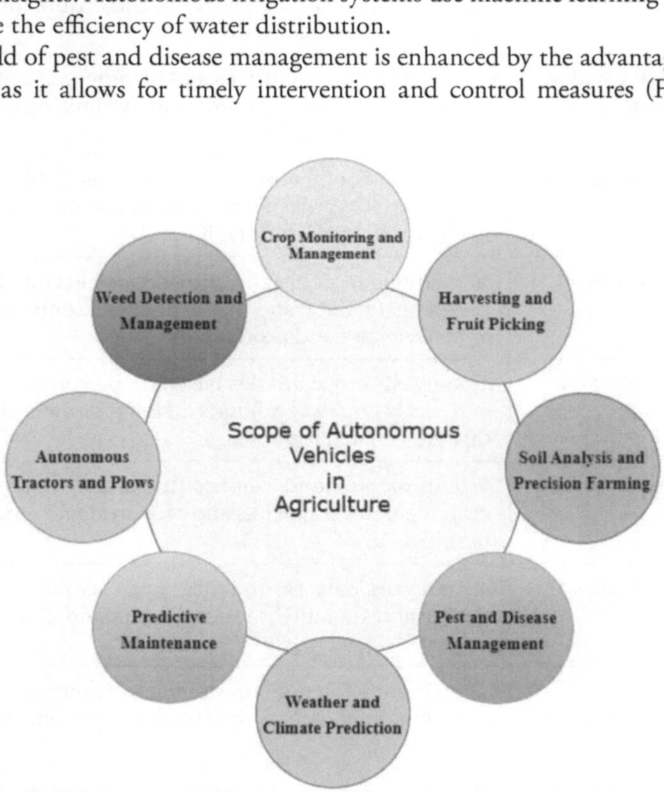

Figure 1.8 Applications of autonomous vehicles using ML in agriculture.

Similarly, the use of ML-based classification techniques improves the efficiency of crop sorting and grading processes. The integration of ML technology into autonomous agricultural vehicles has the potential to significantly improve efficiency, save resources, boost crop production, and promote sustainable practices within the agricultural sector. All the discussed approaches are shown in Figure 1.1. These advancements are expected to deliver substantial benefits for farmers and the environment alike.

1.8.7 Healthcare

The field of ML has shown significant promise within the healthcare domain, and it is plausible to extend these applications to provide advantages in the agricultural sector. ML is used in the healthcare industry for several purposes, including illness identification, health surveillance, predictive analysis, and precision treatment. Likewise, within the field of agriculture, ML might be effectively used to oversee the well-being of crops and animals, anticipate occurrences of diseases, and enhance the allocation of resources via the utilization of predictive models [16].

The use of genomic analysis is of paramount importance in the field of healthcare as it facilitates the comprehension of disease susceptibility. Furthermore, this analytical approach may be effectively employed to augment breeding programs and bolster the resistance of crops. The use of image analysis, remote monitoring, and natural language processing techniques, which are often employed in the healthcare sector, may be applied to the examination of crop and environmental data, the surveillance of agricultural conditions, and the processing of textual agricultural information.

By using machine learning techniques from the healthcare sector, the agricultural industry has the potential to adopt a more data-centric approach, leading to enhanced precision and efficiency. This, in turn, may significantly contribute to heightened production and sustainability in farming methodologies.

1.8.8 Finance

Machine learning applications from the financial sector can be tailored to benefit agriculture by enhancing financial management and risk mitigation for farmers [17]. These applications include accurate credit scoring and risk assessment, precise crop insurance underwriting, informed market predictions, streamlined supply chain financing, and fraud detection. By leveraging these tools, farmers gain access to better credit options, fair insurance coverage, improved market timing, and more efficient financial decision support. These adaptations enhance financial resilience, aid in resource allocation, and optimize investment strategies, ultimately promoting sustainable farming practices [Table 1.4].

Table 1.4 The Adaptation of ML Applications from the Financial Sector to Benefit Agriculture, Enhancing Financial Access, Risk Management, and Decision-Making Processes in Farming.

ML Application in Finance	*Adaptation for Agriculture*
Credit Scoring and Risk Assessment	Accurate assessment of farmers' creditworthiness for improved access to capital.
Crop Insurance Underwriting	Customized insurance coverage, reducing financial risks from crop loss.
Algorithmic Trading for Commodities	Informed selling decisions based on commodity price trend predictions.
Supply Chain Financing	Streamlined payment processes in agricultural supply chains.
Fraud Detection and Risk Management	Protection against financial fraud for farmers and financial institutions.
Portfolio Management	Optimization of resource allocation across different farming operations.
Market Predictions	Real-time insights for adjusting planting and harvesting strategies.
Financial Decision Support	Informed financial decisions and cost-effective farming practices.
Risk Hedging and Derivatives Trading	Management of price risks associated with agricultural commodities.

1.9 Opportunities for Applying ML in Agriculture

Despite the barriers that prevented the agriculture sector from more widely implementing artificial intelligence, it has recently adopted ML-driven initiatives at an increasingly rapid rate. Future seismic developments in the agricultural industry are predicted by trends in academic publications, the drive to found new Ag-tech companies, and the reorientation of agricultural enterprises towards ML approaches. The size of the worldwide market for agricultural analytics is predicted to increase by 75% between 2020 and 2025.

There are four key drivers for the broad-scale application of ML in agricultural activities:

1.9.1 Global Food Demand

The main issue is this explanation. For an obvious reason – namely, the rising need for food on a worldwide scale – agricultural practices are evolving and will continue to do so. The UN projects a 9.8 billion-person world population by 2050, and between 2005 and 2050, global food consumption is expected to increase by 59%–98%. By drawing more water from dwindling supplies, fertilizing the land more, and destroying more forests to make more space for farms, this requirement cannot be met. The production of agriculture is thought to be negatively impacted by climate change globally [18]. The requirement for more food will be met by applying technology in agriculture that increases the effectiveness of sustainable and environmentally friendly methods. Precision farming is the most efficient way to increase productivity, reduce costs, and minimize negative influence on the environment. Precision agriculture requires data collection and data-based agricultural management. This ongoing transformation inevitably creates opportunities for farmers and producers to adopt ML and data-driven methodologies for the obvious reason that since the data has already been collected, what was stopping you from deriving more insights from it? This transformation will take some time to settle, but it has already started and is gaining momentum year after year.

1.9.2 Availability and Affordability of Infrastructures for Data Collection

The amount of data gathered in the agriculture sector has increased by orders of magnitude over the last two decades. Today farmers have access to vast amounts of detailed and precise data at their fingertips on a scale that was unimaginable 20 years ago. Data collection systems are becoming more widely used and accessible every year as a result of advancements in IoT technology. This suggests that growers and farmers may have integrated data collection systems for a fraction of the cost of 15 years ago (taking into account improvements in IoT devices throughout that time).

1.9.3 Competitiveness and Increasing Cost of Labor and Materials

Utilizing data-driven agricultural techniques lowers costs and increases output. On a corporate level, the agriculture industry is undergoing an acknowledged, anticipated shift. The news about every single important agribusiness company (such as Monsanto and Bayer) reveals their investments in ML and that they have already started funding initiatives and programs to make use of ML in farming [19, 20]. A practical use of artificial intelligence technologies is constrained at the grower/farmer level of thinking, showing that its application is still in its infancy in the agricultural sector. Whether it was electricity, internal combustion automobiles, the internet, or smartphones, the history of innovation adoption has a consistent pattern: it is exponential. It represents the fact that technology's use and scope are first poor and expensive but that when it gains traction, it expands exponentially. This trend is still present, especially in the case of small-scale farming, where disruptive technologies may help to close a substantial productivity gap brought on by a lack of suitable equipment.

1.9.4 International/Governmental Initiatives for ML in Agriculture

Data-driven agricultural projects are receiving attention from an increasing number of governments, NGOs, and non-profits on a global scale. Their operations include mostly on open data and infrastructures for data collection but also include education, facilities, grants, and funding. Infrastructures for data gathering are absolutely necessary since they provide the framework for real progress in ML in agriculture. Among useful examples are FAO, GARDIAN, CIMMYT, USDA, and GODAN.

References

1. Trappenberg, T. P. (2019). *Fundamentals of machine learning*. Oxford University Press.
2. Fradkov, A. L. (2020). Early history of machine learning. *IFAC-PapersOnLine*, *53*(2), 1385–1390.
3. Singh, A., Thakur, N., & Sharma, A. (2016, March). A review of supervised machine learning algorithms. In *2016 3rd international conference on computing for sustainable global development (INDIACom)* (pp. 1310–1315). IEEE.
4. Hahne, F., Huber, W., Gentleman, R., Falcon, S., & Carey, V. J. (2008). Unsupervised machine learning. *Bioconductor Case Studies*, 137–157. https://www.huber.embl.de/pub/pdf/HahneHuberGentlemanFalcon2008.pdf
5. Sarker, I. H. (2021). Machine learning: Algorithms, real-world applications and research directions. *SN Computer Science*, *2*(3), 160.
6. Rehman, T. U., Mahmud, M. S., Chang, Y. K., Jin, J., & Shin, J. (2019). Current and future applications of statistical machine learning algorithms for agricultural machine vision systems. *Computers and Electronics in Agriculture*, *156*, 585–605.
7. Simeone, O. (2022). *Machine learning for engineers*. Cambridge university press.
8. Seeniselvi, T., & Nirmala, M. (2019). A survey on data preparation and feature engineering in machine learning. *International Journal of Advanced Research in Computer Engineering & Technology (IJARCET)*, *8*(5), 191–198.
9. Bonaccorso, G. (2017). *Machine learning algorithms*. Packt Publishing Ltd.
10. Sharma, R., Kamble, S. S., Gunasekaran, A., Kumar, V., & Kumar, A. (2020). A systematic literature review on machine learning applications for sustainable agriculture supply chain performance. *Computers & Operations Research*, *119*, 104926.
11. Subeesh, C. R. M. (2021). Automation and digitization of agriculture using artificial intelligence and internet of things. *Artificial Intelligence in Agriculture*, *5*, 278–291.
12. Sharma, A., Jain, A., Gupta, P., & Chowdary, V. (2020). Machine learning applications for precision agriculture: A comprehensive review. *IEEE Access*, *9*, 4843–4873.

13. Javaid, M., Haleem, A., Khan, I. H., & Suman, R. (2023). Understanding the potential applications of artificial intelligence in agriculture sector. *Advanced Agrochem, 2*(1), 15–30.
14. Maffezzoli, F., Ardolino, M., Bacchetti, A., Perona, M., & Renga, F. (2022). Agriculture 4.0: A systematic literature review on the paradigm, technologies and benefits. *Futures*, 142, 102998.
15. Singh, D. K., Sobti, R., Jain, A., Malik, P. K., & Le, D. N. (2022). LoRa based intelligent soil and weather condition monitoring with internet of things for precision agriculture in smart cities. *IET Communications, 16*(5), 604–618.
16. Gera, U. K., Siddarth, D., & Singh, P. (2022). Smart farming: Industry 4.0 in agriculture using artificial intelligence. In *Artificial intelligence for societal development and global well-being* (pp. 211–221). IGI Global.
17. Intelligence, U. A. (2022). Smart farming: Industry 4.0 in agriculture. *Artificial Intelligence for Societal Development and Global Well-Being*, 211. https://www.igi-global.com/chapter/smart-farming/309197
18. Muthoni, F., Thierfelder, C., Mudereri, B., Manda, J., Bekunda, M., & Hoeschle-Zeledon, I. (2021, July). Machine learning model accurately predict maize grain yields in conservation agriculture systems in Southern Africa. In *2021 9th International Conference on Agro-Geoinformatics (Agro-Geoinformatics)* (pp. 1–5). IEEE.
19. Katarya, R., Raturi, A., Mehndiratta, A., & Thapper, A. (2020, February). Impact of machine learning techniques in precision agriculture. In *2020 3rd International Conference on Emerging Technologies in Computer Engineering: Machine Learning and Internet of Things (ICETCE)* (pp. 1–6). IEEE.
20. Nagasubramanian, G., Sakthivel, R. K., Patan, R., Sankayya, M., Daneshmand, M., & Gandomi, A. H. (2021). Ensemble classification and IoT-based pattern recognition for crop disease monitoring system. *IEEE Internet of Things Journal, 8*(16), 12847–12854.

Chapter 2

Introduction to Deep Learning

Neethu Narayanan and Arjun K.P.

The process of deep learning, also referred to as "deeper learning," is a subfield of machine learning that focuses on creating and using neural network simulations inspired by the structure and operation of the human cerebral cortex. It is a subset of artificial intelligence (AI) that has historically flourished due to its amazing capacity for learning and making predictions based on the large quantities of data that is currently available.

Learning techniques are designed to eventually differentiate between supervisors and lower-level attributes, enabling them to effortlessly pick up on data with hierarchical representations. These methods' fundamental components are artificially generated neural networks [1], sometimes known as deep cognitive networks. They are created by making up different hyperlinking layers of synthetic nerve cells. A neuron gets input from the surrounding environment and interprets it using mathematical information, followed by passing its findings to the layer of axons beneath it. The network develops the ability to spot patterns and make predictions as a result of the aforementioned procedure.

Among its primary benefits is that it reduces the need for tedious feature engineering by proactively acquiring how to represent features from raw data. It is used in tasks like speech detection, natural language [2] processing, and image recognition.

Deep learning involves training [3] neural networks with labeled datasets to modify internal parameters through optimization algorithms like stochastic gradient descent (SGD), achieving groundbreaking results in various domains but facing challenges such as the need for large labeled datasets and computational resources, as well as overfitting issues. Ongoing research aims to address these challenges and advance the field.

 DOI: 10.1201/9781003504900-2

2.1 Overview of Artificial Intelligence

AI is a big field in computer science that focuses on making smart machines that can do tasks as humans do. The primary objective is to create computer systems that possess the ability to perceive, reason, learn, and make informed decisions, all with the ultimate aim of achieving specific goals. Figure 2.1 shows comprehensive coverage of AI and its subfields.

AI encompasses a diverse range of techniques, algorithms, and methodologies that enable machines to mimic human-like intelligence. These include:

Machine Learning: A specific branch of AI that deals with the advancement of algorithms that can learn patterns from data and make predictions or decisions without the need for explicit programming. This approach allows systems to enhance their performance through experience and adapt to new information efficiently.

Deep Learning: Deep learning uses multilayered artificial neural networks (ANNs) to learn hierarchical data representations, achieving impressive results in image recognition, speech processing, and more.

Natural Language Processing (NLP) [2]**:** This study area is all about making computers understand, interpret, and create human language. It includes tasks like translating languages, analysing feelings, recognizing speech, and making chatbots.

Computer Vision: This field revolves around educating machines to understand visual data, including images and videos, and covers various tasks such as object recognition, image classification [4], image generation, and video analysis.

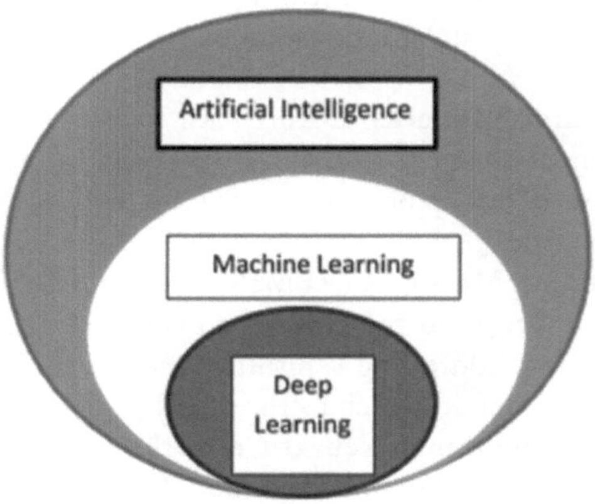

Figure 2.1 Comprehensive coverage of Artificial Intelligence (AI) and its subfields

Robotics: Combines AI with physical systems to create intelligent machines capable of performing physical tasks, perceiving their environment, making decisions based on sensory inputs, and manipulating objects.

Expert Systems: AI systems that utilize knowledge and rules from human experts in specific domains to provide intelligent advice or make decisions, emulating human decision-making abilities.

Reinforcement Learning [5]**:** AI agents learn to make decisions by interacting with an environment, using rewards or penalties to maximize cumulative rewards through trial and error.

AI finds wide-ranging applications across industries such as healthcare, finance, manufacturing, transportation, and gaming, automating tasks, assisting in decision-making, improving efficiency, and enhancing user experiences. However, the advancement of AI also raises ethical and societal concerns, including potential job displacement, privacy issues, algorithmic bias, and impacts on social interactions. Responsible development and deployment of AI systems are essential, considering their potential effects on individuals and society as a whole. As AI progresses, researchers and practitioners strive to create increasingly intelligent and capable machines, pushing the boundaries of AI and exploring new frontiers in the field.

2.2 Evolution of Deep Learning

Deep learning has witnessed significant evolution and advancement over the years, with breakthroughs in algorithms, models, hardware, and data availability. Figure 2.2 depicts the evolution of neural networking.

Deep learning, a subset of machine learning, has experienced significant milestones and developments throughout its evolution:

Emergence of Neural Networks: Dating back to the 1940s, neural networks serve as the foundation of deep learning. The 1980s saw notable progress with the introduction of backpropagation, enabling efficient training of multilayer neural networks [3, 6].

Restricted Boltzmann Machines (RBMs): In the mid-2000s, RBMs played a crucial role in training deep neural networks (DNNs). They allowed for layer-by-layer pretraining, addressing vanishing gradient challenges and facilitating the initialization of deep architectures [7].

ImageNet and Convolutional Neural Networks (CNNs) [4]: In 2012, the ImageNet Large Scale Visual Recognition Challenge (ILSVRC) demonstrated the potential of deep learning. The AlexNet CNN achieved remarkable results, surpassing traditional computer vision techniques and sparking renewed interest in the field.

Neural Networks (1980s)
Restricted Boltzmann Machines and Pretraining (mid-2000s)
ImageNet and CNNs (2012)
Advances in Architectures: VGGNet, GoogLeNet, ResNet
Breakthroughs in NLP: RNNs, LSTM, and Transformers
Generative Models: GANs and VAEs
Transfer Learning and Pretrained Models
Hardware Advancements
Expansion to Various Domains: Speech, Recommendation, Autonomous Driving, Healthcare
Continued Research and Innovation

Figure 2.2 Evolution of neural network

Advancements in Architectures [8]**:** Over time, deep learning architectures have become more sophisticated and effective. Models like VGGNet, GoogLeNet (Inception), and ResNet introduced deeper and more complex structures, leading to improved performance in image classification [4] and computer vision tasks.

Breakthroughs in Natural Language Processing (NLP): Deep learning [9] made big progress in NLP with the introduction of recurrent neural networks (RNNs) and Long Short-Term Memory (LSTM) architecture. It changed tasks like translating languages, recognizing speech, and understanding feelings.

Generative Models: In 2014, new models called generative models like Generative Adversarial Networks (GANs) and Variational Autoencoders (VAEs) [10] came into existence. These models made it possible to create realistic images, text, and data, which opened up new opportunities for generating data.

Transfer Learning [11] **and Pretrained Models:** Leveraging pretraining on vast datasets followed by fine-tuning, models like GPT and BERT achieved state-of-the-art results in NLP tasks, benefiting from extensive text data.

Hardware Advancements: The availability of powerful hardware, such as GPUs and specialized AI accelerators, significantly contributed to deep learning's progress, allowing for faster training and inference times.

Expansion to Various Domains: Deep learning extended its impact beyond computer vision and NLP to domains like speech recognition, recommendation systems, autonomous driving, healthcare diagnostics, and drug discovery, addressing real-world challenges.

Continued Research and Innovation: Deep learning remains a vibrant area of research and innovation, exploring novel architectures [8], optimization techniques [12], regularization methods, and interpretability approaches to enhance model performance and understanding.

The evolution of deep learning has revolutionized AI, achieving remarkable accomplishments and transforming diverse industries. With ongoing advancements and interdisciplinary collaboration, deep learning continues to push the boundaries of AI capabilities.

2.3 Principles of Deep Learning

Deep Learning is a special part of machine learning. It's about teaching ANNs with many layers to understand complex patterns and information from big datasets. This approach has gained prominence due to its ability to handle vast amounts of data and effectively solve complex tasks across diverse domains. Key aspects of Deep Learning encompass its capability to automatically learn hierarchical features and its remarkable success in various applications, such as computer vision, NLP [2], and speech recognition.

Neural Networks: Deep Learning depends on neural networks, which are models inspired by the human brain's structure and how it works. These networks have connected layers of artificial neurons that handle and send information, helping them learn from data and do complicated tasks. The way these layers are arranged lets Deep Learning models get complex patterns and information from the input data. This makes them really good at tasks like recognizing images, understanding natural language, and making decisions.

Deep Neural Networks (DNNs): DNNs are a type of neural network that have many hidden layers between the input and output layers. These hidden layers are very important because they help the network to find more complicated features from the input data. As the data goes through these layers, the network gets better at recognizing more complex patterns and connections in the data. DNNs are very powerful models that can handle difficult tasks in many different areas, like recognizing images, understanding natural language, and other things.

Activation Functions: Activation functions are very important in neural networks because they make the model's computations non-linear. This is necessary for the network to learn and understand complicated relationships in the data.

Some common activation functions are Rectified Linear Unit (ReLU), sigmoid, and hyperbolic tangent (tanh). By using these activation functions, neural networks can find complex patterns in the data and solve different tasks in many different areas.

Backpropagation: Backpropagation is a vital algorithm in training DNNs. It adjusts network weights step by step, reducing errors and improving predictions using feedback from training data. This helps deep learning succeed in various applications.

Convolutional Neural Networks (CNNs): CNNs are specific types of neural networks tailored for image recognition and computer vision tasks. These networks are well-suited for such applications due to their utilization of convolutional layers, which play a crucial role in extracting relevant features from the input images. By employing convolutional layers, CNNs can efficiently process and analyse visual data, making them highly effective in tasks like object detection, image classification [4], and other computer vision-related endeavors.

Recurrent Neural Networks (RNNs): RNNs are designed for sequential data processing, making them suitable for tasks involving time series or natural language.

Deep Learning Frameworks: Frameworks like TensorFlow, PyTorch, and Keras provide APIs, pre-built layers, and optimization algorithms to facilitate Deep Learning model development.

Applications of Deep Learning: Deep Learning has revolutionized domains such as computer vision, NLP, speech recognition, recommendation systems, and autonomous vehicles.

Understanding these foundational concepts is essential for exploring advanced techniques and applications of Deep Learning, enabling the leveraging of DNNs' power to solve complex problems and extract valuable insights from data.

2.3.1 Artificial Neural Networks

ANNs mimic the brain's neuron network to create computational models. These networks form the foundation of machine learning, as they are specifically designed to learn from input data and utilize that knowledge to make predictions or decisions. By simulating the brain's neural connections and processing, ANNs are powerful tools capable of solving complex problems across various domains.

Key aspects of ANNs:

Neurons: Artificial neurons, also called nodes or units, are fundamental in ANNs. They take input signals, perform calculations, and generate output signals.

Layers: Neurons are arranged in layers in ANNs, such as the input layer, hidden layers, and output layer. Each layer has multiple neurons for computations.

Connections and Weights: Neurons in different layers are connected through weighted connections. Each connection has a weight that determines the importance of the signal passing through it.

Activation Function: In an ANN, each neuron typically undergoes a crucial step where it applies an activation function to the weighted sum of its inputs. The introduction of activation functions is vital because it introduces non-linearity to the model. This non-linearity enables ANNs to effectively capture and learn complex relationships present within the data they process. Among the commonly used activation functions are sigmoid, tanh, and ReLU, each playing a vital role in enhancing the network's capacity to handle diverse and intricate data patterns.

Forward Propagation: In forward propagation, data goes through the network from the input layer to the output layer. Neurons calculate their outputs using weighted inputs and activation functions.

Backpropagation: Backpropagation is used to train ANNs. It calculates the error gradient with respect to weights and adjusts them to improve the network.

Learning Algorithms: ANNs can be trained with different algorithms, like gradient descent and SGD. These algorithms update the weights using calculated gradients during backpropagation.

Deep Neural Networks: DNNs are ANNs with many hidden layers. Deep Learning uses DNNs to learn and get complex patterns and information from big datasets.

ANNs find applications in various domains like image and speech recognition, language understanding, pattern recognition, and user recommendations. They are versatile and can learn from data to make predictions, classify objects, or solve difficult problems. Advancements in computing power and training techniques have made ANNs a powerful tool in machine learning and artificial intelligence.

2.3.2 Activation Functions

Deep Learning requires the use of activation functions. To aid neural networks in learning complex patterns and making precise predictions, they incorporate non-linearity. Following are a few typical activation functions and their uses:

- **ReLU (Rectified Linear Unit)** [4]: Employed to introduce non-linearity, ReLU is frequently used in neural networks with hidden layers for computer vision and image classification tasks.
- **Sigmoid:** Sigmoid maps input values to a range between 0 and 1, making it suited for predictions that utilize probability. It is frequently used in binary classification applications.

- **Tanh (Hyperbolic Tangent)** [13]: Similar to sigmoid, tanh distributes output values more evenly by mapping input values to a range between -1 and 1.
- **Softmax:** This algorithm transforms a vector of random real values into a probability distribution [14], and it is mostly employed at the final product layer for problems involving multiple classes.

ReLU (Rectified Linear Unit):

$$\text{Function: } f(x) = max(0, x)$$

Example Application: Utilizing CNNs as an example for image classification. Due to its computational effectiveness and capability to properly handle picture data, ReLU has been routinely incorporated into CNN systems [6].

Sigmoid:

$$\text{Function: } f(x) = 1/\left(1 + e \wedge (-x)\right)$$

Example Application: Binary categorization issues are one example of application. The output layer of a neural network frequently employs the sigmoid activation function to generate probabilities, making it suited for tasks like sentiment analysis or estimating the likelihood of an event occurring [15].

Tanh (Hyperbolic tangent):

$$\text{Function: } f(x) = \left(e \wedge x - e \wedge (-x)\right)/\left(e \wedge x + e \wedge (-x)\right)$$

Example Application: RNNs for the analysis of natural language. Tanh activation function, which may detect values that are both positive and negative, is widely used in RNNs to model sequential input, including text [16].

Softmax:

$$\text{Function: } f(x\_i) = e \wedge (x\_i)/\left(sum\left(e \wedge (x\_j)\right) \textit{ for } j \textit{ in } 1 \textit{ to } n\right)$$

Example Application: Multiclass classification problems [17]. Softmax activation function is typically used in the output layer of a neural network to convert raw scores into class probabilities, allowing the model to make predictions across multiple classes.

Leaky ReLU:

$$\text{Function: } f(x) = max(\alpha x, x),$$

where α is a small positive constant (e.g., 0.01)

Example Application: DNNs with a large number of layers. Leaky ReLU [18] helps mitigate the issue of "dying ReLU" [19] by allowing small negative values, preventing certain neurons from becoming inactive during training.

These are some activation functions used in Deep Learning [20]. The one you choose depends on the problem and network you have. Trying different activation functions is normal to get the best results in applications like recognizing images, understanding natural language, recognizing speech, and more.

2.3.3 Backpropagation Algorithm

Training DNNs involves initializing weights, performing forward and backward propagation, updating weights using an optimization algorithm [21], and evaluating performance on validation data. Backpropagation is crucial as it enables the network to learn from training data and improve predictions iteratively. Hyperparameters are fine-tuned to optimize performance.

2.3.4 Training Deep Neural Networks (DNNs)

It involves optimizing the network's parameters [22], such as weights and biases, using training data to minimize loss or error. The steps for training DNNs are as follows:

Data Preparation: Preprocess the training data, including normalization, scaling, and data augmentation, to make it suitable for training. Split the data into training and validation sets to monitor the network's performance during training.

Network Architecture: Describe the DNN structure [23], including layer count, layer types, and neurons per layer. Select appropriate activation functions for each layer based on the problem and network needs.

Initialization: Initialize the weights and biases of the network using random initialization or specific techniques like Xavier or He initialization.

Loss Function: Select an appropriate loss function based on the problem being solved, such as Mean Squared Error (MSE) [24] for regression tasks or Cross-Entropy Loss for classification tasks.

Optimization Algorithm: Choose an optimization algorithm like SGD, Adam, or RMSprop to update the network's parameters based on computed gradients. Configure hyperparameters like learning rate, momentum, and others for the optimization algorithm.

Mini-Batch Training: Divide the training data into mini-batches to train the network iteratively. Feed a mini-batch through the network, compute the loss, and backpropagate the gradients to update the parameters. Repeat this process for each mini-batch in the training set.

Iterative Training: Repeat the mini-batch training process [25] for a certain number of iterations or epochs. An epoch refers to a complete pass through the entire training data. Monitor the loss and accuracy on the validation set to check for overfitting or underfitting.

Hyperparameter Tuning: Modify network hyperparameters like learning rate, batch size, and regularization techniques (e.g., dropout, L1/L2 regularization) based on validation results. Fine-tune the hyperparameters to improve the network's generalization and performance.

Evaluation [26]: Assess the trained network on a different test set to understand its performance on new data. Calculate metrics like accuracy, precision, recall, or F1 score to evaluate network performance.

Repeat and Refine: If the network's performance is unsatisfactory, iterate through the steps, making changes to the architecture, hyperparameters, or data preprocessing techniques. Continue refining the network until the desired performance is achieved.

Training DNNs [27] requires significant computational resources and time, especially for large-scale networks and complex problems. It often involves experimenting with different architectures, activation functions, optimization algorithms, and hyperparameters to find the best configuration for the given task. Additionally, regularization techniques, early stopping, and model checkpoints can be employed to improve training efficiency and prevent overfitting [28].

2.3.5 Regularization Techniques Regularization

Techniques are essential in Deep Learning to prevent overfitting, in some cases, the model works well on the training data but struggles to adapt to new, unseen data. Here are some commonly used regularization techniques in Deep Learning:

L1 and L2 Regularization (Weight Decay): L1 regularization adds a penalty to the loss function proportional to the absolute values of the weights. L2 regularization, also known as weight decay, adds a penalty proportional to the squared values of the weights. These techniques encourage the model to use smaller weights, reducing the complexity of the model and preventing overfitting.

Dropout: Dropout randomly sets a fraction of the neurons in a layer to zero during training. This technique forces the network to learn more robust and redundant features by preventing the co-adaptation of neurons. Dropout acts as a form of ensemble learning and helps in regularizing the model by reducing over-reliance on specific neurons.

Batch Normalization: Batch Normalization [29] normalizes layer inputs by subtracting the batch mean and dividing by the batch standard deviation. It helps in stabilizing the network's training process, allowing for higher learning rates and faster convergence. Batch Normalization also acts as a regularizer, reducing the effect of internal covariate shift and reducing the need for excessive regularization techniques.

Early Stopping: Early Stopping is a technique where training is stopped early based on the performance on a validation set. It monitors the validation loss or accuracy and stops training when the performance starts deteriorating. Early Stopping helps prevent overfitting by finding the optimal point where the model generalizes well without sacrificing performance.

Data Augmentation [30]: Data Augmentation creates new training samples by applying random transformations to existing data. Techniques like rotations, translations, flips, and zooming increase training set size and diversity, helping the model generalize better by reducing overfitting.

Dropout Regularization [31]: Dropout Regularization combines the dropout technique with weight scaling during training. It helps in preventing co-adaptation of neurons and reducing overfitting by randomly dropping out neurons and scaling the remaining weights.

These regularization techniques are commonly used in Deep Learning to improve model generalization and prevent overfitting. The choice of regularization techniques depends on the specific problem, network architecture, and the amount of available data. By applying appropriate regularization techniques, models can achieve better performance on unseen data and improve their ability to generalize well.

2.3.6 Optimization Algorithms Optimization Algorithms

It plays a crucial role in training DNNs by iteratively updating the network's parameters to minimize the loss function. Here are some commonly used optimization algorithms in Deep Learning:

Gradient Descent (GD) is a crucial optimization technique widely employed in Deep Learning. Its primary purpose is to fine-tune the parameters (weights and biases) of the network by iteratively adjusting them in the direction opposite to the gradients of the loss function concerning those parameters. This process aids in minimizing the loss function, guiding the model toward the optimal solution. The learning rate plays a significant role in GD [32], as it determines the magnitude of parameter updates in each iteration. By appropriately setting the learning rate, one can control the convergence speed and ensure efficient and accurate convergence toward the best possible solution.

SGD [33] represents a variant of the standard Gradient Descent algorithm. In SGD, the model's parameters are updated by computing the gradients of the loss function on a randomly selected subset (mini-batch) of the training data, rather than using the entire dataset. This approach significantly reduces the computational burden, accelerating the convergence process. Although SGD introduces more noise into the parameter updates due to the random mini-batch selection, this characteristic allows it to escape shallow local minima and enhances the model's ability to generalize well on unseen data. As a result, SGD is a popular optimization choice in training deep learning models, providing a balance between computational efficiency and improved performance.

Mini-Batch Gradient Descent balances between Gradient Descent and SGD. It divides training data into smaller batches and updates the model parameters using gradients computed from these mini-batches. By doing so, it combines the benefits of both Gradient Descent, which uses the entire dataset for each update but can be computationally expensive, and SGD, which updates the model after processing each individual data point, introducing high variance. Mini-Batch Gradient Descent offers a more efficient and stable optimization approach, making it widely used in training deep learning models.

2.3.6.1 Example Application: Image Classification

CNNs can be applied to categorize images into predetermined groups, which is the aim of the undertaking.

2.3.6.2 Algorithm Steps in Training a CNN

- Data Preprocessing,
- Obtain a labeled dataset of images for training and testing,
- Preprocess the images by resizing them to a consistent size,
- Normalizing pixel values and applying data augmentation techniques like rotation
- Flipping to increase the training data's diversity.

2.4 Architecture Design of Deep Learning Framework

Create the CNN architecture by deciding the number and layout of convolutional layers, pooling layers, and fully connected layers. Decide on the activation functions, kernel sizes, and the number of filters for each convolutional layer.

2.4.1 Convolution Neural Network (CNN)

To improve a model's generalization ability, regularization techniques like dropout and batch normalization are used. The steps involve forward propagation, classification, loss calculation, backpropagation and parameter updates, training loop, evaluation, and prediction. CNNs can accurately classify images with proper design, data, and hyperparameter tuning.

2.4.2 Recurrent Neural Networks (RNNs)

RNNs are designed for sequential data like text, speech, and time series. Unlike regular neural networks, RNNs have recurrent connections, allowing them to remember and understand relationships across time steps. This unique architecture enables RNNs to grasp temporal dependencies and patterns in the data they process. RNNs find valuable applications in tasks like language modeling, sentiment analysis, and machine translation in NLP. During RNN training, the algorithm iterates through the sequential data, updates the network's parameters, and optimizes it to predict the next element in the sequence accurately based on the previous context.

Example Application: Language Modeling

RNNs can be used for language modeling, where the goal is to predict the next word in a sequence of words given the previous context.

2.4.2.1 Algorithm Steps in Training an RNN

- Data Preprocessing:
  Obtain a large text dataset.
  Convert text into numerical representations.
  Split the dataset into training and validation sets.
- Architecture Design:
  Choose the RNN cell type (e.g., Vanilla RNN, LSTM, GRU).
  Decide the number of hidden layers and their sizes.
  Select the activation function for the RNN cell.
- Forward Propagation:
  Initialize the hidden state.
  Feed the input sequence [14] token by token, updating the hidden state.
  The hidden state captures sequential dependencies.
- Prediction and Loss Calculation:
  Predict the next word's probability distribution for each time step.
  Calculate the loss using Cross-Entropy Loss.
- Backpropagation Through Time (BPTT):
  Unroll the RNN for a fixed number of time steps.
  Perform BPTT to compute gradients.
  Update parameters using an optimization algorithm like Gradient Descent or Adam.
- Training Loop:
  Iterate through the training dataset, updating parameters using BPTT for each sequence.
  Repeat for a fixed number of epochs or until convergence.
- Evaluation:
  Evaluate the trained RNN on a validation dataset.
  Measure metrics like perplexity or accuracy for performance assessment.
  Adjust the model or hyperparameters based on the evaluation.

- Generation:
  Use the trained RNN to generate new text by providing an initial seed. Sample from predicted probabilities iteratively to generate a sequence.

Training an RNN for language modeling allows it to generate coherent text. Model performance depends on the architecture, data quality, hyperparameters, and the optimization algorithm used.

2.4.3 Generative Models: Variational Autoencoders (VAEs) and Generative Adversarial Networks (GANs)

The technique of deep learning uses generative models, such as VAEs [10] and adaptive adversarial networks (GANs) [34], extensively to generate fresh data sets that resemble the training data. These models work by identifying underlying patterns in the incoming data and using this data to produce fresh, believable examples. In contrast to adaptive adversarial networks, which employ a competitive arrangement between a generator and a tool for discrimination, VAEs use an encoder-decoder layout. Practitioners, as well as researchers, can produce synthetic data useful for a variety of applications, including data supplementation, the creation of original content, or the assembly of knowledge for particular tasks by utilizing these modeling techniques

2.4.3.1 Variational Autoencoders (VAEs)

Two well-liked generative models with different strategies for producing fresh data samples are GANs and VAEs.

In order to develop a meaningful representation of input data in a lower-dimensional latent space, VAEs combine autoencoder and probabilistic modeling techniques. They are made up of an encoder network that converts the input data into a latent space that is symbolized by mean and variance. The original input data is then recreated by the decoder network using samples from the latent space. A loss function with a reconstruction loss and a regularization term based on the Kullback-Leibler (KL) divergence is optimized using VAEs during training. VAEs can produce fresh samples while preserving the consistency of the data by sampling from the learned latent space and decoding the representations. In contrast, GANs involve two neural networks: the generator and the discriminator. The generator generates new samples, such as images or text, using either random noise or a latent space as input. The discriminator's task is to differentiate between real training data samples and those produced by the generator. During training, the generator and discriminator engage in a competitive process, where the generator aims to improve its generated samples to deceive the discriminator, while the discriminator learns to better distinguish between real and generated samples. The iterative training process uses a min-max objective function, where the generator minimizes the discriminator's ability to differentiate, and the discriminator maximizes its discriminative power.

Both VAEs and GANs find applications in generative modeling tasks in various domains. VAEs excel in image generation, text generation, and anomaly detection by encoding data into a meaningful latent space and decoding it to reconstruct original data. On the other hand, GANs are widely used for image synthesis, style transfer, data augmentation, and generating realistic images from random noise.

The selection between VAEs and GANs depends on specific task requirements and the desired qualities of the generated data. For image generation, VAEs offer more control over the generated images and data continuity, while GANs provide a higher level of realism and diversity in the generated samples.

In conclusion, both VAEs and GANs have significantly contributed to the field of generative modeling and offer unique strengths for different use cases. Their diverse applications continue to advance the capabilities of generative models in creating high-quality and lifelike data samples.

2.4.4 Transformer Model

Transformer models have had a profound influence not only in NLP but also in various other domains of deep learning. Below are examples of how transformer models are applied in different areas:

2.4.5 Natural Language Processing (NLP)

Transformer models have achieved significant success in various applications, including machine translation, text generation, named entity recognition (NER), and sentiment analysis. They excel in handling long-range dependencies and capturing contextual information, leading to improved performance in these tasks. Transformers [35] are widely used for tasks like language modeling, generating news articles, code writing, dialogue systems, NER, and sentiment classification in social media posts and reviews.

2.5 Computer Vision

Image Recognition: Transformers [35] can be utilized in image recognition tasks by treating images as sequences of patches or regions. This approach allows the models to focus on relevant regions and capture spatial dependencies, resulting in improved image classification and object recognition accuracy.

Image Generation: Transformers [35] have been successfully applied to image generation tasks, where they can generate realistic and high-quality images from textual descriptions or generate image captions based on a given input. These models excel in producing visually coherent and contextually relevant images.

2.5.1 Reinforcement Learning

Reinforcement Learning: Transformers [35] have been integrated into reinforcement learning frameworks to handle high-dimensional and sequential data in reinforcement learning tasks. The models can learn to make optimal decisions in dynamic environments, such as playing games or controlling robots.

Transformer models have shown their adaptability and effectiveness in multiple areas of deep learning, displaying their capacity to capture intricate patterns and relationships in data. Their self-attention mechanism and parallel processing make them especially suitable for dealing with long-range dependencies and capturing global context, leading to their success in diverse applications.

2.5.2 Deep Reinforcement Learning

Deep Reinforcement Learning (DRL) is a distinct and specialized field within the realm of deep learning. It merges reinforcement learning (RL) methodologies with DNNs, empowering agents to acquire complex behaviors and make informed decisions in dynamic and ever-changing environments. In essence, DRL provides a powerful framework for training intelligent agents that can learn from their experiences, optimize their actions, and adapt to various scenarios effectively. Here's a concise overview of DRL.

RL is a fundamental learning technique wherein an agent engages with an environment, acquiring optimal actions through feedback in the form of rewards or penalties. By engaging in trial-and-error, the agent takes actions, receives corresponding rewards, and adapts its policy to maximize the cumulative rewards achieved over time. This iterative process allows the agent to learn from experience and develop strategies that lead to more favorable outcomes within the given environment.

2.5.3 Deep Q-Network (DQN)

DQN is a fundamental DRL algorithm that merges Q-learning with DNNs.

Using a DNN known as the Q-network, DQN estimates Q-values for different actions given the current state.

Training involves minimizing the difference between predicted and target Q-values, leveraging techniques like the Bellman equation.

DQN is well-suited for handling high-dimensional input spaces, like images, using CNNs.

2.5.4 Policy Gradient Methods

Policy Gradient methods [34] directly learn a policy function mapping states to actions without explicitly estimating Q-values.

These methods update the policy parameters via gradient ascent based on expected cumulative rewards.

The policy is usually represented by a DNN (policy network) and updates employ algorithms like REINFORCE.

2.5.5 Proximal Policy Optimization (PPO)

PPO is a popular DRL algorithm that combines value-based and policy-based ideas.

It aims to find a policy that maximizes expected rewards while ensuring stable and efficient learning.

PPO employs a surrogate objective function and includes a clipped surrogate objective to prevent large policy updates.

2.5.6 Actor-Critic Methods

Actor-Critic methods benefit from both policy-based and value-based approaches.

They maintain two networks: an actor-network learning the policy and a critic network estimating the value function.

The actor-network provides actions based on the policy, while the critic network offers feedback on state-action pair values.

By leveraging value estimation and policy improvement, these methods achieve more stable and efficient learning.

2.5.7 Applications of DRL

DRL has excelled in game playing, achieving impressive success in complex games like Go, chess, and video games, surpassing human-level performance.

In robotics, DRL has been instrumental in training robots to perform intricate tasks, such as grasping objects, locomotion, and navigation in challenging environments.

Autonomous systems, including vehicles and drones, benefit from DRL in learning decision-making policies for navigation and control.

DRL has demonstrated its prowess in learning complex behaviors and intelligent decision-making across various domains. With the power of DNNs, DRL continues to make significant strides in solving challenging problems and holds great promise for diverse real-world applications.

2.5.8 Deep Learning for Computer Vision

Applying DNNs to analyse visual data, such as photos and videos, in order to extract useful information and reach conclusions is known as "deep learning for computer vision" [13]. In computer vision tasks, CNNs are the essential architecture. They are suitable for applications like image classification, object detection, segmentation, and visual recognition because they learn hierarchical representations from raw pixel data.

2.5.9 Image Classification

Deep learning models are capable of correctly classifying photos into several groups or classes.

On benchmark image classification datasets like ImageNet, models like AlexNet, VGGNet, and ResNet demonstrate amazing precision.

Object detection is the process of locating and recognizing various items present in an image.

The use of deep learning methods has substantially enhanced object detection tasks, enabling uses like self-driving automobiles and image search. Examples include Faster R-CNN, YOLO, and SSD [36].

2.5.10 Image Segmentation

To precisely localize objects, image segmentation seeks to give labels to image pixels.

Modern efficiency for image segmentation is achieved using deep learning models like U-Net, Mask R-CNN, and DeepLab, and these models are advantageous for medical image analysis and video comprehension.

Advanced generative models have been utilized for tasks such as image synthesis, super-resolution, along with inpainting. Examples include VAEs and GANs.

These models can create brand-new photos that closely resemble the training data or enhance the quality of current images.

2.5.10.1 Transfer Learning and Pre-trained Models

Transfer learning entails using pre-trained deep learning models, trained on extensive datasets, to improve performance on specific computer vision tasks with limited training data.

Models like ResNet, Inception, and MobileNet are often used as feature extractors or fine-tuned for specific tasks.

Deep learning has pushed the boundaries of computer vision, achieving remarkable results in diverse applications such as autonomous vehicles, robotics, medical imaging, surveillance, and more. The process of image classification using deep learning includes important stages such as data collection and preparation, model selection, model training, hyperparameter tuning, model evaluation, and model deployment. These steps may vary depending on the dataset, chosen model, and problem domain. Deep learning frameworks like TensorFlow and PyTorch offer libraries and tools that facilitate the implementation of these steps.

2.5.10.2 Object Detection and Localization

Object detection and localization in computer vision involve using deep learning, specifically CNNs, to identify and locate objects in images or videos.

The process includes dataset preparation, selecting a suitable model architecture (e.g., Faster R-CNN, YOLO, and SSD), training the model, post-processing to refine predictions, and evaluating its performance. Deep learning has greatly improved the accuracy and efficiency of object detection, with applications in autonomous driving, surveillance, augmented reality, and more.

These have numerous applications, including autonomous driving, surveillance systems, object tracking, augmented reality, and image search. Deep learning has significantly improved the accuracy and efficiency of these tasks, enabling computers to detect and localize objects in complex visual scenes with high precision.

2.5.11 Face Recognition

Face recognition is an advanced computer vision technology that uses deep learning, specifically CNNs, to identify and authenticate individuals based on their facial characteristics. The process involves collecting a diverse dataset of labeled face images, preprocessing them for consistency, and training a CNN-based model with techniques like triplet loss. The trained model extracts high-dimensional feature representations (face embeddings) from images, enabling face verification and identification tasks. Face recognition has various applications, including identity verification, access control, surveillance, and personalized services, thanks to the impressive accuracy and robustness of deep learning-based models.

Face recognition has diverse applications, including identity verification, access control, surveillance systems, people counting, and personalized services. Deep learning-based face recognition models have achieved impressive accuracy and robustness, making them valuable tools in various industries and sectors.

2.5.11.1 Image Generation and Style Transfer

The creation and alteration of images based on desired styles or patterns are made possible by the exciting deep-learning applications of image generation and style transfer. An overview of deep learning-based image production and style transfer is provided below:

2.5.12 Image Production

For the creation of images, deep generative models like VAEs and GANs are frequently employed.

By selecting samples from this learned space, VAEs can create new samples by learning a latent space representation of the images.

A discriminator network separates between real and created images in GANs, which are made up of a generator network that learns to create realistic images.

Style Translation:

Style transfer entails merging the content of one image with the style of another image to generate a novel image.

Deep learning techniques like Neural Style Transfer use pre-trained CNNs, such as VGGNet, to extract content and style features from input images.

The content features capture the objects and structure of the content image, while the style features represent the textures, colors, and patterns of the style image.

By optimizing the input image's pixels, the content features are preserved while adopting the style features.

Training and Transfer:

Deep generative models are trained on a sizable dataset of photos to generate new images that are similar to the distribution of the training data.

In order to ensure that the generated images resemble the original input images, VAEs optimize a reconstruction loss during training.

By recognizing its created images as actual, the network of generators tries to trick the discriminator into believing it during adversarial training of the generator and discriminant networks in GANs.

Deep learning has significantly advanced image generation and style transfer, providing powerful tools for creative expression and image manipulation. The development of deep generative models and style transfer algorithms has opened up exciting possibilities in the field of computer vision and digital art.

2.6 Deep Learning for Natural Language Processing

2.6.1 Text Classification and Sentiment Analysis [37]

NLP tasks that require classifying text documents into predetermined categories or figuring out the sentiment conveyed in the text are known as text classification [38] and sentiment analysis. Deep learning models performed remarkably well on these tests. Word embeddings, text preprocessing, model selection, training, evaluation, and inference are all steps in the process. For word embeddings, methods such as Word2Vec, GloVe, or FastText are used, and models such as RNNs, LSTMs, CNNs, or transformer-based models like BERT are frequently used. As a result, the trained model can categorize fresh text documents or assess sentiment instantly, which makes these tasks extremely valuable for a variety of applications.

Text classification [38] and sentiment analysis have extensive uses, such as document categorization, spam detection, sentiment monitoring, customer feedback analysis, and social media sentiment analysis. Deep learning models have significantly enhanced the accuracy and efficiency of these tasks, allowing for better comprehension and analysis of textual data.

2.6.2 Language Translation

Language translation in deep learning involves using neural machine translation (NMT) models trained on parallel text data. The process includes data collection, preprocessing, and dividing the data into training and validation sets. Model architecture selection is crucial, with options like RNNs and Transformers widely used for NMT tasks. Training the model involves minimizing a loss function to improve translation accuracy. Hyperparameter tuning is performed to optimize model performance. Evaluation of a test set is essential to assess translation quality and generalization. After training, the model can be used for language translation [39] by processing the input through the network. Iterative refinement may be done to improve the model continuously.

2.6.3 Named Entity Recognition

NER in NLP involves identifying and categorizing named entities in text, such as names of people, places, organizations, and dates. NER has greatly benefited from deep learning [13] approaches using neural network designs. The process includes data collection with annotated named entities, cleaning and preprocessing the text data, selecting a suitable neural network architecture (e.g., RNNs or Transformers), representing input text as vectors, and assigning numerical labels to entity classes for training and evaluation. The model is trained to predict named entity labels using word representations and minimizing a loss function. Hyperparameter tuning optimizes model performance, and evaluation on a separate dataset measures accuracy using metrics like precision and recall. After training, the model can perform NER on a new text, generating accurately predicted entity labels. Continuous improvement can be achieved through iterative refinement, incorporating diverse data, and employing advanced techniques like BiLSTMs, CRFs, or BERT models based on the specific use case and available resources.

2.6.4 Text Generation

Text generation, also known as language generation or natural language generation (NLG), involves producing text from a machine or artificial intelligence system. Deep learning approaches, including RNNs and transformers, are commonly used for text generation tasks. The process includes data collection, cleansing, tokenization, selecting the right deep learning architecture [40] (e.g., RNNs or Transformers), and preparing the training data. The model is then trained to generate text based on learned patterns and structures from the data. Hyperparameter tuning and evaluation ensure optimal performance. After training, the model can generate text based on a prompt or input. Continuous improvement can be achieved through iterative refinement. Post-processing and ethical considerations are important for producing coherent, meaningful, and contextually appropriate text.

It's important to note that generating coherent, meaningful, and contextually appropriate text is a challenging task, and the generated output may still require post-processing or filtering to ensure quality. Additionally, ethical considerations, such as avoiding biased or inappropriate content, should be taken into account during the next-generation process.

2.7 Application of Deep Learning Framework Applied for Healthcare

Deep learning has numerous applications in healthcare, leveraging its ability to process complex data and extract meaningful patterns. Some key applications of deep learning [13] in healthcare include:

2.7.1 Medical Image Analysis and Diagnosis

Deep learning has made significant advancements in analysing medical images like X-rays, CT scans, and MRI scans. It enables automated image interpretation, aiding in disease diagnosis and treatment decisions. The process involves data collection, preprocessing, model selection, training, and clinical validation.

2.7.2 Disease Prediction and Prognosis

Deep learning models analyse patient data to predict the likelihood of developing diseases or assess the progression of existing conditions. By considering factors like genetics, lifestyle, and clinical history, these models offer personalized insights to guide treatment decisions and improve patient outcomes.

2.7.3 Drug Discovery and Personalized Medicine

Deep learning accelerates drug discovery by predicting drug-target interactions, designing novel drug compounds, and optimizing existing drugs. It also enables personalized treatment predictions based on patient-specific factors, leading to targeted therapies and better treatment responses.

2.7.4 Electronic Health Records (EHR) Analysis

Deep learning helps extract insights from EHR data, such as patient demographics, medical history, diagnoses, and medications. It enables descriptive analysis, predictive modeling, and clinical research, aiding in health policy decisions, quality improvement, and personalized patient care.

2.8 Autonomous Vehicles and Self-Driving Cars

Deep learning plays a crucial role in the development of autonomous vehicles. It facilitates perception and sensing through various sensors, decision-making, planning for safe navigation, and control and actuation for precise movements. Safety, redundancy, and legal considerations are paramount in autonomous vehicle deployment.

While these applications offer immense potential for improving healthcare, responsible use, clinical validation, and adherence to privacy regulations are essential to ensure patient safety and the ethical deployment of AI in healthcare. As technology continues to advance, the integration of deep learning in healthcare is expected to revolutionize patient care and medical research.

2.8.1 Robotics and Manipulation

Robotics and manipulation involve designing and controlling robots to interact with the physical world. Key aspects include robotic manipulators with various end-effectors, perception using sensors and computer vision, planning for grasping and manipulation, control systems, object recognition, human-robot interaction, diverse applications, and challenges like uncertainty and ethical considerations.

2.8.2 Unmanned Aerial Vehicles (UAVs)

UAVs, or drones, are aircraft operated without human pilots. Some key aspects are as follows:

Components and Configurations: UAVs consist of airframes, propulsion systems, sensors, onboard computers, and communication systems. They can be fixed-wing or multirotor designs.

Remote Control and Autonomy: UAVs can be remotely controlled or operate autonomously based on pre-programmed instructions or intelligent algorithms.

Applications: UAVs are used in aerial photography, agriculture, infrastructure inspection, disaster management, delivery services, and more.

Flight Control and Navigation: Control systems manage UAV stability, maneuverability, and navigation. Advanced techniques include GPS-based navigation and obstacle avoidance.

Safety and Regulations: Regulations govern UAV operations to ensure safety and privacy compliance.

Technological Advancements: UAV technology advances in battery life, miniaturization, communication, and intelligent algorithms.

Challenges: Challenges include airspace integration, collision risk, communication reliability, and public acceptance.

2.8.3 Intelligent Assistants and Smart Devices

Intelligent assistants and smart devices use AI and machine learning for convenience and automation. Some key aspects are as follows:

Intelligent Assistants: Virtual assistants use NLP and machine learning to understand and respond to user queries, perform tasks, and control smart devices.
Voice User Interface: Voice commands enable users to interact with intelligent assistants through natural language.
Smart Devices and IoT: Everyday objects with internet connectivity and AI capabilities can automate tasks and communicate with other devices.
Home Automation and Smart Homes: Intelligent assistants enable control and automation of home appliances and systems for convenience.
Personalization and Context Awareness: AI enables personalized responses and context-aware interactions.
Privacy and Data Security: Privacy regulations and security measures safeguard user data.
Expansion of Capabilities: Integration with third-party applications extends the functionality of intelligent assistants and smart devices.

Intelligent assistants and smart devices offer convenience and automation, with the potential for further advancements in AI integration and seamless user experiences.

2.9 Challenges and Future Directions in Deep Learning

Deep learning has made remarkable progress in various domains, but it faces critical challenges and holds potential for future developments. Some key aspects are as follows::

2.9.1 Model Interpretability and Explainability

Model interpretability and explainability present challenges due to the complexity of deep learning models, operating as black boxes, making it difficult to trace their decision-making process. Striking a balance between interpretability and performance, lacking standardized evaluation metrics, scalability, and domain-specific interpretability are some hurdles. Additionally, addressing the ethical concerns of unbiased and fair explanations is essential.

2.9.2 Ethical Considerations in Deep Learning

Ethical considerations in deep learning are crucial to ensure responsible AI deployment. Key concerns include biases and fairness, transparency and explainability, privacy and data protection, accountability and responsibility, dual-use dilemma, equity and access, social impact, and human-AI collaboration. Collaboration among stakeholders is necessary to address these ethical challenges and foster responsible AI development and use.

Addressing these challenges and ethical concerns requires interdisciplinary research, responsible AI practices, and regulatory frameworks to promote trust and accountability in the application of deep learning technologies.

2.9.3 Adversarial Attacks and Robustness

Adversarial attacks and robustness are significant challenges in deep learning, vital to ensuring the reliability and security of AI systems. Here are three key aspects related to these challenges:

Adversarial Attacks: Adversarial attacks involve manipulating input data to mislead deep learning models. These attacks exploit model vulnerabilities by introducing imperceptible perturbations that lead to incorrect predictions. Examples of such methods include the Fast Gradient Sign Method (FGSM), Projected Gradient Descent (PGD), and Carlini and Wagner attacks. The transferability property allows attacks crafted for one model to fool others, raising concerns about model robustness.

Robustness and Defense Mechanisms: Robustness refers to a model's ability to maintain performance in the presence of adversarial examples. Defense mechanisms include adversarial training, where models learn from adversarial examples, and certified defenses, providing provable guarantees of robustness. Gradient masking obscures gradients to hinder attacks but may have limitations.

Transferability and Generalization of Attacks: Adversarial attacks generalize across models and datasets, posing challenges to model robustness. Universal adversarial perturbations are input-agnostic perturbations that cause misclassification across various inputs.

Addressing these challenges requires a combination of techniques like adversarial training and certified defenses. Research efforts aim to develop more robust deep learning models capable of withstanding adversarial attacks in real-world scenarios.

2.9.4 Continual Learning and Transfer Learning

Continual learning is a method where models learn from a continuous flow of data over time. The challenge is to avoid catastrophic forgetting, where learning new data leads to forgetting previous knowledge. Techniques like regularization, replay, task-specific architectures, and memory-based approaches help preserve old knowledge while learning new tasks.

On the other hand, transfer learning leverages knowledge from one task or domain to improve performance on another. Models are pretrained on related tasks or datasets and then fine-tuned for the target task. It helps in situations with limited data, enabling efficient problem-solving and superior performance. Domain adaptation is a subset of transfer learning that aligns representations between different domains. Both continual learning and transfer learning are vital in machine learning for adaptation, knowledge retention, and generalization across tasks or domains.

2.10 Conclusion

2.10.1 Summary of Deep Learning Algorithms and Applications

Deep learning algorithms form a specialized branch within machine learning, drawing inspiration from the intricate workings of the human brain. These algorithms employ ANNs to acquire hierarchical representations of data. By emulating the brain's architecture and learning mechanisms, deep learning models are capable of automatically extracting and comprehending complex patterns and features from the input data. This ability makes deep learning a powerful tool for solving intricate tasks across various domains, such as image recognition, NLP, and many others. Common algorithms include:

Feedforward Neural Networks (FNN): Used for image classification, speech recognition, and language processing.
Convolutional Neural Networks (CNN): Ideal for image-related tasks like classification [37] and object detection.
RNNs: Suitable for sequential data, like language modeling and sentiment analysis.
LSTM Networks: A type of RNN used for text generation and time series analysis.
GANs: Used in image synthesis and data augmentation.
Reinforcement Learning (RL): Used in game playing, robotics, and autonomous driving.

Applications of deep learning include computer vision, NLP, speech processing, healthcare, autonomous systems, and finance. They have transformed image recognition, language translation [39], and medical diagnosis, among others.

2.10.2 Future Trends and Directions in Deep Learning

Explainable Complexity and Interpretability: Deep learning models are often perceived as "black boxes" due to their complex and intricate nature, making it challenging to understand the decision-making process within them. This lack of transparency and interpretability poses a significant concern, as it becomes difficult to grasp the underlying reasons behind the model's choices or predictions. As a result, there is a growing need for developing methods and techniques to make deep learning models [9] more explainable and interpretable, enabling users to gain insights into the factors influencing their outputs and fostering trust in their use across various applications. Research has been investigating techniques that will enhance the interpretability and explicability of models using deep learning, enabling consumers to comprehend why the algorithm made a particular forecast or its conclusion.

Applying Transfer Learning and Pretrained Models: Transfer learning is the procedure of adopting the comprehension obtained from prior a deep learning algorithm on a big dataset to a different task or area with little labeled data. Pretrained neural networks have demonstrated astounding performance in a variety of NLP applications, including OpenAI's GPT (Generative Pretrained Processor).

Procedural Modeling: Generative theories, such as variable autoencoders (VAEs) and generative opponent networks (GANs), have drawn a lot of interest. The above instances of Continual Learning and Lifelong Learning: Traditional deep learning models require large amounts of labeled data for training. Continual learning aims to develop models that can learn from a stream of data over time without forgetting previously learned information. Lifelong learning extends this idea further by enabling models to learn and accumulate knowledge throughout their operational lifespan.

Innovative and Efficient Systems: Models using deep learning need to be deployed on devices that are resource-limited like handheld devices, Internet of Things (IoT) equipment, and servers at the edge as they become progressively complicated and resource-intensive. A significant topic is the creation of powerful deep-learning algorithms that can function with restricted computer power.

Multidisciplinary Neural: Techniques based on deep learning are used to mix information from several modalities, including text, images, videos, and audio. Medical Care, human-computer, and audiovisual analysis are among the many of the areas that can benefit from learning in multiple formats.

Ethical Considerations: As the use of deep learning has spread across a variety of businesses, prejudice, equality, anonymity, and openness are growing as important ethical issues. To guarantee responsible growth, practitioners as well as researchers are constantly tackling these challenges.

References

1. Yang, Menglin, et al. "Hyperbolic graph neural networks: a review of methods and applications." arXiv preprint arXiv:2202.13852 (2022).
2. McCann, Bryan, et al. "The natural language decathlon: Multitask learning as question answering." arXiv preprint arXiv:1806.08730 (2018).

3. Smith, Leslie N. "Cyclical learning rates for training neural networks." Proceedings of the 2017 IEEE winter conference on applications of computer vision (WACV). 2017.
4. Krizhevsky, Alex, Ilya Sutskever, and Geoffrey E. Hinton. "Imagenet classification with deep convolutional neural networks." Advances in neural information processing systems 25.2 (2012). https://doi.org/10.1145/3065386
5. Silver, David, et al. "Mastering chess and shogi by self-play with a general reinforcement learning algorithm." arXiv preprint arXiv:1712.01815 (2017).
6. Vinyals, Oriol, et al. "Show and tell: A neural image caption generator." Proceedings of the IEEE conference on computer vision and pattern recognition. 2015.
7. Feder, Amir, et al. "Causal inference in natural language processing: Estimation, prediction, interpretation and beyond." Transactions of the Association for Computational Linguistics 10 (2022): 1138–1158.
8. Zoph, Barret, et al. "Learning transferable architectures for scalable image recognition." Proceedings of the IEEE conference on computer vision and pattern recognition. 2018.
9. Finn, Chelsea, Ian Goodfellow, and Sergey Levine. "Unsupervised learning for physical interaction through video prediction." Advances in Neural Information Processing Systems 29.8 (2016). https://proceedings.neurips.cc/paper_files/paper/2016/file/d9d4f495e875a2e075a1a4a6e1b9770f-Paper.pdf
10. Liang, Dawen, et al. "Variational autoencoders for collaborative filtering." Proceedings of the 2018 world wide web conference. 2018.
11. Oktay, Ozan, et al. "Attention u-net: Learning where to look for the pancreas." arXiv preprint arXiv:1804.03999 (2018).
12. Raffel, Colin, et al. "Exploring the limits of transfer learning with a unified text-to-text transformer." The Journal of Machine Learning Research 21.1 (2020): 5485–5551.
13. Mnih, Volodymyr, et al. "Playing atari with deep reinforcement learning." arXiv preprint arXiv:1312.5602 (2013).
14. Sutskever, Ilya, Oriol Vinyals, and Quoc V. Le. "Sequence to sequence learning with neural networks." Advances in neural information processing systems 27 (2014). https://www.researchgate.net/publication/326851379_Intrusion_Prediction_with_System-call_Sequence-to-Sequence_Mod
15. Vaswani, Ashish, et al. "Attention is all you need." Advances in Neural Information Processing Systems 30 (2017): 15. https://arxiv.org/abs/1706.03762
16. Devlin, Jacob, et al. "Bert: Pre-training of deep bidirectional transformers for language understanding." arXiv preprint arXiv:1810.04805 (2018).
17. Goodfellow, Ian, et al. "Generative adversarial nets." Advances in Neural Information Processing Systems 27 (2014): 9. https://proceedings.neurips.cc/paper_files/paper/2014/file/5ca3e9b122f61f8f06494c97b1afccf3-Paper.pdf
18. Radford, Alec, Luke Metz, and Soumith Chintala. "Unsupervised representation learning with deep convolutional generative adversarial networks." arXiv preprint arXiv:1511.06434 (2015).
19. He, Kaiming, et al. "Deep residual learning for image recognition." Proceedings of the IEEE conference on computer vision and pattern recognition. 2016.
20. Oord, Aaron van den, et al. "Wavenet: A generative model for raw audio." arXiv preprint arXiv:1609.03499 (2016).
21. Christiano, Paul F., et al. "Deep reinforcement learning from human preferences." Advances in Neural Information Processing systems 30 (2017). https://proceedings.neurips.cc/paper_files/paper/2017/file/d5e2c0adad503c91f91df240d0cd4e49-Paper.pdf

22. Alammar, Jay. "The illustrated transformer." The Illustrated Transformer–Jay Alammar–Visualizing Machine Learning One Concept at a Time 27 (2018). https://jalammar.github.io/illustrated-transformer/
23. He, Kaiming, et al. "Mask R-CNN." Proceedings of the IEEE international conference on computer vision. 2017.
24. Arık, Sercan Ö., et al. "Deep voice: Real-time neural text-to-speech." International conference on machine learning. PMLR, 2017.
25. Chen, Ricky TQ, et al. "Neural ordinary differential equations." Advances in Neural Information Processing Systems 31 (2018). https://arxiv.org/abs/1806.07366
26. Brock, Andrew, Jeff Donahue, and Karen Simonyan. "Large scale GAN training for high fidelity natural image synthesis." arXiv preprint arXiv:1809.11096 (2018).
27. Brown, Tom, et al. "Language models are few-shot learners." Advances in Neural Information Processing Systems 33 (2020): 1877–1901.
28. Chen, Mark, et al. "Generative pretraining from pixels." International conference on machine learning. PMLR, 2020.
29. Wolf, Thomas, et al. "Transformers: State-of-the-art natural language processing." Proceedings of the 2020 conference on empirical methods in natural language processing: system demonstrations. 2020.
30. Liu, Xiao, et al. "Self-supervised learning: Generative or contrastive." IEEE Transactions on Knowledge and Data Engineering 35.1 (2021): 857–876.
31. Kaplan, Jared, et al. "Scaling laws for neural language models." arXiv preprint arXiv:2001.08361 (2020).
32. Berthelot, David, et al. "Understanding and improving interpolation in autoencoders via an adversarial regularizer." arXiv preprint arXiv:1807.07543 (2018).
33. Davison, Joe. "Zero-shot learning in modern nlp." https://joeddav.github.io/blog/2020/05/29/ZSL.html (2020).
34. LeCun, Yann, et al. "Gradient-based learning applied to document recognition." Proceedings of the IEEE 86.11 (1998): 2278–2324.
35. Tunstall, Lewis, Leandro Von Werra, and Thomas Wolf. Natural language processing with transformers. "O'Reilly Media, Inc.", 2022.
36. Chen, Liang-Chieh, et al. "DeepLab: Semantic image segmentation with deep convolutional nets, atrous convolution, and fully connected CRFs." IEEE Transactions on Pattern Analysis and Machine Intelligence 40.4 (2017): 834–848.
37. Qi, Charles R., et al. "Pointnet: Deep learning on point sets for 3d classification and segmentation." Proceedings of the IEEE conference on computer vision and pattern recognition. 2017.
38. Lea, Colin, et al. "Temporal convolutional networks: A unified approach to action segmentation." Computer Vision–ECCV 2016 Workshops: Amsterdam, The Netherlands, October 8-10 and 15-16, 2016, Proceedings, Part III 14. Springer International Publishing, 2016.
39. Huang, Xun, et al. "Multimodal unsupervised image-to-image translation." Proceedings of the European conference on computer vision (ECCV). 2018.
40. Barret, Zoph, and V. Le Quoc. "Neural architecture search with reinforcement learning." International Conference on Learning Representations. 1. 2017.

Chapter 3

Role of AI in the Field of Precision Agriculture

Ayushi Bhavesh Tanna, Raashi Samir Jatakia, and Gajanan Dattatraya Nagare

3.1 Introduction to Precision Agriculture

3.1.1 Concept

One of the oldest and most important sectors is agriculture. For the vast majority of people on earth, it serves as their primary source of sustenance. The global food market currently faces immense challenges, which are predicted to grow drastically over the next decade. With adequate incentives and funds, much could potentially be accomplished immediately with present technologies and expertise. To tackle future concerns, more cutting-edge discoveries would be necessary. [1] Furthermore, agricultural intensification has had a negative environmental impact over the past years. The overall efficiency declines, affecting the natural resources, declining farm incomes, absence of an eco-regional strategy, declining and divided property holdings, agricultural trade liberalization, inadequate job prospects in the non-agricultural sector, and universal climate change have all become significant challenges in agricultural progression and advancement. As a result, the widespread use of more modern technological developments is viewed as one path to increasing agricultural output in the decades ahead. Rather than managing an entire field based on some hypothetical average environment that may not exist anywhere in the field, precision farming identifies site-specific requirements within the field and modifies the management methods accordingly. Obtaining varying yields throughout the landscape is well established by farmers.

DOI: 10.1201/9781003504900-3

The observed differences can be attributed to management strategies, soil qualities, and/or environmental factors.[2]

The level of understanding of the landscape can be challenging to maintain due to the enormous size and changes caused by annual adjustments in lease arrangements in the farm area. PA has the ability to streamline and automate data collection and processing. It enables management decisions to be made and implemented swiftly on tiny portions throughout bigger fields.

PA, also known as precision farming or smart farming, has the prospect to contribute to the broader objective of meeting rising food demand while ensuring primary production sustainability using a more precise and resource-efficient approach to production management. It is a management strategy that accumulates, processes, and analyses temporal, spatial, and individual data and integrates it with other information to support management decisions based on estimated variability in order to improve resource use efficiency, quality, efficiency, revenue, and the long-term viability of agricultural production.[3] PA is an ensemble of technologies that work together for optimal resource utilization. It uses accurate amounts of supplies to achieve greater average yields than traditional growing techniques.

Precision farming is a method that uses site-specific nutrient management, remote sensing, a global information system, a global positioning system, and variable rate application to precisely regulate production intake. Managing agricultural production inputs such as water, seed, fertilizer, and so on in order to maximize output, quality, profit, reduce waste, and become more sustainable. Precision farming aims to increase the accuracy of their applications by matching agricultural inputs and procedures to crop and agro-climatic conditions. Nevertheless, precision farming technology has enabled more than merely crop and fruit growing; farmers engaged in livestock husbandry are now reaping the benefits.[1]

3.1.2 What Is Precision Farming

PA, also known as precision farming, is a method of increasing average yields by using precise amounts of inputs in comparison to traditional cultivation techniques. It is a comprehensive management strategy that focuses on intensive data gathering and processing to guide targeted efforts that increase the efficiency, production, as well as sustainability of agricultural operations by integrating information, technology, and management are all critical components. The idea of crop management for specific sites is centred on recognizing and monitoring temporal as well as spatial variability. Precision crop farming relies heavily on the "sensing" component of the idea.[4]

Sensors installed in fields and crops are beginning to offer precise data points on soil conditions, as well as accurate data on climate, fertilizer specifications, water availability, and outbreaks of insects. Aerial photos obtained by non-aviation vehicles, humans, or drones can be used to monitor fields, notify farmers of crop maturity or potential challenges, and offer warning of variations from predicted growth rates or quality. Satellites can be utilized as well for precision crop farming, allowing for the identification of significant modifications in the field through the use of satellite photography. [2]

Typical farming processes necessitated the extensive use of agricultural inputs such as automation, pesticides, water, and various other inputs, resulting in severe environmental effects such as contamination of the environment through greenhouse gas emissions. Thus, precision farming is an appealing idea, and its concepts naturally lead to the belief that farming inputs may be used more effectively, resulting in higher profitability and less ecologically problematic output. Precision farming, particularly for small farmers in developing nations, presents the potential for significant yield growth with minimum external input consumption. Precision farming seeks to lower a farmer's expenses by reducing the demand for fertilizer, pesticides, and herbicides.[5] Hence, it is an integrated system designed to optimize production, improve quality, and optimize chemical consumption. Lowering health risks, preserving energy, and safeguarding the environment.[4]

3.2 Importance and Benefits of Precision Agriculture

One pursuing precision farming should be acquainted with the technology tools available in order to successfully acquire and use information. These resources cover hardware, software, and optimal practices.

3.2.1 Methods Used (Software and Sensors)

- Global Positioning System (GPS)

  The Global Positioning System (GPS) is a satellite navigation system that employs a network of 24 satellites to enable users to capture positional data (latitude, longitude, and elevation). These satellites provide signals that enable GPS receivers to compute their position with an accuracy of 100 to 0.01 metres. This information is delivered in real time, which means that it is provided while the vehicle is in motion. Those farmers who possess accurate location information at all times may determine the exact location of field information such as soil type, insect prevalence, weed invasion, water holes, boundaries, and challenges.[6] The technology enables farmers to properly identify field regions based on performance specifications and previous input applications, allowing inputs (seeds, fertilizers, insecticides, herbicides, and irrigation water) to be provided to each specific field. Users are able to return to specific sites to collect data or remedy specific areas using GPS receivers, which are often carried or placed on equipment.[7]

- Sensor technologies

  To detect humidity, vegetation, temperature, form, framework, physical character, nutrient level, vapour, air, and so on, several technologies such as electromagnetism, electrical conductivity, photoelectricity, and ultrasound are employed. Remote sensing data is used to distinguish crop species, identify stressful situations, locate pests and weeds, and evaluate drought, soil, and plant parameters. Sensors allow vast volumes of data to be collected without the need for testing in laboratories. Sensor technology is an important component of PA technology, and its use to provide information on soil quality and plant fertility/water status is well established. Scanning the field using soil-evident electrical conductivity (ECa) sensors, which capture data continuously as they travel across the field surface, is one of the most frequent ways for assessing soil heterogeneity. These sensors are an excellent starting point for site-specific management since ECa is highly responsive to variations in soil texture and salinity levels.
- Geographic information system (GIS)

  This technology consists of hardware, software, and techniques for assembling, storing, retrieving, and analysing feature attributes and location information in order to build maps. GIS gathers information in a single location so that it may be extrapolated as needed. GIS maps differ from typical maps in that they include a great deal of information (for example, yield, soil survey maps, rainfall, crops, soil nutrient levels, and pests). GIS is a form of digitized map, but its primary role is to analyse traits and geography via the use of statistics and spatial techniques. You may overlay data on field topography, soil types, surface drainage, subsurface drainage, soil testing, irrigation, chemical application rates, and crop yield in an agricultural GIS database.[6] This information, once analysed, is used to understand the relations amongst the numerous components impacting a crop in a given location. Aside from data storage and display, the GIS could also be used to analyse current and potential management scenarios by merging and modifying data layers.
- Grid soil sampling and variable-rate fertilizer (VRT) application

  Variable-rate technologies (VRTs) are automated and can be used in various farming activities. Fertilizers, pesticides, lime, gypsum, irrigation water, and other necessary inputs may be administered at varied rates across a field depending on the soil type identified on a soil map using VRT systems. VRT and variable-rate application (VRA) can range from basic flow rate control to more complicated rate, chemical mix, and application pattern management. GIS-derived information can be used to control operations like as seeding, fertilizer and pesticide implementation, herbicide picking and distribution at a variable rate in the appropriate place at the right time. Grid soil sampling is similar to soil testing but increases the sampling intensity. Soil samples collected in a systematic grid offer location information, allowing data to be easily demonstrated. An application map, which is a map of nutrient needs, is created using grid soil sampling. Samples can be collected from many different parts of a field that have the same yield, the colour of the soil, and so on and hence fall into the same region. Grid soil specimens are tested in the laboratory, and crop nutrient needs for each soil sample are computed. After that, the fertilizer distribution map is constructed using all of the soil samples. The application map is uploaded to a computer that is installed on a variable-rate fertilizer delivery. Using the application map and a GPS receiver, the computer instructs a commodity-delivery controller to change the dosage and/or type of fertilizer substance according to the requirements on the application map.[6]

- Crop management

  Satellite data helps farmers understand how variations in soil circumstances and terrain affect crop production in fields.[7] Farmers may therefore precisely regulate production parameters including seeds, fertilizers, pesticides, herbicides, and water control to improve productivity and efficiency.
- Precision irrigation in pressurized structures

  GPS-based sprinkler irrigation controllers are currently being made for commercial usage. Aside from movement control, wireless transmission and technologies for sensors are being developed for tracking soil and environmental variables, as well as irrigation machine operational parameters (e.g., flow and pressure), in order to improve the distribution of water efficiency and crop utilization.[6]
- Software

  Precision agricultural technologies tend to necessitate the employment of software to perform multiple functions, including display-controller interaction, statistics layer mapping, data analysis and interpretation before and after processing, farm reporting of inputs per field, and many more. The most frequent include software for creating maps (e.g., yield, soil); software for filtering acquired data; software for creating variable rate administration maps (e.g., for fertilizer, chemicals); software for merging various maps; and software for providing expert geostatistical features.[7] There are a few firms that operate globally and provide combined software features for creating various sorts of maps, statistical evaluation tools, and record-keeping. Machinery firms supply yield meters and sells software for creating yield maps, while fertilizer companies sell software for creating variable rate application maps. Some programmes are highly difficult for farmers to operate and rather pricey, while others are simpler and cheaper with lesser possibilities. The products are more user-friendly and provide the farmer with more possibilities. Even so, data sharing amongst farmers, cooperatives, and consultants, remains a challenge. Overlaying maps, primarily soil and yield maps, has proven to be a difficult task thus far.

- Yield monitor

  Yield monitoring is comprised of several parts. They typically include a data collection device, a user interface, and an operating computer installed in the combined box to manage the combination and connection of these components.[7] Grain flow mass or volume (grain inflow sensors), divider speed, surface speed, and grain are all monitored by sensors. Grain flow mass or volume (grain inflow sensors), divider speed, surface speed, and grain are all monitored by sensors. Grain yield monitors measure and record grain flow in real time. Yield data is essential when evaluating the impacts of controlled inputs like as fertilizer, lime, seed, herbicides, and cultural techniques such as tillage and irrigation. The weather may have a major impact on yield numbers from one year.

3.2.2 Parameters Monitored by Sensors

Figure 3.1 depicts the 8 important agricultural parameters monitored by AI, namely, pest protection, decision-making, irrigation, weather forecasting, harvesting, weed protection, intelligent spraying, and soil management.

3.2.3 Tasks for Which Guidance Is Provided

After accumulating the data, the collected data can be used to provide information about:

i. Optimal harvesting time and conditions
ii. Optimal planting times
iii. Crop rotation
iv. Pest and weed protection.
v. Soil management

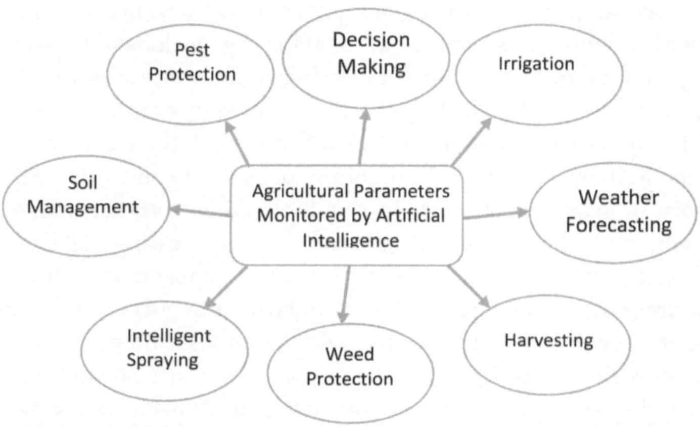

Figure 3.1 Essential parameters monitored by AI.

3.2.4 Improvements Brought by Precision Agriculture

- Fostering a new business in agriculture and related areas that capitalizes on variability can be a very inventive and long-term method to increasing agricultural output and contributing to food security.
- It lowers the rate of erosion, runoff, and sedimentation, that is crucial for the health and long-term viability of marine ecosystems and water supplies. These processes are linked and, if not effectively controlled, can have negative consequences on both the environment and human activity.[8]
- Reduced irrigation water application, nutrient leaching, and deep percolation are essential components of sustainable agriculture and water resource management.
- Reducing chemical doses with Variable Rate Application (VRA) technology is a method of applying chemicals such as fertilizers, insecticides, and herbicides in agricultural activities in a focused and effective manner.
- Reducing cultivation costs through SSCM strategies entails customizing agricultural activities to the unique demands and conditions of various sites within a field. This method optimizes resource consumption while reducing needless costs.
- Improving input productivity through site-specific management involves modifying the utilization of resources like fertilizers, herbicides, water, and energy to meet individual demands and circumstances of different places within a field. This method improves resource usage, eliminates waste, and increases total agricultural output.[9]
- Proper planning of the land can help avoid soil and environmental degradation from the start. It entails zoning and regulating land to prevent the excessive use of fertilizers, pesticides, and water. This not only enhances resource efficiency but also reduces the danger of fertilizer and chemical runoff into bodies of water.[8]
- Reducing nitrogen fertilizer application, which is essential to prevent nitrate pollution in groundwater and nitrous oxide emissions into the atmosphere. Nitrate pollution in groundwater can affect the quality of drinkable water and aquatic ecosystems, and nitrous oxide is a harmful greenhouse gas that contributes to climate change.

3.3 Artificial Intelligence and Machine Learning in Farming

3.3.1 Introduction to Artificial Intelligence and Machine Learning in Agriculture

AI and ML developments have aided robotics, healthcare, ML, and numerous other industries. ML, a branch of AI that focuses on schooling, is a useful strategy that provides better yield prediction based on multiple traits. ML can identify links and trends in datasets and uncover information.[10] Various AI technologies have been proposed to alleviate existing agricultural challenges and increase output. The models must be trained using datasets that describe the outcomes based on previous experience. The predictive model is constructed with numerous features, and as a result, crop yield prediction is one of the most challenging issues in PA, with several models presented and verified thus far. This challenge demands the usage of a large number of datasets since agricultural output is affected by numerous factors such as climate, weather, soil, fertilizer use, and seed type.[11,12] This recommends that crop production prediction is not a straightforward process but rather a series of complex procedures. Crop yield prediction initiatives can now effectively forecast actual yield, although improved yield prediction performance based on historical data is still required. ML decision is an important tool for agricultural yield prediction, including recommendations on which crops to cultivate and what to do during the crop's growth season. Multiple ML techniques have been used to aid agricultural production prediction studies. Another significant AI technique is artificial neural networks (ANN).[11] These models are made up of connected nodes that perform functions similar to those of our human brain.

3.3.2 Process and Working of AI and ML in Agriculture

AI monitors agriculture factors. Agriculture remains to be a very labour-intensive industry, so a labour shortage is expected and is being addressed with self-driving tractors, smart irrigation, spraying, and fertilizing systems, and AI-based harvesting robots. As illustrated in Figure 3.2, AI and ML have made huge amount of progress in the field of farming. Software companies find it tough to help farmers grasp the complete AI system. In agriculture, AI is utilized for field harvesting, health monitoring, pest management, and deficiency detection. ML and AI are being used in agriculture and are replacing archaic forecasting and intelligence approaches.[13] AI makes agriculture more adaptive by introducing cutting-edge

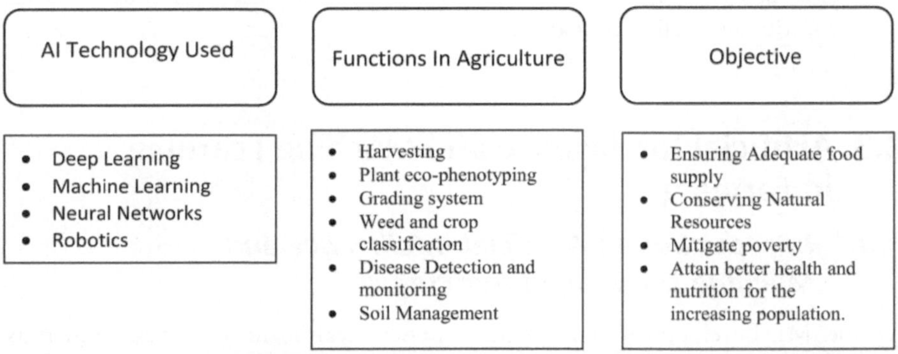

Figure 3.2 Progress of AI and ML in Agriculture.

technologies to the field. Biosensors have even enabled the monitoring of soil moisture and fertility. Instead of employing simple linear regression models, raw data and alternative ways are collected.

Past weather trends with unpredictable patterns may be assessed and predicted using neural networks. As a consequence, for key commodities like rice, wheat, and maize, AI can be used to sow the seeds at the ideal time because they primarily require substantial rains to thrive and are often farmed. Farms generate thousands of data points every day on temperature, soil, water consumption, weather, and other aspects. This data is used in real-time by AI and ML models to gain insightful knowledge, such as when to sow seeds, which crops to choose, which hybrid seeds to choose for greater yields, and so on. AI systems aid PA by increasing harvest accuracy and quality. AI technology aids in the detection of diseases, pests, and nutritional deficiencies in farms. AI sensors can identify and target weeds before determining the best herbicide to use in the area. As a result, herbicide use is reduced, resulting in cost savings.[13]

Several technological firms have developed robots that employ computer vision and AI to correctly monitor and spray weeds. AI has made advances in a wide range of specialties, including businesses, healthcare, architecture, marketing, and finance. Devices based on modern technologies like as AI, ML, and the IoT are being utilized to irrigate, secure, and fertilize crops. Crops must be irrigated at the correct times to obtain the desired yield.[14] Large-scale watering by a handful of farmers became a complicated task due to a lack of resources. Intelligent irrigation devices are now marketed. These systems are composed of two parts: an ML model that estimates the quantity of moisture in the soil and hardware implementation utilizing IoT devices. ML's monitoring systems for the crops and soil, remote sensing, satellite imagery, drones, and precision technologies all benefit from this application. Autonomous robots are being designed to perform traditionally labour-intensive tasks, like as agricultural harvesting, more rapidly and substantially than traditional human labourers. Nano-based sensing mechanisms and smart sensors are being produced and tested for the early and exact detection of diseases, allergies, chemicals, and contaminants in foods, plant and animal production systems, water, and soil.[14]

Farmers can finally make calculated choices in real time, using AI to solve a long-standing issue. Farmers can forecast weather conditions ahead of time, allowing them to select the optimal time to seed their crops. Another issue that AI has dealt with is water use.[14] AI has enabled the assessment of hybrid seeds and their yields prior to planting them in the fields, reducing the possibility of unsuccessful harvests and enhancing output. Modern agriculture technology lets farmers acquire data that they may use to track and enhance their crops. It is also frequently revised with changing environmental and ecological conditions. Emerging agricultural technologies suggest that technical improvements are critical to the agriculture sector today. Crop cultivation can benefit from data acquired from sensors, drones, and satellites. The data acquired may subsequently be analysed using AI in agriculture, helping farmers to make more informed decisions. AI may be used in phenotyping to assess the biomass and features of plants.[13]

AI technology could eventually be developed to detect the particular causes of these illnesses by analysing changes in plant biomass and external influences to uncover similar trends amongst affected crops. The detection of contaminants and pests in crops can help farmers increase their total productivity. This technology can help farmers overcome various difficulties, including climate change, water scarcity, pests, and so on. The adoption of modern agricultural technology helps meet agricultural sustainability goals.[6] AI technology is supporting numerous sectors in enhancing efficiency and output. AI in agriculture is aiding farmers in minimizing negative environmental consequences while enhancing efficiency. Hence, the agriculture industry has actively accepted AI.

3.4 Implementation of Artificial Intelligence in Precision Farming

Nowadays, AI has been rapidly evolving in terms of data management, data procurement, and its ability to mimic human intelligence. These factors make it an explicit choice for implementation in PA or Precision Farming. It can be widely utilized for the development and upgradation of remote sensing technologies for soil and crop health analysis, pest detection, disease prediction, weather forecast, and many more parameters that play a major role in productive harvest. The tendency of AI to "train" itself to new inputs and supervise vast amounts of complex data can be very beneficial and bring changes in modern agriculture.

3.4.1 Weather Prediction

The use of AI in weather prediction aids in providing farmers with valuable information about the weather forecast, cloud patterns, humidity, and many more. Satellites, ground stations, and drones help in collecting real-time weather data by taking multispectral images.[15] The ability of Machine Learning Models for predicting future events from past inputs can bring revolutionary changes, especially in the aspects of weather prediction. Machine Learning Algorithms, Deep Learning Architectures, and Neural Network architectures can train themselves to automatically detect weather patterns such as cloud formation, temperature changes, storm waves, etc. from the images captured by satellites.[15] This allows farmers to analyse better and makes decision-making easier for them. As a result, healthier crops can be harvested since the information provided by AI technology not only detects weather forecasts and climatic conditions but also guides farmers on the ideal time for seed plantation, tilling, harvesting, and various other activities which are highly dependent on weather conditions surrounding the agricultural field.

3.4.2 Soil Management

Soil plays a major role in the performance of crop health, productive yield and in providing a good growing environment for the crops. Different crops require different soil characteristics in terms of nutrients, humidity, nitrogen level, and overall chemical make-up. Farms equipped with soil sensors, IoT devices, and soil analysis drones collect real-time data from the soil and monitor its nutrition level, soil quality, and soil moisture.[16] By analysing this data acquired from sensors and satellite images, soil faults and deficits can be detected. ML algorithms have the ability to structuralize the raw data acquired from the soil and display its predominant characteristics on mobile applications accessible to ground stations and farmers. This helps farmers in deciding the amount and type of fertilizer to be added to the soil for improved crop yield. The use of AI and IoT sensors can extensively collect large amounts of data from the soil, measure its essential features and analyse this data to provide farmers with precise guidance and predictions.[15] Along with the measurement of soil characteristics, this technology also aids in maintaining a balance on the soil properties which benefits crop yield as well as soil fertility, preventing barren and infertile lands.

3.4.3 Data Collection and Analysis

Before AI was introduced to the agricultural industry, farmers used traditional methods in fields like manual detection of diseases, pests, soil quality, yield prediction, and weather conditions. These traditional methods involved the use of field cameras, taking aerial images, and satellite photography which has now been replaced by the IoT, drones, ML, and Wireless Sensor Networks.[17] The volume of data collected by soil sensors, IoT devices, satellite and drone images, and other remote sensing technologies is vast and very complex to analyse manually or by using technologies that easily overload with data. Robots and intelligent sensors integrated in the agricultural field collect data from the soil, agricultural vehicles, and multispectral images obtained from unmanned aircraft systems and drones hovering above the field. This data can be transmitted to AI algorithms for data structuralizing, detection, and analysis using various sources of information.[18] AI algorithms are capable of using the inputs obtained from sensors as experiences and learning from these experiences (inputs) to give future predictions and outputs. This computational technology is flexible to the amount and types of parameters that can be analysed by it. The collection of real-time data makes the agriculture system more productive and precise.[19] Fundamentally, these algorithms tend to procure and fuse data and extract valuable information from it for various precision agricultural processes. Big Data Analytics also plays a major role in the collection and analysis of vast volumes of soil sensor data and soil monitoring. This system provides guidance to farmers regarding fertilizer requirements, optimal harvesting times, optimal planting times, and smart decision-making services using decision support systems (DDS).

3.4.4 Decision Making

Over the last few decades, DSS using AI have been progressively increased for usage in precision farming. A variety of sensors are used in PA that provide precise data extracted from soil, crop, and weather, which leads to the development of an accurate and precise DDS. Application of ML models, working primarily on training data to perform tasks, has a wide range of scope for accurate and informed decision-making. These models are applied to sensor data to provide improved solutions and help farmers in making effective decisions. There are an abundant number of ML models and algorithms developed for this purpose. Two of them are decision trees and ANNs.[20]

Decision trees are classification models mapped into a tree-like architecture which branches out its data into subsets as well as organizes a tree graph simultaneously. In this model, the internal nodes are represented as a pairwise comparison, each branch is represented as the comparison outcome and the leaf nodes represent the decision taken after computing the attributes.[21]

ANN can function based on human brain intelligence such as cognition, learning ability, decision-making, and adjustments in response to surroundings. ANN mimics the structure of the interconnected data transmitting system of the human brain, by arranging processing units interconnected to each other in an organized and constant pattern. The use of AI in decision-making is very beneficial to the agricultural sector as it allows farmers to make informed and smart decisions, reducing the rate of crop wastage and unnecessary expenses.[21]

3.4.5 Disease and Pest Detection

AI assists farmers in disease and pest detection by using predictive analysis from the data acquired by the soil sensors. Pest detection and disease prediction are essential for the prevention of crop loss or damage in the future. Traditional methods for disease and pest detection require a lot of expertise knowledge and is time-consuming, which is not convenient for farmers.[4] The ML model that can be used for this task is Convolutional Neural Network (CNN) and the data collection process can be done using drones. Hyperspectral images are also distinctively effective in pest detection and plant disease monitoring.[2] GPS-guided drones and Hyperspectral Remote Sensing help farmers obtain precise data for smart decision-making as well as for the controlled use of pesticides and fertilizers in specific fields.[22]

3.4.6 Weed Protection

Weeds are one of the major problems encountered by farmers on a daily basis as they compete with significant crops over light, water, and nutrients which affect crop yield. The traditional methods used by farmers include manual removal of weeds or the use of herbicides and weedicides on the targeted field area. However, these methods cause inconvenience in terms of extensive use of labour, environmental damage, and these methods are ineffectual on herbicide-resistant weeds.[23]

Thus, application of AI, ML, and DL techniques allow farmers to control weeds efficiently. AI uses automated and robotic techniques [24] to distinguish plants from weeds by image analysis. The images are obtained from Unmanned Aircraft Systems (UAS) that fly over a large scouting area of the field. After analysis of data collected from the spatial images, farmers are provided with precise information about weed growth prediction which helps them in making analysed decisions as well as carry out automated weeding by spot spraying on targeted weed areas.[23]

3.4.7 Labour and Equipment Management

Labour and agricultural equipment consume a significant amount of cost for their services. Keeping a track of labour working on the field and regular maintenance of equipment is time intensive and requires attention in case of manual supervision. GPS-enabled mobile applications can help in monitoring labour movements, the amount of work done by them, and the areas of field covered. This data can make the workflow efficient and ensure timely task completion. Agricultural equipment is considered as an investment but if not maintained and serviced regularly, it can turn into a bane. Thus, drone imagery can help farmers keep track of and decide on equipment placement across the field and automated record-keeping software enables up-to-date reminders for different equipment maintenance.

3.5 Benefits of Artificial Intelligence in Precision Agriculture by Case Studies

Implementation of AI has brought revolutionary changes globally, which prompted various industries to adopt this technology for their tasks. Agricultural sector has the maximum benefit from the development of AI-driven technologies as it helps farmers and industries with better crop yield, smart decision-making, and better soil health. Some of these implementations have been briefed below in the form of case studies

3.5.1 Case Study on Use of Data-Based Decisions in Farming

The European project Internet of Food and Farm 2020 (IoF2020) implemented the QUHOMA platform which is a smart irrigation solution to explore the potential of IoT in the field of agriculture, especially for automated irrigation. The study was reliable on the daily accumulated data from the sensors integrated in the soil which brought up some limitations such as loss of communication with ground sensors, sensor failures, etc. [19] The project implemented AI-based predictive analytics that trained the model by combining the real-time data with the public data. The irrigation schedule was set up in an organized manner by the use of a Complex Event Processing (CEP) algorithm. The project also had a back-up for situations where the sensors might lose connection for a longer period of time. The back-up was to use the previously accumulated irrigation scheduling data and historical data stored in the sensor to regularize irrigation schedule. In case of inaccurate sensor data, the historical data used by predictive analytics detected abnormalities in the sensor data values and used this data for maintenance of the targeted sensors.[19]

3.5.2 Case Study on Automation Impact

Simultaneous Localization and Mapping (SLAM)[19] is a mapping system which can be used to provide an automated vehicle or robot with information about its location, the surrounding map, and the obstacles or crops in its path for safe navigation. A SLAM system can be integrated with such data and programmed according to the sensors present on board such as ultrasonic devices and cameras that determine the location of the vehicle as well as the crop rows and trees in its navigation path. [19] The French-German MOVEON project researchers have used a visual SLAM system integrated with a single camera with DL techniques and multi-view-stereo (MVS) algorithms. It has brought forth some advancements in the current autonomous robots and vehicles such as repetitive patterns, blunt movements due to uneven roads, and the need for detailed reconstructive information. The Rosaria Dataset [25] has been used for this experiment consisting of six sequences of soybean field whose images were captured by a front stereo camera the images consisted of burnt field, similar patterns of field and many more to explore the abilities of the system.[19]

3.5.3 Case Study on Use of Drones in Fields

In an experiment performed by two researchers from Shibaura Institute of Technology in Japan, they aimed at early detection of plant diseases using drones and AI technology to avoid crop wastage and prevent the rice from severe infection.[26] They used drones, GPS sensor and cameras for data acquisition by flying over the entire field and scouting the rice field. They used the GPS data to get the latitude and longitude format of their location and used an HD camera to capture images in RGB colour format. These images were sent to the ground station where the images were processed, analysed, and then classified by using Support Vector Machine (SVM). The entire flowwork of this technique is shown in Figure 3.3. The images were classified into four different types of diseases based on their colour and shape.[26]

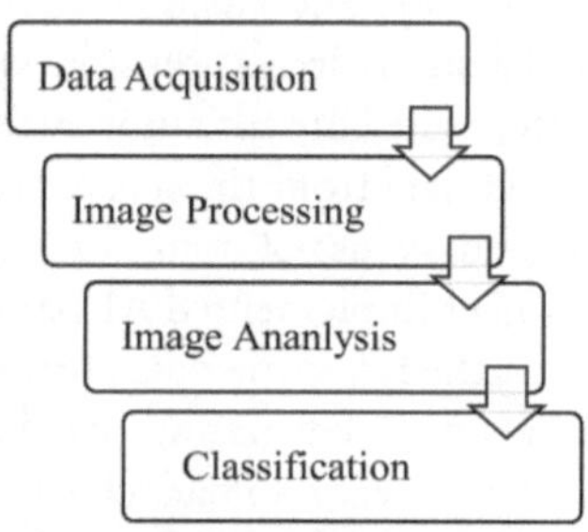

Figure 3.3 Framework of the detection technique. [26]

Key Activities	Key Resources
• Data collection • Data analysis • Machine design • Software Development	• Robotic based see and spray machine • Software • Camera

Figure 3.4 Key activities and key resources of Blue River Technology. [27]

3.5.4 Case Study on Blue River Technology

Blue River technology is a US-Based Company working in the agricultural sector, mainly on agricultural technology. It develops innovative technologies for improvements in the agricultural industry and for the betterment of the farmers, as shown in Figure 3.4.[27] It also provides effective solutions to farmers' problems with minimized inconvenience to farmers. The business model of Blue River technology includes SWOT analysis and PESTLE analysis. SWOT stands for Strength, Weakness, Opportunity, Threats and PESTLE stands for Political, Economic, Social, Technological, Legal, Environmental factors. The company has made excellent advancements in the integration of AI technology in the Indian Farming Industry.[27]

3.6 Challenges for Artificial Intelligence in Precision Agriculture

AI, ML, and Deep Learning algorithms are widely beneficial for crop cultivation and various other tasks essential for a good yield and better lifestyle for farmers. However, there exist some factors that do not allow a widespread expansion of the awareness of these technologies around the world, especially in rural areas. Despite the significant number of benefits of AI in agriculture, there are some aspects which do not allow people to adapt to these technologies fully and create a barrier between technology and farmers.

3.6.1 Lack of Time and Technical Skills in Farmers

Majority of farmers lack the ability to adapt to new technology and time to learn technical skills for investigation of AI solutions by themselves. Farmers living in rural areas lack the proper technical knowledge and digital skills to understand the scope of AI in this industry. Due to a lack of knowledge of such technologies, there is no awareness among their community about the evolution of such revolutionary technologies in the world. Expertise and appropriate technical skills are mandatory to be able to operate and maintain AI-driven technologies and in rural areas, it is difficult to have access to training and support systems to understand these algorithms.[28]

3.6.2 Cost and Accessibility

One of the biggest challenges faced by farmers is the cost of AI technology and equipment. The high cost of the entire AI system is not affordable to farmers located in rural areas of developing countries which builds a great barrier for the farmers with limited resources.[29] AI does prove to be cost-effective in the long run; however, it does demand a large investment which cannot be affordable for farmers. Implementing AI in small-scale farms might be difficult for the time being but as the technology develops, the cost might reduce, and the technology might be accessible to small-scale farmers as well.

3.6.3 Data Management

The data collected from the sensors and IoT devices might create errors due to faulty sensors or obstructions in general. But in some cases, some environmental factors create non-fixable errors which are not in our control. This questions the ability of IoT devices to procure accurate and precise data from the sensors and it still requires a vast number of developments.[30] The usability of AI technology in agriculture has a breaking point beyond which the agronomists and data scientists might not have obtained strong results. There is more room for the development in this subject which is why data management will not be a limitation in the long run.

3.6.4 Acceptance and Adoption

Another factor that causes obstruction in awareness of ML and Deep Learning systems is the willingness of people to resist the change. Traditional methods of farming are passed down from generations to generations which resists people to adapt to new technologies.[31] Even though this technology can bring improvements in crop yields, reduce unnecessary costs, and improve efficiency, the awareness of such technologies needs to be increased in order to increase the use of AI in agriculture.[2]

Despite these challenges, AI remains to be the fastest developing technology in the world till date and the agricultural companies who have embraced these technological changes continue to see refinements in their yield and crop health and have greatly benefited from it.

Glossary

GPS	Global Positioning System
ECa	Electrical Conductivity sensors
GIS	Geographic Information System
VRT	Variable-Rate Technologies
VRA	Variable-Rate Applications
AI	Artificial Intelligence
ML	Machine Learning
ANN	Artificial Neural Network
IoT	Internet of Things
DL	Deep Learning
DSS	Decision Support System
CNN	Convolutional Neural Network
UAS	Unmanned Aircraft System
IoF2020	Internet of Food and Farm 2020
QUHOMA	QUalitative HOrticulture MArketplace
CEP	Complex Event Processing
SLAM	Simultaneous Localization and Mapping
MVS	Multi View Stereo

References

1. D. K. Shannon, D. E. Clay, and K. A. Sudduth, "An Introduction to Precision Agriculture," ASA, CSSA, and SSSA Books, pp. 1–12, Jun. 2018, doi: 10.2134/precisionagbasics.2016.0084
2. P. Singh et al., "Hyperspectral remote sensing in precision agriculture: Present status, challenges, and future trends," Hyperspectral Remote Sensing, pp. 121–146, 2020, doi:10.1016/b978-0-08-102894-0.00009-7
3. N. M. Patil, "Precision agriculture: Smart farming taken to the next level," SSRN Electronic Journal, 2021.
4. M. Javaid, A. Haleem, I. H. Khan, and R. Suman, "Understanding the potential applications of artificial intelligence in agriculture sector," Advanced Agrochem, vol. 2, no. 1, pp. 15–30, 2023, doi: 10.1016/j.aac.2022.10.001
5. A. Monteiro, S. Santos, and P. Gonçalves, "Precision agriculture for crop and livestock farming—brief review," Animals, vol. 11, no. 8, p. 2345, 2021, doi: 10.3390/ani11082345.
6. V. Hakkim, E. Joseph, A. Gokul, and K. Mufeedha, "Precision farming: The future of Indian agriculture," Journal of Applied Biology and Biotechnology, pp. 068–072, 2016, doi: 10.7324/JABB.2016.40609
7. U. Shafi, R. Mumtaz, J. García-Nieto, S. A. Hassan, S. A. R. Zaidi, and N. Iqbal, "Precision agriculture techniques and practices: From considerations to applications," Sensors, vol. 19, no. 17, p. 3796, 2019, doi: 10.3390/s19173796

8. B. Nowak, "precision agriculture: Where do we stand? A review of the adoption of precision agriculture technologies on field crops farms in developed countries," Agricultural Research, vol. 10, no. 4, pp. 515–522, 2021, doi: 10.1007/s40003-021-00539-x
9. P. Mondal, and V. K. Tewari, "Present status of precision farming: A review," International Journal of Agricultural Research, vol. 2, no. 1, pp. 1–10, 2006, doi: 10.3923/ijar.2007.1.10
10. Y. Ampatzidis, "Applications of artificial intelligence for precision agriculture," EDIS, vol. 2018, no. 6, Dec. 2018, doi: 10.32473/edis-ae529-2018
11. V. Sachithra, and L. D. C. S. Subhashini, "How artificial intelligence uses to achieve the agriculture sustainability: Systematic review," Artificial Intelligence in Agriculture, vol. 8, pp. 46–59, 2023, doi: 10.1016/j.aiia.2023.04.002
12. G. Nagasubramanian, R. K. Sakthivel, R. Patan, M. Sankayya, M. Daneshmand, and A. H. Gandomi, "Ensemble classification and IoT-based pattern recognition for crop disease monitoring system," IEEE Internet of Things Journal, vol. 8, no. 16, pp. 12847–12854, 2021.
13. T. Talaviya, D. Shah, N. Patel, H. Yagnik, and M. Shah, "Implementation of artificial intelligence in agriculture for optimisation of irrigation and application of pesticides and herbicides," Artificial Intelligence in Agriculture, vol. 4, pp. 58–73, 2020, doi: 10.1016/j.aiia.2020.04.002
14. R. Bongiovanni, and J. Lowenberg-Deboer, "Precision agriculture and sustainability," Precision Agriculture, vol. 5, no. 4, pp. 359–387, 2004, doi: 10.1023/b:prag.0000040806.39604.aa
15. C. E. Hachimi, S. Belaqziz, S. Khabba, B. Sebbar, D. Dhiba, and A. Chehbouni, "Smart weather data management based on artificial intelligence and big data analytics for precision agriculture," Agriculture, vol. 13, no. 1, p. 95, 2022, doi: 10.3390/agriculture13010095
16. T. van Klompenburg, A. Kassahun, and C. Catal, "Crop yield prediction using machine learning: A systematic literature review," Computers and Electronics in Agriculture, vol. 177, p. 105709, 2020, doi: 10.1016/j.compag.2020.105709
17. R. Akhter, and S. A. Sofi, "Precision agriculture using IoT data analytics and machine learning," Journal of King Saud University - Computer and Information Sciences, vol. 34, no. 8, pp. 5602–5618, 2022, doi: 10.1016/j.jksuci.2021.05.013
18. J. Backman, M. Koistinen, and A. Ronkainen, "Agricultural process data as a source for knowledge: Perspective on artificial intelligence," Smart Agricultural Technology, vol. 5, p. 100254, 2023, doi: 10.1016/j.atech.2023.100254
19. M. T. Linaza et al., "Data-driven artificial intelligence applications for sustainable precision agriculture," Agronomy, vol. 11, no. 6, p. 1227, Jun. 2021, doi: 10.3390/agronomy11061227
20. F. Balducci, D. Impedovo, and G. Pirlo, "Machine learning applications on agricultural datasets for smart farm enhancement," Machines, vol. 6, no. 3, p. 38, 2018, doi: 10.3390/machines6030038
21. W. Yong et al., "Smart sensors from ground to cloud and web intelligence," IFAC-PapersOnLine, vol. 51, no. 17, pp. 31–38, 2018, doi: 10.1016/j.ifacol.2018.08.057
22. N. Kitpo, and M. Inoue, "Early Rice Disease Detection and Position Mapping System using Drone and IoT Architecture," 2018 12th South East Asian Technical University Consortium (SEATUC), Mar. 2018, doi: 10.1109/seatuc.2018.8788863
23. Ghatrehsamani et al., "Artificial intelligence tools and techniques to combat herbicide resistant weeds—a review," Sustainability, vol. 15, no. 3, p. 1843, Jan. 2023, doi:10.3390/su15031843

24. F. A. Auat Cheein, and R. Carelli, "agricultural robotics: Unmanned robotic service units in agricultural tasks," IEEE Industrial Electronics Magazine, vol. 7, no. 3, pp. 48–58, 2013, doi: 10.1109/mie.2013.2252957
25. T. Pire, M. Mujica, J. Civera, and E. Kofman, "The Rosario dataset: Multisensor data for localization and mapping in agricultural environments," The International Journal of Robotics Research, vol. 38, no. 6, pp. 633–641, 2019, doi: 10.1177/0278364919841437
26. R. Chin, C. Catal, and A. Kassahun, "Plant disease detection using drones in precision agriculture," Precision Agriculture, vol. 24, no. 5, pp. 1663–1682, 2023, doi: 10.1007/s11119-023-10014-y
27. S. Panpatte, and C. Ganeshkumar, "Artificial Intelligence in Agriculture Sector: Case Study of Blue River Technology," Proceedings of the Second International Conference on Information Management and Machine Intelligence, pp. 147–153, 2021, doi:10.1007/978-981-15-9689-6_17
28. U. A. Bhatti, and M. Masud, "Investigating AI-based smart precision agriculture techniques," Frontiers Research Topics, 2023, doi: 10.3389/fpls.2023.1237783
29. J. Su, X. Zhu, S. Li, and W.-H. Chen, "AI meets UAVs: A survey on AI empowered UAV perception systems for precision agriculture," Neurocomputing, vol. 518, pp. 242–270, 2023, doi: 10.1016/j.neucom.2022.11.020
30. J. P. Fulton, and K. Port, "Precision Agriculture Data Management," ASA, CSSA, and SSSA Books, pp. 169–187, Jun. 2018, doi: 10.2134/precisionagbasics.2016.0095
31. C. S. Alexander, M. Yarborough, and A. Smith, "Who is responsible for 'responsible AI'?: Navigating challenges to build trust in AI agriculture and food system technology," Precision Agriculture, Aug. 2023, doi: 10.1007/s11119-023-10063-3

Chapter 4

How AI Contributes to Precision Agriculture

Eugene Berna I., Vijay K., J. Jeyalakshmi, and Prithi Samuel

4.1 Introduction

Precision agriculture is not immune to the rise of artificial intelligence (AI) as a game-changing technology. Precision agriculture, often known as smart farming, is the application of contemporary technologies to optimize agricultural practices and increase resource efficiency while minimizing environmental impact. By enabling data-driven decision-making, AI plays a crucial role in developing these achievements. Precision agriculture [2, 10, 14] is based on the principle of treating each portion of a field as an independent entity with distinct features. Traditional agricultural practices may apply comparable treatments to entire fields, resulting in the waste of resources such as water, fertilizers, and pesticides [5]. Farmers can tailor their operations based on real-time and historical data when AI is integrated into precision agriculture, resulting in greater crop output, quality, and sustainability.

Precision agriculture has profited immensely from the application of AI, as shown in Figure 4.1, a revolutionary approach to modern farming. This collaboration has resulted in significant changes in agricultural practices, helping farmers to boost yields, reduce resource waste, and make more informed decisions [7, 9]. The bond between AI and precision agriculture addresses the complexities and restrictions of modern food production by integrating data-driven insights, predictive analytics, and automation. Precision agriculture, as opposed to treating entire fields consistently, is the meticulous administration of agricultural practices inside fields at a micro-level. This level of precision is achieved by the collection, analysis, and use of large amounts

 DOI: 10.1201/9781003504900-4

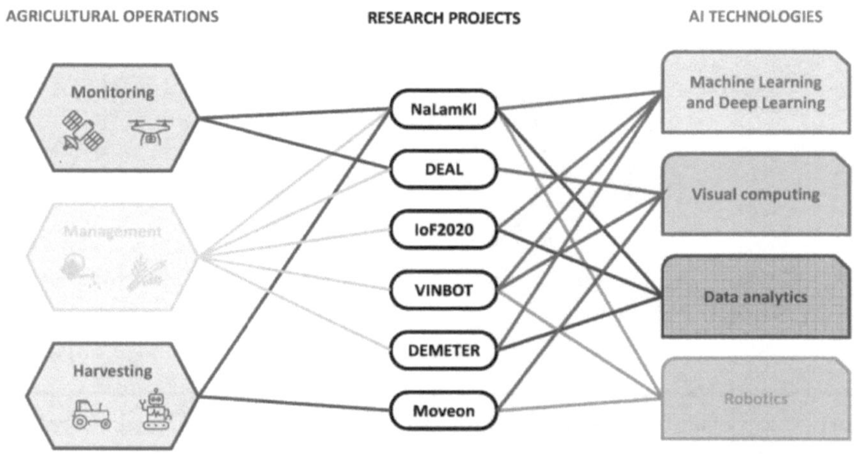

Figure 4.1 Precision agriculture with artificial intelligence.

of data. AI technologies provide the computational power required to efficiently process enormous amounts of data and derive meaningful insights [8].

Potential benefits and drawbacks of using information and communication technologies in conventional farming are discussed. In-depth discussions on robotics, Internet of Things (IoT) gadgets, and ML difficulties, as well as ML, AI, and sensor applications in farming, are included [1]. Drones are also the subject of study for the purposes of crop monitoring and management. When relevant, we also highlight state-of-the-art global farming systems and technologies based on the IoT. We analyse the most recent publications in each of our fields of study. Based on these findings, we make some predictions for the future of AI and identify some of the most critical areas for future AI research in agriculture.

Climate change, population increase, competition for farmland from biofuels, and degrading soil quality are all factors that undermine the world's ability to provide food for its people [3]. We need to explore radical new techniques and bring in knowledge from other fields, such as materials science and computer science if we want to discover answers that will persist. Future food security could be improved with the use of nanotechnology, AI, and precision agriculture, where farmers respond to in-season changes in crop development in real time. Existing nutrient cycling and crop productivity models must be integrated with nano-informatics methodologies for optimal targeting, absorption, delivery, nutrient capture, and long-term effects on soil microbial populations.

4.1.1 Precision Agriculture (PA)

AI technologies are underutilized in agriculture due in large part to a lack of publicly available information about their potential uses. Therefore, many research efforts have focused on documenting the most widespread AI tools

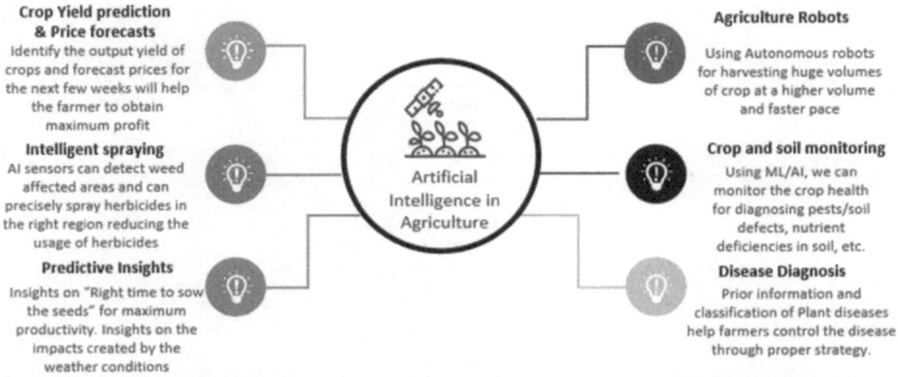

Figure 4.2 Visualization of various agricultural activities and AI-related technologies.

for PA and the various ways in which they have been applied. Linaza et al. [11] conducted a comprehensive analysis of research into the application of ML in agriculture. Management of crops, water, and soil are the three most important areas of use. They argued that the introduction of AI algorithms to Farm Management Systems will considerably help agriculture thanks to ML technology. This would allow for the development of real-time decision support systems (DSS).

Finally, the agriculture economy, which covers a lot of ground, is a major beneficiary of AI and robotics. Agricultural production and management have benefited greatly from robotics, with autonomous robots weeding, watering, and monitoring individual plants. Robotic agricultural occupations are discussed in detail by Zhang et al. [10], who provide a comprehensive overview of contemporary robotic grippers, grasping, and sensor-based control techniques. The standards for agricultural gripping are higher than those in the manufacturing sector since most food and agricultural products are perishable, easily damaged, sticky, or slippery in Figure 4.2.

4.2 Smart Farming and Precision Agriculture

In their response, Bacco and coworkers noted that by keeping tabs on weather patterns, soil conditions, moisture levels [15], and other variables, "smart agriculture" is able to solve multiple issues related to crop production. Improve crop yields and aid management decisions with the help of modern sensors and data analysis tools through the practise of "precision agriculture" (PA). To "increase

Figure 4.3 Smart farming prototype.

output, decrease labour time, and ensure effective fertiliser and irrigation management," PA is a new concept that has been embraced globally (Mulla, 2013). To boost agricultural resource utilization, crop yields, and quality, it makes extensive use of data and information. In order to increase crop management practises and yields while minimizing the use of costly inputs like fertilizer, water, and pesticides, precision agriculture uses data from multiple sources, as shown in Figure 4.3.

In order to protect their crops from incursions and keep tabs on their progress in real time, farmers can implement a crop management approach known as "smart agriculture." Sensors that measure chemical concentrations, pH levels, wind speed and direction, precipitation amounts, soil temperatures and humidity levels, and even sound waves are employed in today's farms. The capabilities and hardware requirements of a sensor are highly context-dependent. Wireless technologies, data analytics, big data capabilities, and image processing systems are all ways in which

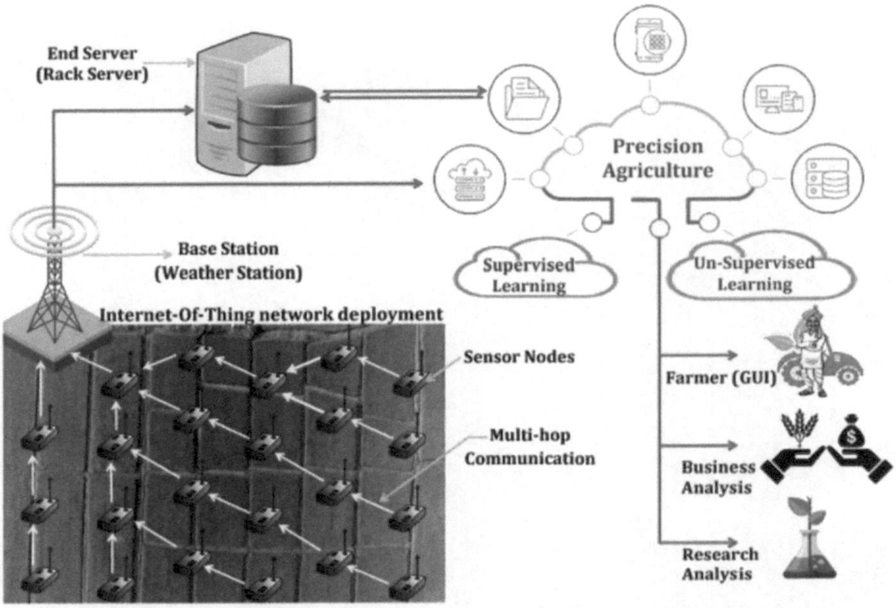

Figure 4.4 Overview of the primary elements in smart and precision agriculture.

the IoT might improve smart farm management [1, 6]. Precision agriculture's core elements are shown in Figure 4.4.

Utilizing data science and the IoT to examine, foresee, and arrange farming activities in light of their physical setting [13]. Sensors can be used to measure a wide range of factors, including soil temperature, moisture, foliage, daylight levels, and even orientation. The cornerstones of smart agriculture (SA) are the use of digital technologies for tasks including harvesting crops, storing data, sharing information, making decisions, and analysing results.

Among the many data-gathering and communication technologies utilized in this field, sensor networks are among the most common. Environment data is essential for precision farming. They serve a dual purpose in crop management and farm organization by exploring different approaches. Gathering this data represents a fresh input with the potential to greatly increase agricultural output [1, 26]. Figure 4.5 depicts the global infrastructure of IoT farm applications.

4.3 The Role of AI and IoT in Agriculture

Figure 4.6 depicts the many branches of AI. These branches include machine learning, deep learning, natural language processing (NLP), swarm intelligence (SI), expert systems, fuzzy logic, and computer vision. This area of study can be useful

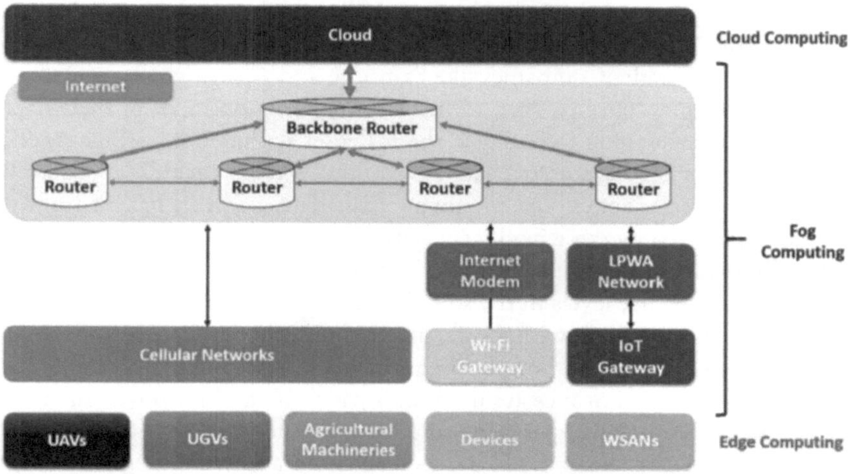

Figure 4.5 Global IoT framework in automated and smart agricultural practices.

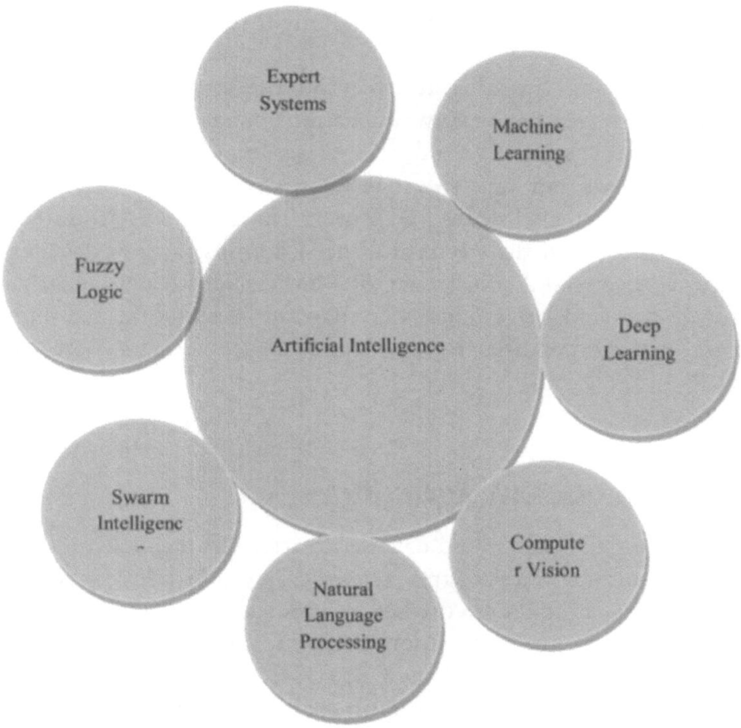

Figure 4.6 AI techniques.

in many contexts. Research into developing intelligent AI programmers is active in many industries, including medicine, agriculture, banking, the economy, technology, business, and trade. Companies like Samsung and Apple, which specialize in consumer electronics, have pledged to include this function in all of their future releases. Another emerging technology, the IoT, uses the internet to link together different types of smart sensors and gadgets. Multiple sectors, including solar power plants, farms, disaster zones, and the manufacturing sector, can benefit from the information gathered by these intelligent sensors.

In a smart farming system built on the IoT, sensors track things like fertilizer levels and moisture levels in the soil. Researchers are looking into using ML algorithms to figure out how much fertilizer should be applied to soils before planting crops [2]. The use of drones in agriculture is having a profound effect. These camera-equipped drones are used for a wide range of purposes, including crop and field monitoring, pesticide application, and drip irrigation. Drone-collected imagery can be analysed by deep learning and computer vision algorithms to detect plant diseases and weeds at any stage of the crop's growth. After that, these unmanned aircraft are used to apply herbicides to weeds and diseased crops. The ever-changing nature of the weather is the primary source of anxiety for farmers. Drone irrigation is an effective AI-powered system that can be easily taught on weather patterns and might considerably reduce farmers' water challenges.

Drones are having a profound effect on the agricultural sector. These camera-equipped drones have many applications in agriculture, including crop monitoring, pesticide spraying, and drip irrigation. Using DL and computer vision algorithms, we can analyse drone images collected at various stages of a crop's growth to detect diseases and weeds. The chemicals are subsequently sprayed by the drones on the weeds and the sick crops [6]. For decades, farmers have fretted over the inconsistency of Mother Nature. Drone irrigation is a powerful AI-powered irrigation device that can be easily trained on weather patterns and has the potential to significantly lessen farmers' water concerns [22, 23].

4.4 AI Techniques in Agriculture

It entails employing technology and data-driven strategies, as shown in Figure 4.7, to optimize agricultural practices and increase crop yield while reducing resource waste [4]. Several techniques are used in various stages of precision agriculture. In precision agriculture, the cultivation phase refers to the time when crops are actively produced and managed in the field. During this phase, numerous techniques and technologies are used to optimize crop growth, monitor field conditions, and make educated decisions in order to achieve improved yields and resource efficiency.

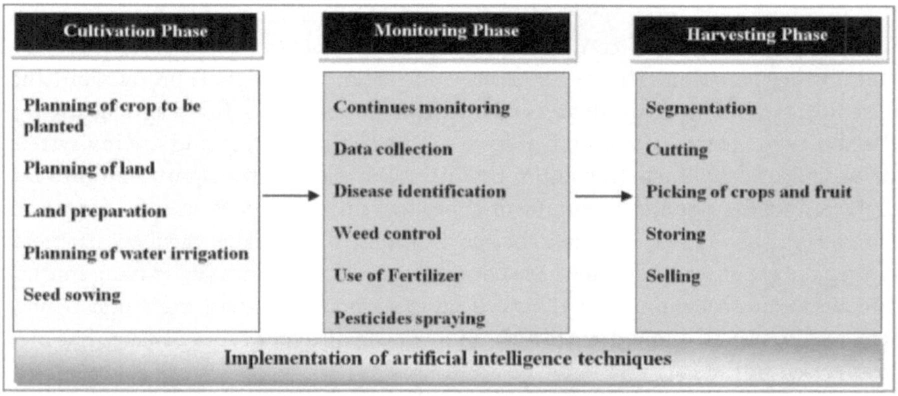

Figure 4.7 Implementation of AI techniques.

4.4.1 Soil Care

Soil is crucial to agricultural success [13] since it provides plants with their initial exposure to water, nitrogen, phosphorus, potassium, and proteins. Using compost and manure, which improve soil porosity and aggregation, and switching to a less destructive kind of tillage can both improve soil quality. Effective management of soil may mitigate the hazards posed by contaminants and diseases transmitted via the soil [13]. AI-generated soil maps are useful for visualizing subsurface soil composition and landscape interactions [14].

4.4.2 Weed Control

Without proper weed management, dry bean and maize harvests might drop by as much as half, while wheat yields could drop by as much as 48 percent due to weed competition. Weeds compete with crops for water, nutrients, and sunlight [13], despite the fact that some weeds are harmful and may represent a threat to public health. Weeds can be controlled with spraying, however using too much of it can be dangerous to humans and contaminate natural areas. In order to reduce expenses and the likelihood of crop damage, AI weed detection systems have been designed and tested in lab settings [19].

4.4.3 Use of Internet of Things Technology

The "Internet of Things" is the global network of computers, machinery, and other items that can be individually identified and which can send and receive data and commands. Therefore, neither a human nor a computer can act as a go-between.

The IoT builds on prior technological advancements including wireless sensor networks (WSNs), cloud computing, and radio frequency identification (RFID). Monitoring, precision farming, tracking and tracing, greenhouse production, and agricultural machinery are all areas where the IoT could be useful in agriculture [5]. The ability to receive data is crucial for a supply chain tracking and tracing system for agricultural products (the entire product life cycle, the transportation process, etc.), store it for a specific amount of time, transmit it, process it, and output it. Agricultural businesses can make better decisions, find reliable business partners, and reduce expenses when they have complete product histories [9]. Having mutual trust between a seller and a buyer is crucial to a smooth business exchange. Sensor data, audio, visual, and video are only a few of the different types of data that may be analysed in the IoT.

4.5 Insect Infestations and Grain Illnesses

Grain production can be negatively impacted by a number of unfavourable situations, such as disease epidemics or pest and bug infestations. Market prices could drop if crop growth or grain quality are negatively impacted. Successful disease control, essential for profitable and sustainable agriculture, depends on early identification and accurate diagnosis of infections in grain crops [6]. When dealing with a pest infestation, for example, prompt identification is essential for delivering the appropriate reaction and implementing the necessary control measures. Visual inspection is the norm when diagnosing plant diseases, yet this method is problematic due to its difficulty and subjectivity. Techniques for automating plant disease identification utilizing computer vision and AI have been proposed in this regard. Automatic disease detection using images relies heavily on identifying the most useful distinguishing characteristics for accurate recognition. Classification algorithms can thus be used to determine if a given image depicts a healthy or unhealthy subject.

4.5.1 Detection of Bakanae in Rice Crops

Chung and co-workers developed a system to identify healthy and sick Bakanae seedlings at the three-week mark. The rice (*Oryza sativa* L.) disease bakanae is caused by the fungus *Fusarium fujikuroi*. Some of the morphological abnormalities noticed in infected plants include stem elongation, stunting, an increased angle between the leaves and the stem, and a change in leaf colour (the leaves turn yellow). The illness begins in the fields, where it infects the grains and spreads to the seeds before multiplying throughout the crop during storage. Depending on the level of resistance in the plant, the symptoms can be complicated and difficult to predict. By taking this tack, they complicate computer vision recognition.

Figure 4.8 Detection of bakanae disease.

The procedure described here makes use of scanned plant pictures in Figure 4.8. The plants' morphological and colour characteristics are measured, and their anatomical locations are recognized mechanically. The only part that can't be automatically identified is the base of the stem.

4.5.2 Leaf Illnesses in Soybeans

Shrivastava and co-workers investigated soybean foliar diseases to see if computer vision technologies are useful for recognizing and classifying foliar diseases in Figure 4.9.

Figure 4.9 Leaf disease in soybeans.

In our research, we paid special attention to the following: histogram, CCV (Colour Coherence Vector), which enables colour-based image comparison; CDH (Colour Difference Histogram), which enables encoding of the images' visual appearance; LBP (local binary pattern) descriptor, which encodes image texture information; SEH (Structure Element Histogram), which encodes both colour and texture information, is similar to the aforementioned but aims to reduce the size of the LBP; and LAP.

4.5.3 Aphids Detection in Wheat

The increase in aphids could threaten this year's wheat harvest. These bugs cause growth stunting and viral transmission when they feast on wheat's phloem sap. Manual labour is still widely used in the counting and identifying processes. Liu and co-workers created the SMH method (Support Vector Machine, Maximally Stable Extremely Regions, and Histograms of Oriented Gradient approach) for identifying aphid populations and types in the field. Included in this method are the use of support vector machines, maximally stable extreme areas, and a histogram of gradients in Figure 4.10.

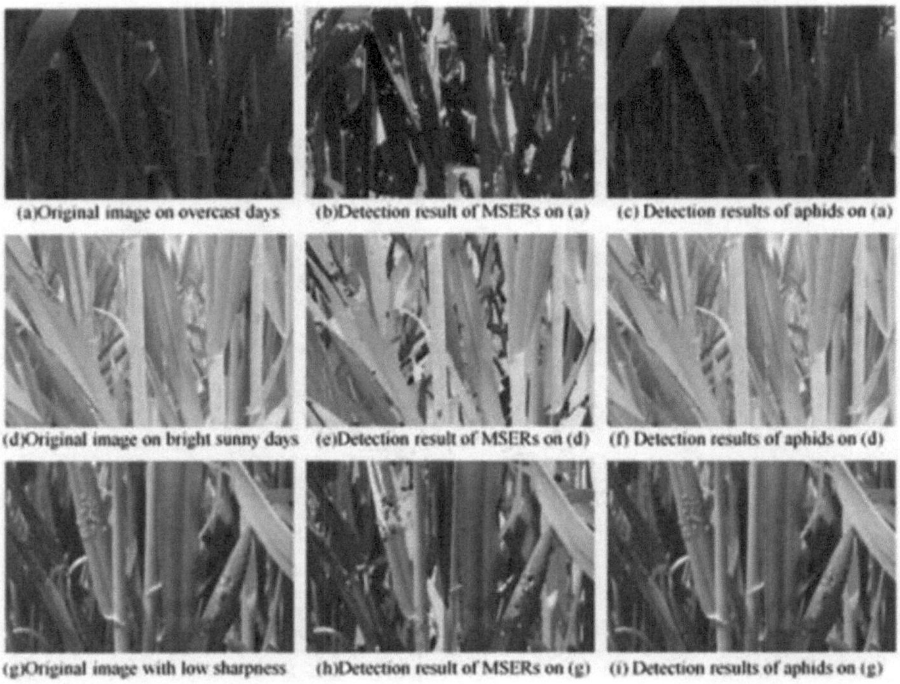

Figure 4.10 Aphids detection in wheat.

Figure 4.11 Plant and weed detection utilizing computer vision and artificial intelligence.

4.6 AI and Object Detection

Figure 4.11 shows how the smart sprayer can tell the difference between a cotton or wheat plant and weeds. When compared to traditional broadcast sprayers, this method can cut pesticide use by as much as 90 percent. Traditional pest management methods employing broadcast sprayers treat the entire field at once, regardless of which areas actually require treatment [7]. Reducing costs, crop loss, pesticide residue, and environmental impact by applying agrochemicals only where pests are actually present.

Another example is Harvest CROO Robotics' (HCR, harvestcroorobotics.com) robotic strawberry harvester. The harvester for strawberries uses AI and machine vision to detect and locate ripe fruit [15]. An autonomous platform (Figure 4.12) navigates through strawberry fields, while the HCR strawberry harvester's many picking robots and support mechanisms (Figure 4.13) shift leaf foliage. Strawberries are a perishable fruit that are typically picked by hand at the market. The high expense of harvesting strawberries and the increased risk of a failed harvest are both exacerbated by labour shortages. Growers are unable

Figure 4.12 HCR harvester main picking component.

Figure 4.13 On a NVidia Jetson TX2 board, real-time citrus detection utilizing YOLO (a real-time AI object detection method).

to pick all of their salable fruit because the number of domestic farmworkers has plummeted dramatically. According to a recent study (Guan, Wu, 2018).

4.6.1 Detection of Pest

The damage that pests may do to a farm is extensive. The Asian citrus psyllid (ACP) is a major problem for Florida's citrus sector because it transmits citrus huanglongbing (HLB). The capacity of Florida's farmers to send fresh fruit across the country and throughout the world is in jeopardy due to bugs and diseases. Visual inspections are the primary method used by growers and inspectors for pest detection in the field. However, inspector availability severely restricts the usefulness, cost, and feasibility of visual detection. Growers need to be able to recognize ACP automatically to make appropriate management decisions and reduce the development of disease, as shown in Figure 4.14.

4.6.2 Data Fusion Using AI for Crop Monitoring

Due to a lack of streamlined data collection and integration, most information about soil moisture, plant health, and insect infestation in agricultural areas must be gathered by hand. However, with the advent of remote sensing for automated

Figure 4.14 AI-based vision system detects and counts ACP.

scanning and data analysis, additional data sources (soil analyses, weather forecasts), and powerful AI algorithms, the application of fertilizers and pesticides has been on the decline in recent years.

4.6.3 Geospatial Monitoring of Agricultural Digital Twins

Because it allows for the establishment of a reliable database for plants and fields, enhances the precision and breadth of crop predictions, and reduces the need for manual effort and monetary investment, a DT [12] is a generic but crucial component of agricultural applications. The DT simulates the real-world system in a controlled environment, allowing for the testing of multiple complex situations at once. The collecting of data on phenotypic traits, plantation inventories, the production of precise weed control maps, and the management of asset evolution across time are all examples of the current limited use cases.

The German DEAL project uses ML and DL-based geospatial AI algorithms to autonomously predict vegetation components from 3D point clouds (Figure 4.15). As a result, a self-sufficient process leading to an agricultural DT can be established. [17].

Figure 4.15 Two examples of vegetation objects created from 3D point cloud analysis.

Geospatial AI uses ML and DL algorithms to "distil" object-based information from 3D point clouds. This has practical applications in areas such as the interpretation of orchards and tree plantations, including the entire spectrum of what is seen as "trees" and "vegetation" (e.g., "crooked trees," "flat bushes," "diseased trees," etc.).

4.6.4 Irrigation Using AI to Save Water

The project used the QUHOMA platform [12], an end application for smart irrigation enabled by FIWARE (see Figure 4.16). Complex event processing (CEP) AI is used to determine the irrigation plan based on the most up-to-date irrigation algorithms. This approach is grounded in the daily collection of sensor data. Therefore, if the data stream was disrupted or imprecise as a result of sensor malfunction or failure and/or loss of connectivity for extended periods of time, computations would diverge from the optimal target.

Figure 4.16 Sensors and QUHOMA platform components.

4.7 Materials and Procedures

4.7.1 Software That Is Hosted on the Cloud

Amazon Web Services (AWS) provides a number of instance types—computers optimized for specific tasks—as well as the ability to create instance images—exact copies of an instance's configuration and software—so that several identical machines can be used. For the purposes of this research, a regular AWS instance (m5a.xlarge) serves as the application's central processing unit (CPU), housing both the website and the GUI. The AWS Command Line Interface (CLI) is used for interprocess communication between the several running instances. The CLI allows a single AWS instance to manage the allocation of resources across many AWS instances through the exchange of cloud-based instructions.

The AWS platform allows users to build instance images, which are copies of an instance's configurations and software that may be used to spin up multiple identical computers and to choose from a wide range of instances, or machines with varied configurations for different purposes. The principal application control machine (website server) in this study's code is a generic AWS instance (m5a.xlarge). The AWS Command Line Interface (CLI) is used for interprocess communication between the several running instances in Figure 4.17. With the AWS Command Line

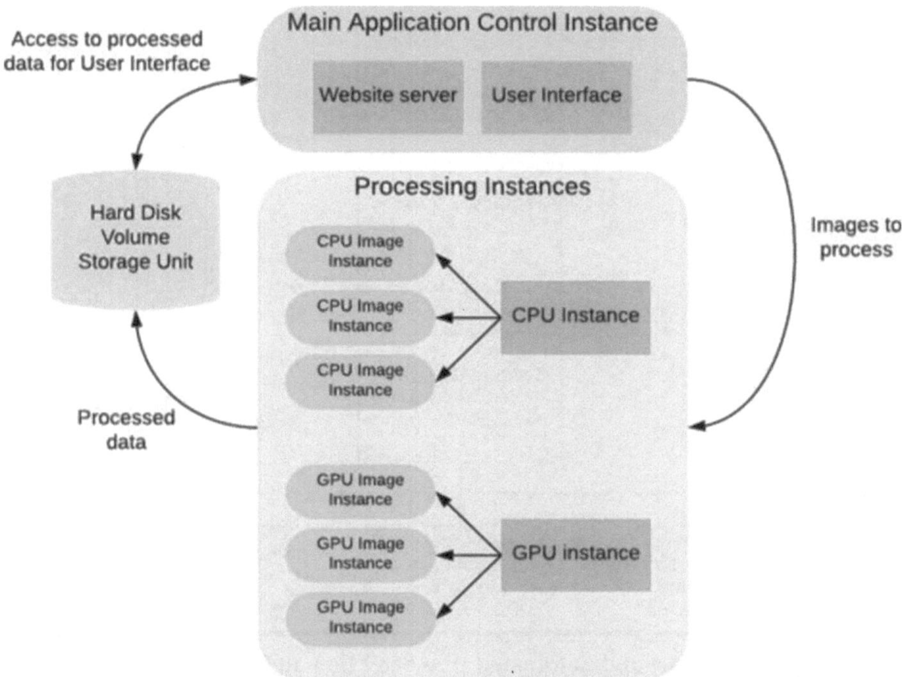

Figure 4.17 The Amazon Web Services (AWS) system workflow.

Interface (CLI), administrators of many cloud instances may manage and control their resources from a single location.

4.7.2 Algorithm for Detecting Trees

Data on (i) individual trees and tree gaps (skips), (ii) tree height, (iii) an estimate of tree canopy area, and (iv) tree health/stress is extracted from the Ortho mosaic using a C-based code (Figure 4.18). This research improves upon the techniques

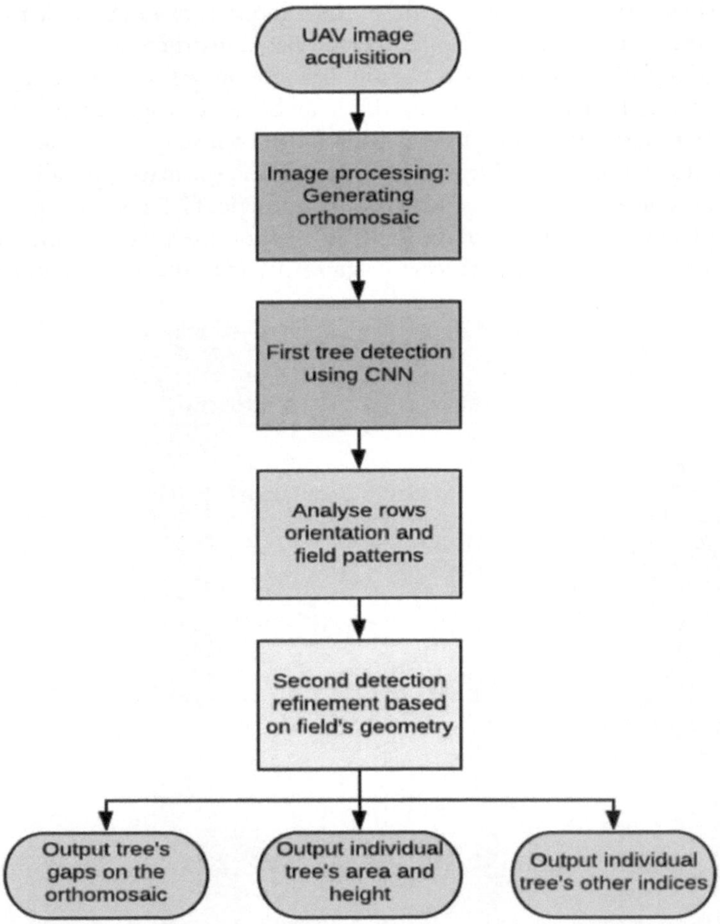

Figure 4.18 The smart and automated tree detection algorithm's workflow.

developed by Ampatzidis and Partel (2019). To sum up, the technique begins with the detection of CNN trees on the map and then uses the processed output to assess row orientation patterns in the field and compute distances between trees and rows. The algorithm predicts where trees might be planted based on the contours of the field and then applies a second classificatory CNN to the places that have already been identified.

Gathering visual data from an orchard using UAVs (or similar aerial or ground platforms) is the first step in this "tree detection" approach (see Figure 4.19). Users can also choose to have their raw UAV photos analysed by a cloud-based engine and converted into an Ortho mosaic map via the Agroview website.

Having the Ortho mosaic hosted online gives the user additional room to make adjustments in the event of a mistake. To begin using the tree detection method, the user must (i) define the search region by drawing field borders around the area

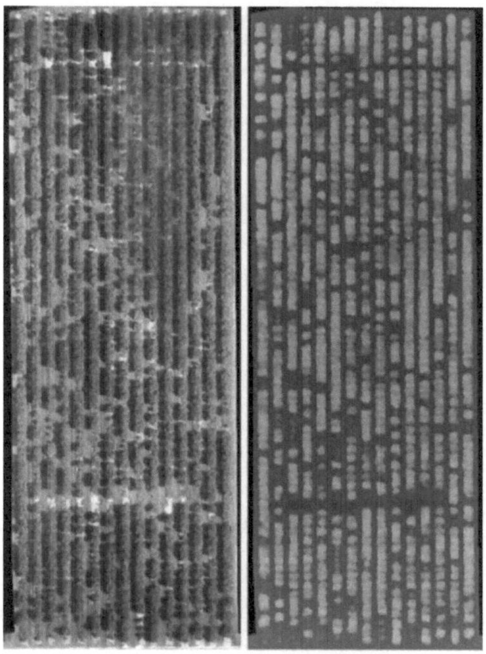

(a) (b)

Figure 4.19 The Open Drone Map (ODM) engine created the following citrus grove blocks: (a) Ortho mosaic map and (b) DSM.

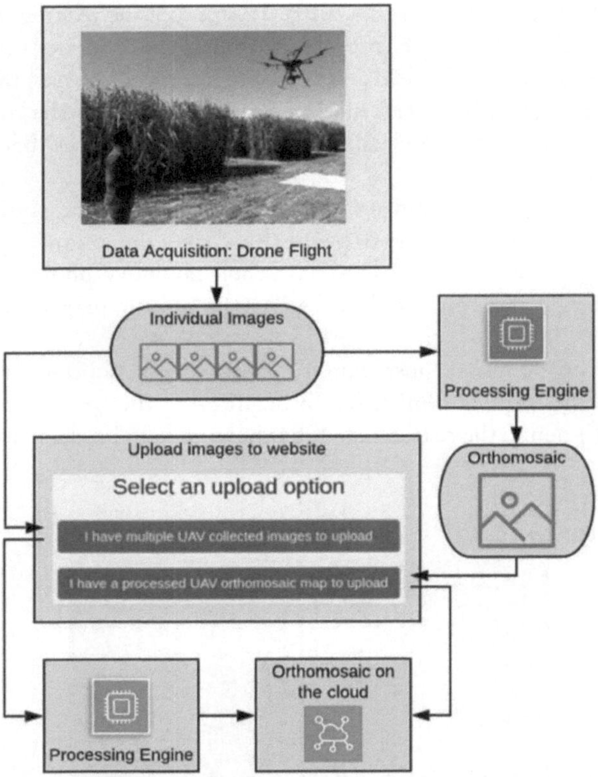

Figure 4.20 Data gathering from UAVs and image upload to the Agroview website workflow.

of interest, as shown in Figure 4.20. This is where the user can instruct the algorithm to ignore data from within the region that is likely to be irrelevant, such as bodies of water or man-made structures (Figure 4.21).

4.8 Vertical Farming Opinions in Support of PA

In Figure 4.22, we can see the proposed layout for the VF model that ushers in the next generation of agricultural technology. VF will thrive thanks to the Intelligent IoT idea, which incorporates AI and ML algorithms with productivity and quality measures. A data-driven approach, intelligent system agriculture, leverages emerging technologies to improve agricultural knowledge systems in light of a plethora of new data points [9]. With the development of data management comes a greater emphasis on data collection in modern agriculture; this helps users make more informed decisions. Many terms have been coined to describe the same concept of data-centric agriculture [16], which resulted from the combination of telematics,

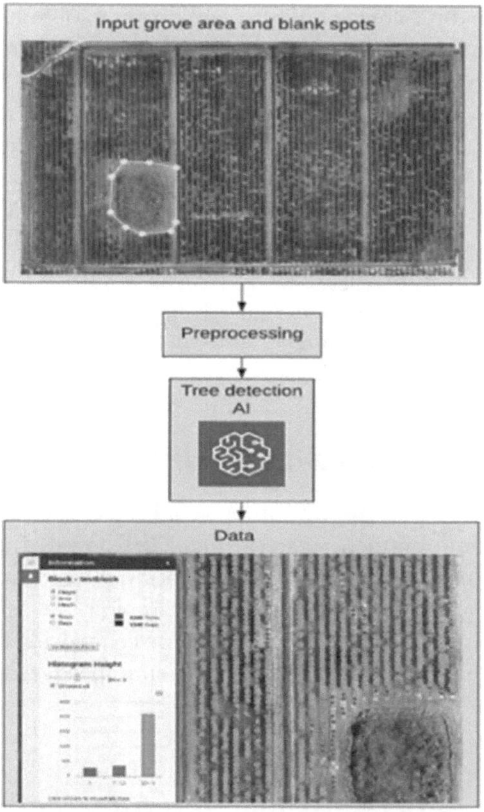

Figure 4.21 Workflow.

Figure 4.22 Smart vertical farming conceptual architecture

data management, and the concept of precision farming to improve crop yields and efficiency. Sensors and other devices made possible by the IoT provide farmers with a comprehensive record of their operations [7].

IoT innovations are a paradigm for Agriculture 4.0 [8] and may be a major reason why agricultures may supply a great quantity of vital information. It's expected that the expansion of the IoT' capabilities would have far-reaching effects on farming [6, 20]. As of late [10], agricultural apps developed by the CGIAR (Consortium of International Agricultural Research Centres) have been utilizing big data. Preprocessing activities and an AI algorithmic approach are used in data-based management in agriculture to ensure that the right data is collected and the right decisions are made [15]. Included in this category are sensor readings, information gathered from the ground around plants, and the final decision-making process.

4.8.1 AI Is Collaborating with Other Emerging Technologies

Smart greenhouses, cattle farms, and fisheries all make use of AI in conjunction with machine learning, big data analytics, and IoT-powered smart edge devices like GPS, drones, sensors, RFID, and LED lighting. The most fascinating aspect is the application of innovative blockchain technology in agricultural settings. Supply chain optimization that helps farmers in general is facilitated by blockchain because of its existing transparency, processing speed, zero-time latency, and increased accountability [18, 21].

4.8.2 Precision Agriculture Solutions from Innominds

Its basic inference engine and AI and machine learning capabilities enable it to collect and analyse large amounts of data from a wide variety of heterogeneous sources, including sensors, drones, and other IoT—enabled connected/edge devices [15]. The number of nodes that iFusion can process in a short amount of time is staggering. To help farmers make quick, educated decisions (decision agriculture), the collected and analysed data is presented in highly visual dashboards as tailored, accurate agriculture reports/analysis. Our state-of-the-art technical solutions have been extremely helpful to businesses in the worldwide food, agriculture, and farming industries. We offer IoT-powered connected smart (edge)devices including smart sensors, drone solutions, monitoring systems, and so on by combining our knowledge of hardware, software/embedded engineering, and the IoT [24, 25].

Figure 4.23 depicts the need for consistent collaboration between industry leaders, technological companies, and agriculture specialists to enhance, innovate, and harness the power of artificial intelligence.

As AI and other emerging technologies have been shown to be a game changer with huge potential to raise farm crop output, governments throughout the world should educate and pressure farmers to use them to boost farm crop productivity.

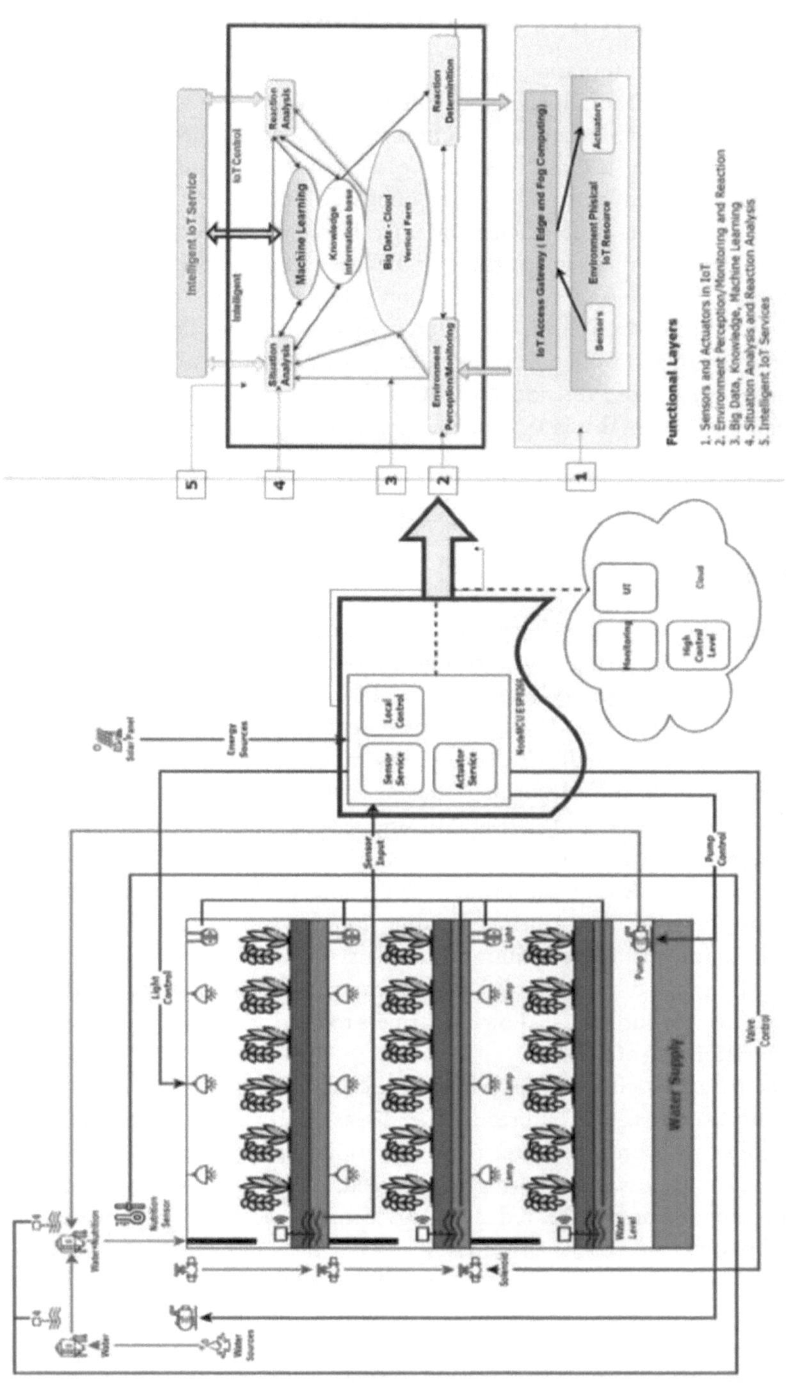

Figure 4.23 Innominds solutions for precision agriculture.

4.8.3 Agriculture's Future Will Be Technology-Enabled

Smart technologies and automation are the future of agriculture, and early adopters will reap the benefits. Although the exact price is yet unknown, we can be confident that as more technology companies enter the market, competition will drive efficiency and cost down, leading to widespread adoption and, ultimately, better agricultural outcomes and increased food productivity for everyone. By making use of their limited lands in a more efficient manner, farmers may be able to not only find lasting solutions to their agricultural challenges but also generate substantial profits. However, this can only come to fruition if leaders in the agricultural business, tech firms, and AI researchers all work together to improve, develop, and harness AI's potential. In addition, governments everywhere should spread the word to farmers about the value of Precision Agriculture by promoting the use of cutting-edge technologies like AI, This is a game-changer with huge potential for increasing crop yields in agriculture

4.9 Conclusion

Precision agriculture, made possible by breakthroughs in AI, has changed modern farming practices. This cutting-edge farming approach makes use of data, automation, and smart technologies to boost farm profits, resource efficiency, and environmental sustainability [5, 8]. AI algorithms process huge amounts of data from various sources, including satellites, drones, sensors, and historical farm data. As a result, farmers may make more informed decisions about planting, irrigation, fertilization, and pest management. As a result, businesses can distribute resources more effectively and decrease waste. Machines and robotics powered by AI are integrated with sensors and cameras that can evaluate soil conditions, monitor crop health, and perform precise tasks like planting, weeding, and harvesting [13]. This reduces the need for human work while simultaneously ensuring consistent, precise operations over a wide range of domains. AI models can forecast weather patterns, bug outbreaks, and disease outbreaks, allowing farmers to take preventive measures and alter their operations accordingly.

This proactive approach lowers crop losses while improving yield. AI models can forecast weather patterns, bug outbreaks, and disease outbreaks, allowing farmers to take preventive measures and alter their operations accordingly. This proactive approach lowers crop losses while improving yield. With AI-driven automation, farmers can optimize water usage, reduce energy consumption, and limit the use of fertilizers and pesticides [15]. Not only does this save waste, but it also lessens agriculture's ecological footprint. AI enables continuous agricultural monitoring via drones and satellites, providing real-time crop health insights. Early discovery of issues such as disease or nutrient deficiencies allows for rapid control, ultimately improving crop quality and productivity [16]. AI enables continuous agricultural monitoring via drones and satellites, providing real-time crop health insights. Early discovery of issues such as disease or nutrient deficiencies allows for rapid control, ultimately improving crop quality and productivity. Farm management software driven by AI makes it easier for farmers to plan, track, and analyze their actions. Inventory management, financial management, and regulatory compliance are all covered.

Gathering and photographing a large number of samples in their natural and dynamic environment is essential for training a vision-based AI system to execute effectively, as this best simulates the settings under which the AI would operate. As more high-quality data is collected, a deep learner's performance usually improves, helping the system to more effectively deal with image problems including inadequate illumination, poor alignment, and erroneous cropping of the object t [7]. When mobile hardware is combined with these AI algorithms and technologies [6], a platform is created that can identify and geolocate pests and diseases and give a prescription map (suited for use with precision equipment) for variable rate pesticide administration. Pesticide applicators will have the resources they need to reduce waste and ecological damage by applying the appropriate amount of pesticide in the appropriate location. Automated harvesting and trimming solutions for fruits and vegetables can also be developed using these technologies [7]. The development of efficient and cost-effective AI-based solutions for precision agriculture is still in its infancy. AI's revolutionary effect on precision agriculture helps feed a growing global population while reducing pollution and wasting fewer natural resources [8]. It is a significant step towards a more sustainable and efficient agriculture industry that can feed the world's population while remaining environmentally responsible.

References

1. Shaikh, Tawseef Ayoub, Rasool, Tabasum and Lone, Faisal Rasheed, "Towards leveraging the role of machine learning and artificial intelligence in precision agriculture and smart farming, computers and electronics in agriculture", Comput. Electron. Agric. 198, 2022, 107119, ISSN 0168-1699, https://doi.org/10.1016/j.compag.2022.107119.
2. Sharma, A., Jain, A., Gupta, P. and Chowdary, V., "Machine learning applications for precision agriculture: A comprehensive review", IEEE Access, 9, 2021, 4843–4873, doi: 10.1109/ACCESS.2020.3048415.
3. Zhang, P., Guo, Z. and Ullah, S. et al., "Nanotechnology and artificial intelligence to enable sustainable and precision agriculture", Nat. Plants 7, 2021, 864–876, https://doi.org/10.1038/s41477-021-00946-6.
4. Wakchaure, Manas, Patle, B.K. and Mahindrakar, A.K., "Application of AI techniques and robotics in agriculture: A review, artificial intelligence in the life sciences", Artif. Intell. Life Sci. 3, 2023, 100057, ISSN 2667-3185, https://doi.org/10.1016/j.ailsci.2023.100057.
5. Oliveira, R.C.d. and Silva, R.D.d.S.e., "Artificial intelligence in agriculture: Benefits, challenges, and trends" Appl. Sci. 13(13), 2023, 7405. https://doi.org/10.3390/app13137405.

6. Patrício, Diego Inácio and Rieder, Rafael, "Computer vision and artificial intelligence in precision agriculture for grain crops: A systematic review, computers and electronics in agriculture", Comput. Electron. Agric. 153, 2018, 69–81, ISSN 0168-1699, https://doi.org/10.1016/j.compag.2018.08.001.
7. Ampatzidis, Yiannis and Gorucu, Serap, "Drone Injuries and Safety Recommendations: AE560/AE560, 06/2021", EDIS 2021(3), 2021, https://doi.org/10.32473/edis-ae560-2021.
8. Ampatzidis, Yiannis, Partel, Victor and Costa, Lucas, "Agroview: Cloud-based application to process, analyze and visualize UAV-collected data for precision agriculture applications utilizing artificial intelligence", Comput. Electron. Agric., 174, 2020, 105457, ISSN 0168-1699, https://doi.org/10.1016/j.compag.2020.105457.
9. Siregar, R.R.A., Seminar, K.B., Wahjuni, S. and Santosa, E. "Vertical farming perspectives in support of precision agriculture using artificial intelligence: A review", Computers 11, 2022, 135. https://doi.org/10.3390/computers11090135.
10. Sishodia, R.P., Ray, R.L. and Singh, S.K, "Applications of remote sensing in precision agriculture: A review", Remote Sens. 12, 2020, 3136. https://doi.org/10.3390/rs12193136.
11. Linaza, M.T., Posada, J., Bund, J., Eisert, P., Quartulli, M., Döllner, J., Pagani, A., G. Olaizola, I., Barriguinha, A. and Moysiadis, T.; et al., "Data-driven artificial intelligence applications for sustainable precision agriculture" Agronomy 11, 2021, 1227. https://doi.org/10.3390/agronomy11061227.
12. Leal Filho, W., Yang, P. and Eustachio, J.H.P.P. et al., "Deploying digitalisation and artificial intelligence in sustainable development research", Environ. Dev. Sustain. 25, 2023, 4957–4988. https://doi.org/10.1007/s10668-022-02252-3.
13. Javaid, Mohd, Haleem, Abid, Haleem Khan, Ibrahim and Suman Rajiv, "Understanding the potential applications of Artificial Intelligence in Agriculture Sector." Advanced Agrochem. 2(1), 2023, 15–30. https://doi.org/10.1016/j.aac.2022.10.001
14. Sharma, Amit, Sharma, Ashutosh, Tselykh, Alexey, Bozhenyuk, Alexander, Choudhury, Tanupriya, Abdu Alomar, Madani and Sánchez-Chero, Manuel, "Artificial intelligence and internet of things oriented sustainable precision farming: Towards modern agriculture", Open Life Sci. 18, 2023, 20220713. doi: 10.1515/biol-2022-0713
15. Sowmia, K. R., Prithi, S., Vijay, K., Eugene Berna, I. and Bhuvaneswaran, B., "Crop Monitoring System with Water Moisture Level using Arduino," In 2023 5th International Conference on Smart Systems and Inventive Technology (ICSSIT), Tirunelveli, India, 2023, pp. 495–498, doi: 10.1109/ICSSIT55814.2023.10060913
16. Ampatzidis, Yiannis, "Applications of artificial intelligence for precision agriculture: Ae529, 12/2018," EDIS 2018(6), 2018, doi:10.32473/edis-ae529-2018
17. Nemenyi, Miklos, "Precision crop production and artificial intelligence–the future of sustainable agriculture," Acta Agraria Debreceniensis 2018, 47–58, doi:10.34101/actaagrar/150/1701
18. Vijay, P., Shanthi D, S., P, R., Ohja, S., N, S. and Nandhiannal, "Grainchain –Agricultural supply chain traceability and management technique for farmers sustainability using blockchain hyper ledger", Int. J. Intell. Syst. Appl. Eng., 2022, 10(3), 141–146.
19. Das, Archana, Das, Saswat and Rathee, Naveen, "Roles of Big Data, Data Science, Artificial Intelligence in Entrepreneurships", In Artificial Intelligence in Entrepreneurships (December 25, 2021). Proceedings of the International Conference on Advances in Management Practices (ICAMP 2021). 2021. doi:10.2139/ssrn.3993704.

20. Adewusi, Adebunmi Okechukwu, Onyeka Franca Asuzu, Temidayo Olorunsogo, Chinwe Iwuanyanwu, Ejuma Adaga and Donald Obinna Daraojimba, "AI in precision agriculture: A review of technologies for sustainable farming practices", World J. Adv. Res. Rev. 21(1), 2024, 2276–2285, https://doi.org/10.30574/wjarr.2024.21.1.0314.
21. Leema, Roselin G., Kiruba, K., Hemalatha, P., Manikandan, J., Madhin, M. and Mohan, Raj S.. "Revolutionizing Secure Commercialization in Agriculture Using Blockchain Technology." In IEEE 2021 International Conference on System, Computation, Automation and Networking (ICSCAN), pp. 1–6, 2021
22. Srinivasan, S.P., Shanthi, D.S. and Anand, A.V., 2017, June. Inventory transparency for agricultural produce through IOT. In IOP Conference Series: Materials Science and Engineering (Vol. 211, No. 1, p. 012009). IOP Publishing.
23. Gnanavel, S., Sreekrishna, M., DuraiMurugan, N., Jaeyalakshmi, M. and Loksharan, S., "The Smart IoT based Automated Irrigation System using Arduino UNO and Soil Moisture Sensor," In 2022 4th International Conference on Smart Systems and Inventive Technology (ICSSIT), Tirunelveli, India, 2022, pp. 188–191, doi: 10.1109/ICSSIT53264.2022.9716368.
24. Murugamani, C., Shitharth, S., Hemalatha, S., Kshirsagar, Pravin R., Riyazuddin, K., Naveed, Quadri Noorulhasan, Islam, Saiful, Mazher Ali, Syed Parween and Batu, Areda, "Machine learning technique for precision agriculture applications in 5G-based internet of things", Wireless Commun. Mobile Comput., 2022, Article ID 6534238, 11 pp., 2022. https://doi.org/10.1155/2022/6534238
25. Nagasubramanian, G., Sakthivel, R. K., Patan, R., Sankayya, M., Daneshmand, M. and Gandomi, A. H, "Ensemble classification and IoT-based pattern recognition for crop disease monitoring system", IEEE Internet Things J., 2021, 8(16), 12847–12854.
26. Mekala, M. S., & Viswanathan, P. A Survey: Smart agriculture IoT with cloud computing. Proceedings of the 2017 International Conference on Microelectronic Devices, Circuits and Systems (ICMDCS), 2017, Vellore, India, pp. 1–7. IEEE, http://dx.doi.org/10.1109/ICMDCS.2017.8211551.

Chapter 5

AI Model Generation Platforms

J. Jeyalakshmi, Sowmia K.R., R. Babu, and Ravikumar S.

5.1 Introduction

Cloud-based artificial intelligence (AI) model generation systems for precision agriculture enable farmers and agricultural experts to create and deploy their own AI models for use in the sector. Common AI resources may be accessed through these hubs. Common precision agricultural AI activities like crop yield prediction, pest detection, and disease diagnosis may be performed quickly and accurately with pre-trained AI models. Datasets used for precision agricultural training, machine learning libraries, and deep learning frameworks are examples of AI development tools and frameworks. Compared to more conventional approaches to building AI models, platform-based AI model development for precision agriculture offers several advantages, including tools for developing and deploying AI models in the cloud. Firstly, they remove the necessity for farmers and agricultural experts to have knowledge of machine learning or AI in order to create and implement AI models. Secondly, they facilitate access to a wealth of AI technologies and resources that would otherwise be too expensive or time-consuming to acquire and administer. Thirdly, they facilitate the deployment of AI models to production with little effort and no requirement for farmers and agricultural specialists to manage their own hardware.

In order to anticipate crop yields, identify pests, and diagnose diseases, farmers can use Agrilytics, a cloud-based platform that uses AI. Access to several AI development tools, datasets for training, and pre-trained models is all available through

DOI: 10.1201/9781003504900-5

Agrilytics. FarmShots is an online service that helps farmers track their crops and anticipate their yields using AI. FarmShots gives you entry to several pre-trained AI models, datasets for training, and AI-building resources. To better anticipate crop yields, regulate nutrients, and combat pests, farmers may make use of Granular, a cloud-based platform that facilitates the creation and deployment of AI models. Granular gives users access to a library stocked with AI development tools, training datasets, and pre-trained models.

Farmers and agricultural experts may enhance their crop yields, decrease their costs, and boost their profitability with the use of AI model generation platforms for precision agriculture. The use of AI model-generating platforms in precision farming.

Precision agriculture AI model-generation systems allow for the creation of several AI models for usage in a wide range of agricultural tasks. To cite just a few potential applications of AI models:

- Estimate harvest success
- Pest and disease detection
- Keep an eye on the crop's condition
- Nutritional management
- Managing pests
- Irrigation optimization
- Schedule changes

Once an AI model has been built, it may be implemented on various gadgets including tractors, mobile phones, and tablets. This facilitates instantaneous access to and utilization of AI models by farmers and agricultural experts in the field.

Precision agriculture can benefit greatly from the use of AI model-generating platforms. Using AI model-generation platforms for precision agriculture has several advantages.

- Improved harvest success
- Saved money
- Revenue growth
- Enhanced long-term viability
- Impact on the environment diminished
- A more secure food supply

Farmers and agricultural experts who want to boost their operations and increase sustainable food production might benefit greatly from using AI model generation platforms for precision agriculture.[1]

The impact of AI on society is growing rapidly. The CEO of Google, Sundar Pichai, has compared the impact of AI to that of electricity and fire, and leading tech companies are spending extensively in the field. Companies that have implemented

AI to enhance their offerings have seen substantial financial benefits. The potential of AI is only now becoming apparent. However, many companies still struggle to see the business potential in AI. It's exceedingly difficult to deploy meaningful AI into production since tools are always evolving, use cases are dispersed across the company, and there are so many of them. When applying machine learning and deep learning to business challenges, data science and engineering teams face a bewildering array of platforms, libraries, and tools.[2] It takes a collaborative effort across departments, strategic investments in technology, and regular re-calibration of processes to evaluate and make decisions about how to apply cutting-edge tools and approaches to design, deploy, and maintain high-performing models.

You can weigh the benefits and drawbacks of using an AI platform to help you build and implement data science and machine learning algorithms, models, and systems. As you want to leverage the potential of AI at scale, we'll go through some of the most prevalent use cases among enterprise organizations and the open-source technologies at your disposal. Finally, we will discuss what to look for in a reputable AI platform and what questions to ask any prospective vendors (Figure 5.1).[3].

The following are leading agricultural research organizations.

- **Food and Agriculture Organization of the United Nations (FAO):** To end world hunger, the United Nations established the FAO. The FAO works towards a world where everyone always has enough to eat so that they can thrive physically, mentally, and socially.
- **World Economic Forum (WEF):** While it lacks legal authority to make decisions on its own, the WEF advocates for policies that will benefit humanity on a worldwide scale. The group relies on dues paid by its members, who include many influential people in business and politics.
- **McKinsey Global Institute (MGI):** They help businesses with everything from developing a global business services and sourcing program to implementing a shared services strategy, establishing a shared services environment, scaling up that environment, and transforming it with cutting-edge digital and analytics tools.

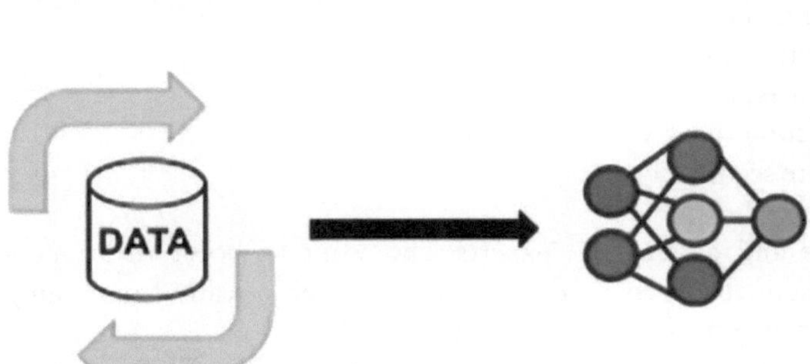

Figure 5.1 Data-Centric AI Model.

- **Industry publications:** Several academic periodicals and magazines have published special issues devoted to the use of AI in the agricultural sector.
- **Academic research:** There is a proliferation of academic research on the topic of using AI in agriculture. There have been several conferences and events where this research has been presented.

The following are a few samples of the information offered by different organizations about the advantages of AI in agriculture:

- **Increased crop yields:** There is evidence that using AI in farming can increase yields by as much as 20%. This is due to the fact that AI is assisting farmers in increasing yields through the improved application of precision farming practices such as irrigation, fertilization, and pest control.
- **Reduced costs:** AI could save the agricultural industry up to 15% in costs by boosting productivity and reducing waste.
- **Increased profits:** AI has the potential to increase agricultural profits by as much as 10 percent. Reasons for this include decreased prices and increased harvests. That's what Figure 5.2 shows.
- **Improved sustainability:** With the help of AI, farmers may be able to reduce their reliance on chemical fertilizers and pesticides, thus making their operations more environmentally friendly. AI has the potential to help farmers in a number of other ways, including water conservation and erosion prevention.
- **Reduced environmental impact:** Through the use of fewer pesticides and fertilizers, farmers' environmental impact can be lessened with the assistance of AI. Water management and soil erosion are two other areas where AI may aid farmers.
- **Improved food safety:** By identifying pests and diseases at an early stage, AI can assist farmers in maintaining a safe food supply and keeping track of their goods throughout the distribution process to guarantee their integrity and quality.

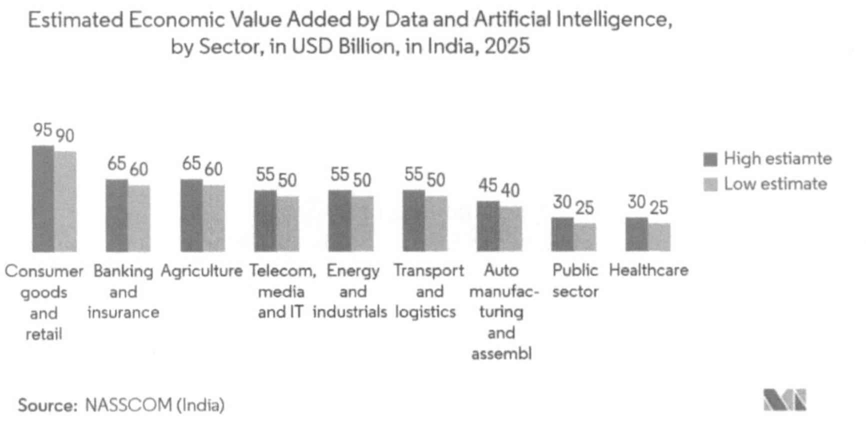

Figure 5.2 Nasscom(India) report on Estimated Economic Value Addition by AI

These broader advantages are in addition to the fact that AI may provide answers to some pressing issues in the agricultural industry. One application of AI is the development of new varieties of disease- and pest-resistant crops. Using AI in new farming techniques could help mitigate the disastrous effects of climate change.[4]

In general, AI has the ability to completely alter the farming sector. AI can help farmers increase their agricultural yields and decrease their operating expenses and carbon footprints.[5]

Here are a few concrete applications of AI in the field:

- Drones with AI cameras can detect the earliest signs of crop damage or disease.
- Tasks that can be dangerous for people or take a long time to complete, such as agricultural harvesting, are increasingly being performed by robots powered by AI.
- With the help of AI, pest- and disease-resistant agricultural varieties are being produced.
- Greenhouse gas emissions are being reduced through the application of AI in new farming practices.

Although AI is still a young technology, it is already making a significant impact in the farming industry. As AI develops, it will become increasingly useful in the agricultural sector.[6]

5.2 Artificial Intelligence and Machine Learning Models in Agriculture

An AI platform is a collection of tools for creating, testing, releasing, and updating machine learning and deep learning models. AI-enhanced machines outperform humans in areas where speed, accuracy, and data analysis are crucial.[7]

Developers, testers, publishers, and upgraders of ML and DL models need access to an AI platform. The phrase "artificial intelligence" (AI) is used to describe a group of technologies that enable machines to evaluate data, make predictions, and take action in ways that rival or exceed those of human beings:

1. combine data science projects and findings
2. streamline ML operations (also known as ML development and production workflows)
3. help data scientists and engineers work together
4. automate some of the processes needed to create AI systems
5. control the implementation of AI models and systems

There are many ways to prepare an AI platform for data science and engineering practitioners.[8]

Many agricultural issues may be addressed with the help of AI models.

Predicting agricultural yields using weather, soil, and previous data may be done with the help of an AI model. Farmers may use this data to refine their strategies for seeding, watering, and fertilising.

Early identification of pests and illnesses can prevent extensive agricultural loss when AI models are deployed. This data can aid farmers in taking preventative measures against pests and illnesses.

Monitoring crop health and spotting issues early may be accomplished with the help of AI models. Farmers may use this knowledge to take preventative actions and reduce the likelihood of crop failure.

AI models can be used to improve nutrition management. Farmers can benefit from this since it can increase their agricultural yields while decreasing their ecological footprint.

- The health of cattle may be monitored and possible issues identified with the use of AI models. Farmers may use this knowledge to make preventative efforts to reduce the number of animal deaths.
- Scheduling and allocating resources are only two examples of how AI models may be utilized to enhance farm management. This can enhance productivity and income for farmers.
- Machine learning methods are often used in the development of agricultural AI models. The algorithms used in machine learning are "trained" using datasets that have the intended result labelled. An AI model for pest identification, for instance, may be taught using a dataset comprising both pest-free and pest-infested plant photos. A trained model may then be used to determine if an unlabelled image includes a pest.

Although still in their infancy, AI models have the potential to completely transform the agricultural sector. AI can help farmers increase food production while decreasing their negative effects on the environment by increasing crop yields, decreasing costs, and increasing sustainability.[9]

Some current applications of AI-based models in the agricultural sector include the following:

- AI algorithms are being used by farmers to forecast crop harvests and pinpoint problem spots in their farms. With this data, they'll be able to refine their strategies for seeding, watering, and fertilizing.
- Preventing severe crop loss by using AI models to identify pests and illnesses as soon as they appear. This data can aid farmers in taking preventative measures against pests and illnesses.
- The well-being of crops and cattle is being tracked using AI algorithms. This knowledge will aid farmers in spotting developing issues and taking corrective action.
- Nutrient management strategies are being improved with the use of AI algorithms. Farmers can benefit from this since it can increase their agricultural yields while decreasing their ecological footprint.

- Scheduling and resource allocation are only two examples of farm management practises that have benefited from the application of AI models. This can enhance productivity and income for farmers.
- The use of AI farming models might significantly improve international food supplies. AI may help ensure that everyone has access to the food they need by assisting farmers in producing more food with less resources.[10]

5.3 AI Model Generation for Precision Agriculture

A methodical approach is required in order to generate an AI model which is generalized irrespective of the deployment industry. Firstly, the issue must be defined, and then pertinent, high-quality data must be collected. After that, preprocess the data and engineer features based on them. To avoid overfitting, select a suitable machine learning or deep learning model, create its architecture, and train it while monitoring its performance. Adjust the hyperparameters as needed to achieve the desired results, then assess the model using the metrics that are most applicable.[11] Once you are pleased with the model, you should deploy it to production settings, regularly check its performance, and retrain whenever necessary. Throughout the entirety of the process, it is essential to prioritize ethical concerns, documentation, scalability, security, legal compliance, and input from users. Keeping up with the latest breakthroughs in AI is absolutely necessary for finding success in this fast-paced sector. Figure 5.3 and Table 5.1 narrate the steps involved in the generation of an AI model for any generic purpose.[8]

The process is narrated and simplified as a series of steps in below Table 5.1.

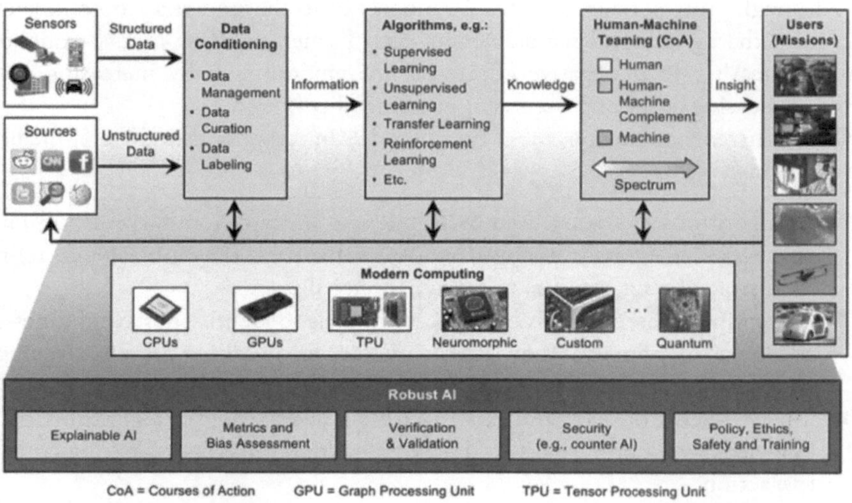

Figure 5.3 AI Model Generation Process

Table 5.1 Steps of Model Development AI

Step	*Description*
1. Define the Problem	Clearly state the problem to be solved and establish objectives.
2. Data Collection	Gather relevant, high-quality data for the AI model.
3. Data Preprocessing	Clean, preprocess, and handle data issues like missing values.
4. Feature Engineering	Select an engineer features from the data for model input.
5. Model Selection	Choose an appropriate machine learning or deep learning algorithm.
6. Model Architecture	Design the neural network architecture for deep learning models.
7. Training	Train the model using training data while monitoring performance.
8. Hyperparameter Tuning	Experiment with hyperparameters to optimize model performance.
9. Evaluation	Assess the model's performance using test data and relevant metrics.
10. Model Deployment	Deploy the model into production environments for practical use.
11. Monitoring & Maintenance	Continuously monitor and retrain the model to keep it up-to-date.
12. Ethical Considerations	Address ethical and fairness issues related to the model.
13. Documentation	Properly document the model, its architecture, and preprocessing steps.
14. Scaling	Scale the model to handle larger datasets or more users if needed.
15. Security	Implement security measures to protect the model and data.
16. Legal Compliance	Ensure compliance with relevant laws and regulations
17. User Feedback	Collect user feedback and use it to improve the model.
18. Incremental Learning	Keep up with the latest AI developments to stay at the cutting edge.

5.4 Platforms for AI Model Generation in Precision Agriculture

The term "artificial intelligence" (AI) now refers to a phenomenon that has permeated nearly every facet of our lives. Increasing numbers of people are employing AI technology in their enterprises and their day-to-day activities. It is essential to have a solid grasp of the various AI platforms that are currently accessible in order to make the most of the potential of AI. Various platforms are narrated in Table 5.2.[9]

Table 5.3 shows the incorporation of platforms for Smart Agriculture and a comparison of their services. The data-centred management of farming using AI

Table 5.2 AI Model Generation Platforms

Platform/Framework	*Description*
TensorFlow	Google's open-source deep learning framework for various AI tasks.
PyTorch	Facebook's open-source deep learning framework with dynamic computation graph.
Keras	High-level neural networks API that runs on top of TensorFlow, Theano, or CNTK.
Scikit-learn	Versatile Python library for traditional machine learning algorithms.
MXNet	Open-source deep learning framework known for efficiency.
Caffe	Deep learning framework favoured for image-related tasks.
Microsoft Azure ML	Cloud-based service for building, training, and deploying machine learning models.
Google Cloud AI Platform	Google's cloud-based platform for machine learning and AI model development.
IBM Watson Studio	IBM's cloud-based platform with AI model development and collaboration tools.
AutoML Platforms	Automated Machine Learning platforms for simplifying model training.
Docker and Kubernetes	Containerization and orchestration tools for deploying AI models in production environments.

Table 5.3 AI Model for Precision Agriculture

Platform	*Key Features*	*Data Sources*	*Customization*	*Integration*	*Industry Focus*
Agrilytics	Provides predictive analytics for crop management and yield optimization. Offers disease and pest detection using AI.	Integrates weather data, satellite imagery, and on-farm sensors.	Customizable AI models for specific crop and field conditions.	Supports integration with third-party farm management software.	Broad range of crops and farming types.
FarmShots	Offers satellite and drone imagery analysis for crop health assessment. - Provides alerts for potential issues in the fields.	Utilizes satellite and drone imagery.	Limited AI model customization compared to some other platforms.	Integrates with various farm management systems and mobile apps.	Focuses on crop health monitoring.
Granular	Offers farm management software with AI-driven insights. - Focuses on financial planning, budgeting, and inventory management.	Utilizes farm data, including yield records, equipment data, and field observations.	Customizable features for farm management.	Integrates with various equipment and software systems.	Broad range of crops with a focus on farm financials.

and drones is shown in Figure 5.4. The model shows the application of technology through the life cycle of plant growth. Also known as precision agriculture, data-driven farming involves the use of information gathering, statistical analysis, and technological advancements to enhance agricultural productivity.

Sensors, drones, and satellites are just some of the tools farmers use to collect and analyse data on things like weather, soil, and crop health. Informed decisions regarding planting, irrigation, and resource allocation may subsequently be made with the use of this processed data.[12, 13] A data-driven approach to farming improves productivity by keeping track of crops in real time and reducing waste.

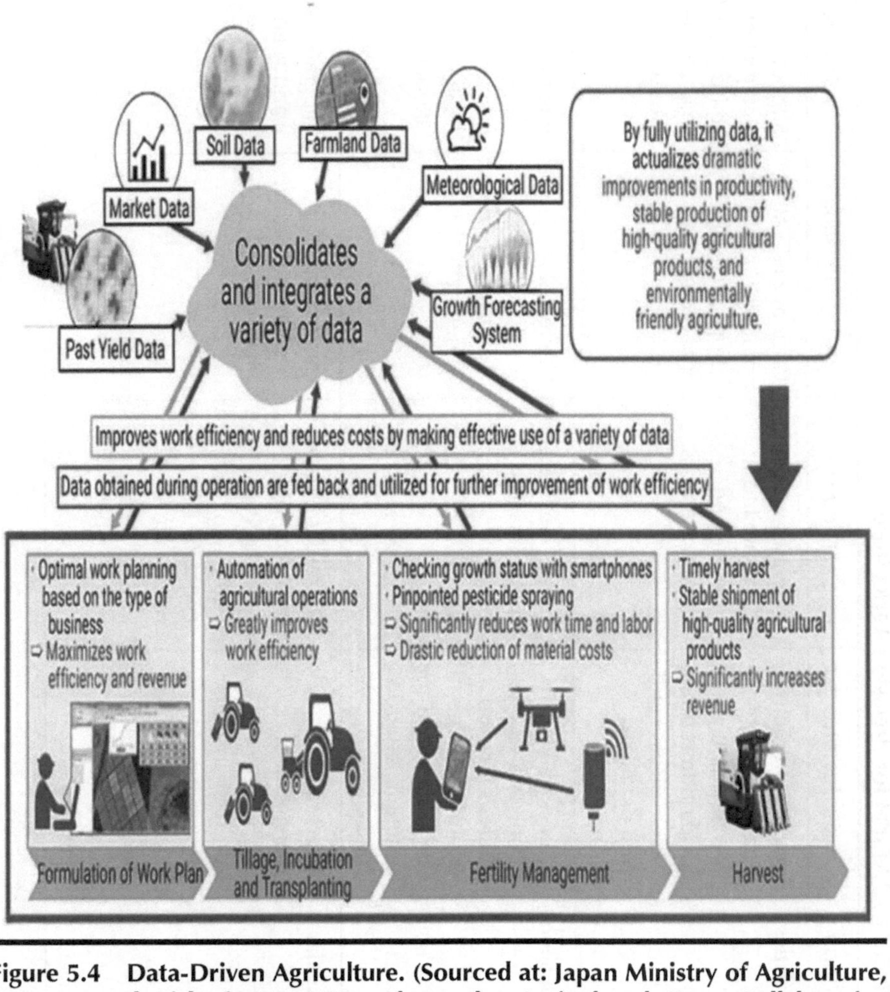

Figure 5.4 Data-Driven Agriculture. (Sourced at: Japan Ministry of Agriculture, Forestry and Fisheries (MAFF), *About the Agricultural Data Collaboration Platform*, January 2021. Abridged translation by Alexander)

It supports automated farming techniques and can be integrated with other farm management programs. A new era of agricultural production and environmental sustainability is being ushered in by this approach because it allows farmers to make decisions based on solid information.[14, 15, 16]

To improve agricultural decision-making, "data-driven agriculture" makes use of a wide variety of data, including that collected by farm sensors, satellite imaging, meteorological data, and market data. It has several uses in agriculture, including crop yield prediction, disease and pest early detection, crop health monitoring, nutrient management optimization, livestock health monitoring, and general farm management enhancement. Farmers may improve them practises across the board by using data to guide their decisions about seeding, watering, and fertilizing, among other things. More accurate and data-informed agricultural practises have the potential to significantly improve crop yields, lower production costs, and improve sustainability in the agricultural sector.[17]

The issues with smart agriculture are presented in Figure 5.5 table. AI opacity, AI unpredictability, Data Validity, and accuracy issues are major challenges in the deployment of AI models and platforms.

Classification	Causes of Risk	Impacts of Risk
Data Accuracy	Insufficient accuracy in data collection by sensors. Example: Deterioration of the soil sensor reduces the measurement accuracy, resulting in data collection with insufficient accuracy.	Analysis is based on inappropriate data so that wrong decisions are made. Example: Fertilizer is not applied when it should be because it is judged that there are still nutrients in the soil. Crop quality and yield are impacted. Example: When pesticides are used under low rainfall conditions, the pesticides are not sprayed because it is determined that soil moisture is still present. Residues of pesticides in crops above the standard values affect the health of crop consumers and can be subject to punishment by regulatory authorities.
Data Validity	The data collected by the sensors is biased. Example: Solar radiation sensors are installed in a limited number of locations so that the data are biased when estimating the amount of solar radiation for the entire cultivated land.	Analysis is based on inappropriate data so that wrong decisions are made. Example: Irrigation (watering) is not implemented when it should be because it is judged that soil moisture is still present. Impacts crop quality and yield.
AI Unpredictability	It is not possible to verify and predict the processing results of the machine learning (ML) engine at the construction stage. Example: When building a function for sorting (grading) crops by image recognition using ML, the correct answer rate does not reach 100 percent, even with a limited amount of evaluation data.	Inappropriately processed outputs result when targets that could not be learned in the training data at the time of construction are input during production operation. Example: AI gives a low grade to crops that are rare (not included in the training data) and have high-quality characteristics. Shipping price is impacted.
AI Opacity	It is difficult to analyze what calculations were made to reach the results of the processing of the ML model. Example: It is not possible to analyze and understand the basis for sorting (grading) of agricultural products from the image recognition and sorting engine using machine learning.	One is unable to account for the rationale for the processing results of the ML engine to affected stakeholders. Example: One is unable to provide an explanatory response to a complaint received from a crop consumer that the quality of the product is poor even though it is a high grade (the response can only be that AI judged it). The reputation of the crop brand is impacted.

Figure 5.5 Examples of Risks in Smart Agriculture. (Sourced at: https://www.isaca.org/resources/isaca-journal/issues/2022/volume-1/data-and-ai-management-in-smart-agriculture-using-soil-and-crop-data)

Agricultural drones, more commonly referred to as "ag drones," have quickly become indispensable equipment for today's farmers. They provide a number of benefits, including crop monitoring, precision farming, aerial mapping, yield prediction, monitoring of animals, and other similar services. Imagery captured by drones at a high resolution enables earlier diagnosis of crop problems and more focused resource applications, which in turn reduces resource waste and increases crop yields. They are helpful in the management of irrigation systems, the control of weeds, and even the collecting of meteorological data. In addition, drones improve farm safety and security, and the data collected by drones interacts with farm management systems, which enables farmers to make more informed decisions. Drones are playing an increasingly important part in today's agriculture because of their ability to cut expenses, boost productivity, and lessen their negative effects on the environment. Figure 5.6 shows the deployment model for AI-integrated drone-supported agriculture platform design.

The ag drones can be deployed for various purposes like crop management, pest management, aerial survey, livestock and irrigation control, and weather and safety

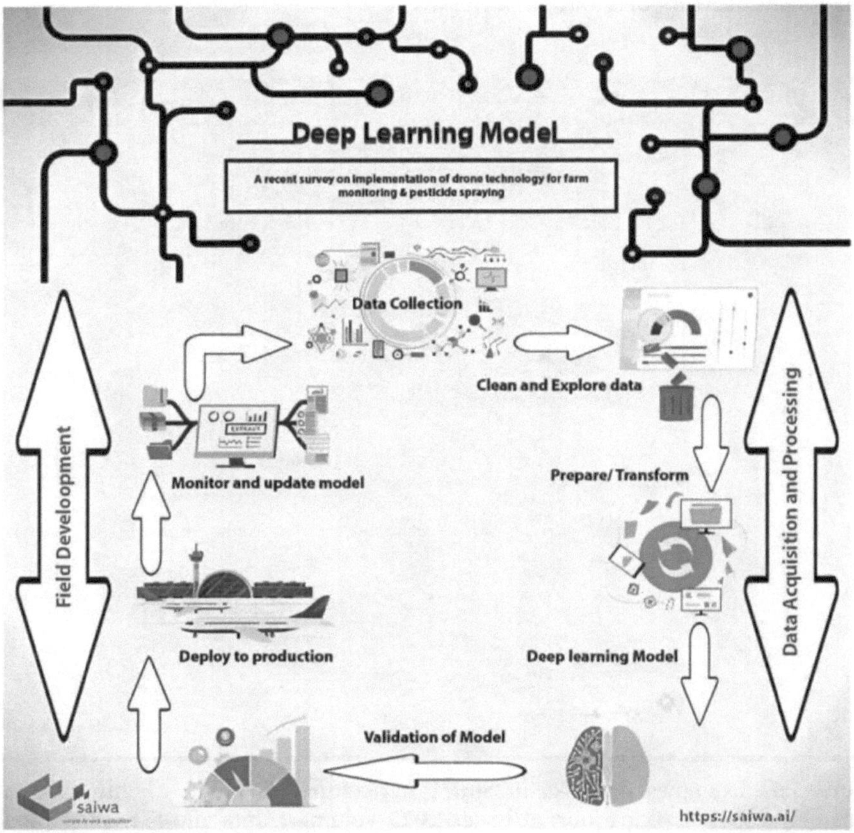

Figure 5.6 Drone Integration in Smart Agriculture (Source at: https://saiwa.ai)

mechanisms. When it comes to security, blockchain-based supply chain models are gaining importance in smart farming.

There are several advantages to be gained by utilizing blockchain and AI technology in conjunction with smart farming. While AI examines the data to get useful insights, blockchain maintains the data's authenticity and protects its confidentiality. It increases the visibility of the supply chain, makes it easier to use smart contracts to automate operations, and lends support to the integration of IoT devices to enable real-time monitoring of crop conditions and management of resources. Precision farming practises, which improve crop yields, reduce environmental impact, and promote sustainability, are made possible by predictive analytics powered by AI and decision-making that is driven by data.[18] This combination strategy improves efficiency and transparency across the agricultural value chain, so contributing to the development of an agricultural sector that is both more productive and less harmful to the environment. The blockchain-based architecture is explained in Figure 5.7, and smart farming architecture is explained in Figure 5.8.

It is anticipated that the market for AI applications in agriculture will considerably expand, reaching $4.7 billion by 2028. The use of AI in farming has several potential benefits, including increased productivity and more environmental

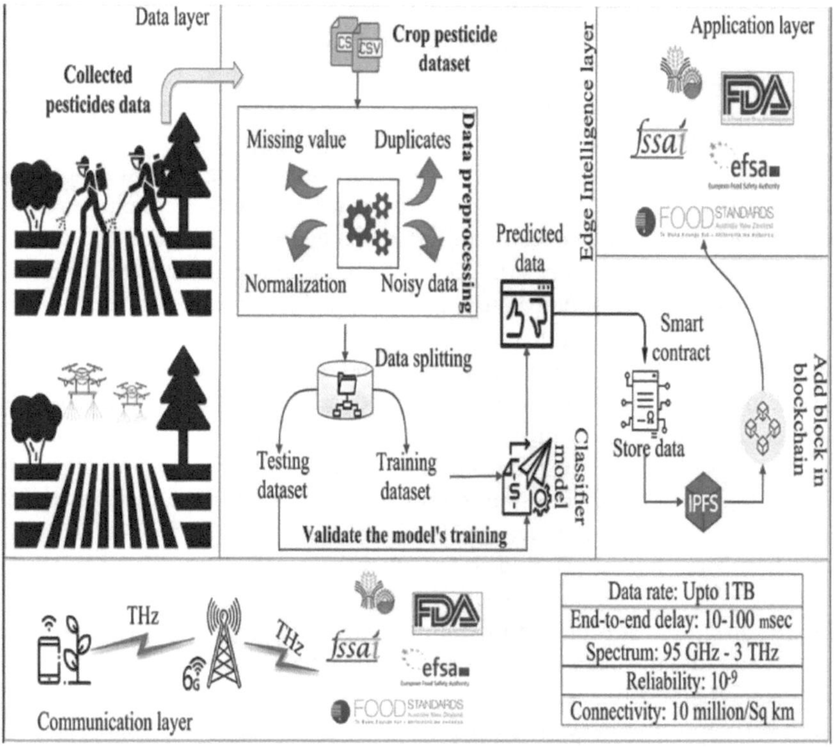

Figure 5.7 Blockchain-based Smart Agriculture Framework[19]

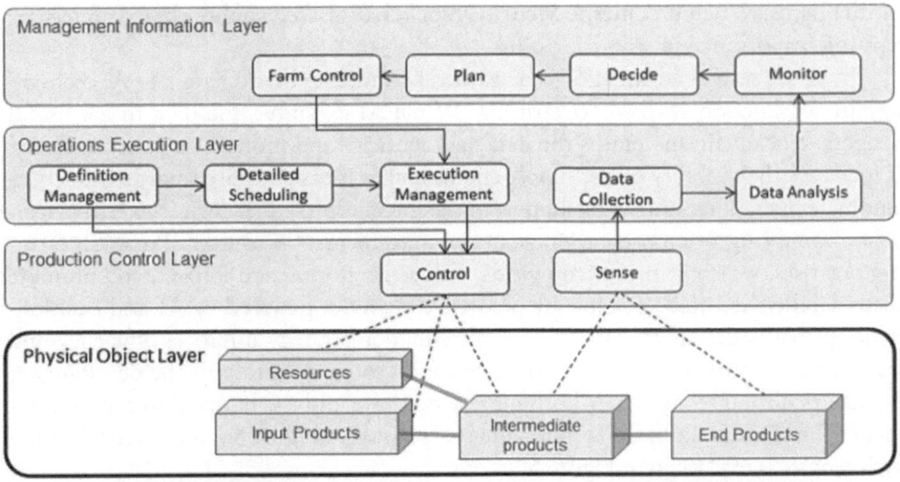

Figure 5.8 Smart Farming Architecture[20]

friendliness. Some of the most important uses of AI in farming are irrigation optimization; leak detection in irrigation systems; crop and soil monitoring; disease and pest detection; livestock health monitoring; pesticide application intelligence; yield mapping; automatic weeding, harvesting, and sorting; and surveillance [21, 22].

Combining AI with other technologies like big data, IoT sensors, and automation helps farmers make better decisions, reduces risk, and improves plant breeding, soil analysis, and crop health checks. The early expenses of adopting AI, reluctance to new technologies, a lack of practical expertise, infrastructural limits, technological limitations, privacy, and security concerns are some of the hurdles that must be overcome. In spite of these obstacles, AI has the ability to completely transform the agricultural sector, making it more productive, sustainable, and efficient in the process.

In the context of "smart agriculture," conventional farming methods are combined with cutting-edge technologies like AI, robots, and the IoT to increase the agricultural sector's value-added potential, quality, and productivity. The research and development of support services for smart agriculture that make use of data analysis and AI is experiencing a boom at the moment. These services gather data about crops, soil, and the habitats around them using sensors. AI then analyses or processes this data to provide assistance with production-related decision-making and tasks (e.g., tilling, sowing, fertilizing, harvesting, and shipping). This indicates that digital technology may be able to aid some of the complex decisions and actions done by savvy producers who have relied on their expertise and intuition up until now. This would make it less difficult to increase the quantity, value, and quality of agricultural products.

Figures 5.9 and 5.10 show the modern AI-based platforms deployed for smart farming with optimized and precise outcomes. In the near future, many such platforms and applications will rise to increase the productivity in agriculture.

Figure 5.9 FarmWise AI: Screenshots

Figure 5.10 Agrimetrics Analytics and Models

5.5 Conclusion

Platforms for generating AI models for agriculture are essential since they help solve a wide range of problems across the ecosystem. They simplify the farming process, which in turn makes supply more widely available and the process is effective. In complicated fields like agriculture involving many parameters to consider, these systems' capacity to maximize resource utilization, improve repeatability, and guarantee scalability is invaluable.[17] The platforms and models for AI deployment in agriculture help with the full AI model lifecycle by easing collaboration, version management, and model deployment. Responsible AI development is encouraged by AI platforms in agriculture because they allow for ethical AI practises, advance security, and help with regulatory compliance. In a nutshell, the goal of these platforms is to increase productivity, spur creativity, and foster ethical AI use cases across a wide range of fields and sectors.

References

1. Reuther, Albert & Michaleas, Peter & Jones, Michael & Gadepally, Vijay & Samsi, Siddharth & Kepner, Jeremy. (2020). Survey of Machine Learning Accelerators, pp. 1–10. 10.1109/HPEC43674.2020.9286149.
2. Y. Ding, J. Wu, Y. Gao, M. Wang and H. K. -H. So, "Model-Platform Optimized Deep Neural Network Accelerator Generation through Mixed-Integer Geometric Programming," 2023 IEEE 31st Annual International Symposium on Field-Programmable Custom Computing Machines (FCCM), Marina Del Rey, CA, USA, 2023, pp. 83–93, doi: 10.1109/FCCM57271.2023.00018.
3. J. Luo, Z. Ni, Z. Tian, X. Xie, X. Zhou and L. Li, "Extending the Control System by Integrating AI Models," 2022 International Conference on Computer Engineering and Artificial Intelligence (ICCEAI), Shijiazhuang, China, 2022, pp. 659–662, doi: 10.1109/ICCEAI55464.2022.00140.
4. L. De Laurentiis et al., "A New User Oriented Platform to Develop AI for the Estimation of Bio-Geophysical Parameters from EO Data," 2021 IEEE International Geoscience and Remote Sensing Symposium IGARSS, Brussels, Belgium, 2021, pp. 262–265, doi: 10.1109/IGARSS47720.2021.9553913.
5. R. Y. Cohen and V. P. Kovacheva (2023). "A methodology for a scalable, collaborative, and resource-efficient platform, MERLIN, to facilitate healthcare AI research," IEEE Journal of Biomedical and Health Informatics, 27(6), 3014–3025,, doi: 10.1109/JBHI.2023.3259395.
6. L. Chen et al., "A Federated Parallel Data Platform for Trustworthy AI," 2021 IEEE 1st International Conference on Digital Twins and Parallel Intelligence (DTPI), Beijing, China, 2021, pp. 344–347, doi: 10.1109/DTPI52967.2021.9540175.
7. S. Jena, A. Raina, H. Raina and C. Nehete, "AI based Dynamic Model for Question Paper Generation and Performance Monitoring," 2022 International Conference on Applied Artificial Intelligence and Computing (ICAAIC), Salem, India, 2022, pp. 16–23, doi: 10.1109/ICAAIC53929.2022.9792657.

8. I. B. K. Y. Utama, D. H. Tran, M. M. Faridh, B. Chung and Y. M. Jang, "Development of an Intelligent IoT Platform for PV Power Plant Monitoring and Control," 2023 Fourteenth International Conference on Ubiquitous and Future Networks (ICUFN), Paris, France, 2023, pp. 690–693, doi: 10.1109/ICUFN57995.2023.10199576.
9. R. Cantoni, J. Essenziale, M. Pezzera, R. Mainetti and N. A. Borghese, "Procedural Constrained Story Generation Based on Propp's and Fabula Models," 2020 IEEE 8th International Conference on Serious Games and Applications for Health (SeGAH), Vancouver, BC, Canada, 2020, pp. 1–8, doi: 10.1109/SeGAH49190.2020.9201843.
10. M. Z. M. Shamim (2022). "TinyML model for classifying hazardous volatile organic compounds using low-power embedded edge sensors: Perfecting factory 5.0 using edge AI," IEEE Sensors Letters, 6(9), 1–4, Art no. 6003204, doi: 10.1109/LSENS.2022.3201398.
11. Y. Feng, S. Vanam, M. Cherukupally, W. Zheng, M. Qiu and H. Chen, "Investigating Code Generation Performance of ChatGPT with Crowdsourcing Social Data," 2023 IEEE 47th Annual Computers, Software, and Applications Conference (COMPSAC), Torino, Italy, 2023, pp. 876–885, doi: 10.1109/COMPSAC57700.2023.00117.
12. S. Liao et al. (2020). "Cognitive popularity based AI service sharing for software-defined information-centric networks," IEEE Transactions on Network Science and Engineering, 7(4), 2126–2136, doi: 10.1109/TNSE.2020.2993457.
13. A. Kelkar and C. Dick, "A GPU Hyperconverged Platform for 5G vRAN and Multi - Access Edge Computing," 2021 IEEE Canadian Conference on Electrical and Computer Engineering (CCECE), ON, Canada, 2021, pp. 1–6, doi: 10.1109/CCECE53047.2021.9569133.
14. P. Rathnayaka, H. Moraliyage, N. Mills, D. De Silva and A. Jennings, "Specialist vs Generalist: A Transformer Architecture for Global Forecasting Energy Time Series," 2022 15th International Conference on Human System Interaction (HSI), Melbourne, Australia, 2022, pp. 1–5, doi: 10.1109/HSI55341.2022.9869463.
15. Ravikumar, S., K. Vijay, B. Bhuvaneswaran, Babu Rajendiran, and S. Vidhya. "Water Quality Monitoring and Notification System Based on Arduino, MATLAB and Visualization Techniques." In Recent Trends in Computational Intelligence and Its Application: Proceedings of the 1st International Conference on Recent Trends in Information Technology and its Application (ICRTITA, 22), p. 431. CRC Press, 2023.
16. Nagasubramanian, G., Sakthivel, R. K., Patan, R., Sankayya, M., Daneshmand, M., & Gandomi, A. H. (2021). Ensemble classification and IoT-based pattern recognition for crop disease monitoring system. IEEE Internet of Things Journal, 8(16), 12847–12854.
17. Y. Liu, X. Liu, J. Guo, R. Lou and Z. Lv, "Digital Twins of Wave Energy Generation Based on Artificial Intelligence," 2022 IEEE Conference on Virtual Reality and 3D User Interfaces Abstracts and Workshops (VRW), Christchurch, New Zealand, 2022, pp. 718–719, doi: 10.1109/VRW55335.2022.00210.
18. Nilesh Kumar Jadav, Tejal Rathod, Rajesh Gupta, Sudeep Tanwar, Neeraj Kumar, Ahmed Alkhayyat (2023). Blockchain and artificial intelligence-empowered smart agriculture framework for maximizing human life expectancy, Computers and Electrical Engineering, 105, 108486, ISSN 0045-7906, https://doi.org/10.1016/j.compeleceng.2022.108486.
19. K. R. Sowmia, S. Prithi, K. Vijay, I. Eugene Berna and B. Bhuvaneswaran, "Crop Monitoring System with Water Moisture Level using Arduino," 2023 5th International Conference on Smart Systems and Inventive Technology (ICSSIT), Tirunelveli, India, 2023, pp. 495–498, doi: 10.1109/ICSSIT55814.2023.10060913.

20. R. Kabrick, J. Leidel and D. Donofrio, "Toward HDL Extensions for Rapid AI/ML Accelerator Generation," 2021 IEEE High Performance Extreme Computing Conference (HPEC), Waltham, MA, USA, 2021, pp. 1–6, doi: 10.1109/HPEC49654.2021.9622832.
21. P. Vijay, S. Shanthi D, R. P, S. Ohja, S. N, and Nandhiannal (2022). Grainchain –agricultural supply chain traceability and management technique for farmers sustainability using blockchain hyper ledger, International Journal of Intelligent Systems and Applications in Engineering, 10(3), 141–146.
22. Cor Verdouw, Harald Sundmaeker, Bedir Tekinerdogan, Davide Conzon, Teodoro Montanaro (2019). Architecture framework of IoT-based food and farm systems: A multiple case study, Computers and Electronics in Agriculture, 165, 104939, ISSN 0168-1699, https://doi.org/10.1016/j.compag.2019.104939.

Chapter 6

AI Deployment Models in Precision Agriculture

J. Jeyalakshmi, K. Vijay, B. Saiganesh, and Eashaan Manohar

6.1 Introduction

As the world's population approaches 10 billion by 2050, the need to expand food output and maximise yields becomes increasingly urgent. Both increasing land usage and transitioning to industrial-scale farming, as well as embracing creative practises and making use of technical advances to boost productivity on current farms, have been proposed as solutions to projected food shortages.

The modern agricultural landscape is changing and expanding in many different ways as a result of the numerous challenges that farmers face in trying to achieve their desired levels of output. These challenges include a lack of available land and labour, as well as the effects of climate change, environmental concerns, and declining soil fertility. The use of manual ploughs and horse-drawn farm equipment has given place to more modern methods. Every harvest time sees the introduction of cutting-edge tools to maximise output and productivity. However, many farmers and agribusinesses, both large and small, fail to make use of AI's potential in the agricultural sector. The applications of AI technologies and areas are shown in Figure 6.1.

Artificial Intelligence (AI) is a fast-developing science that is revolutionising numerous sectors, including the agricultural sector. AI, at its heart, is a technique for utilising computers to interpret enormous amounts of data for the sake of making predictions or choices. It has the potential to increase agricultural output and efficiency, benefiting both farmers and agricultural businesses. It may assist farmers

DOI: 10.1201/9781003504900-6

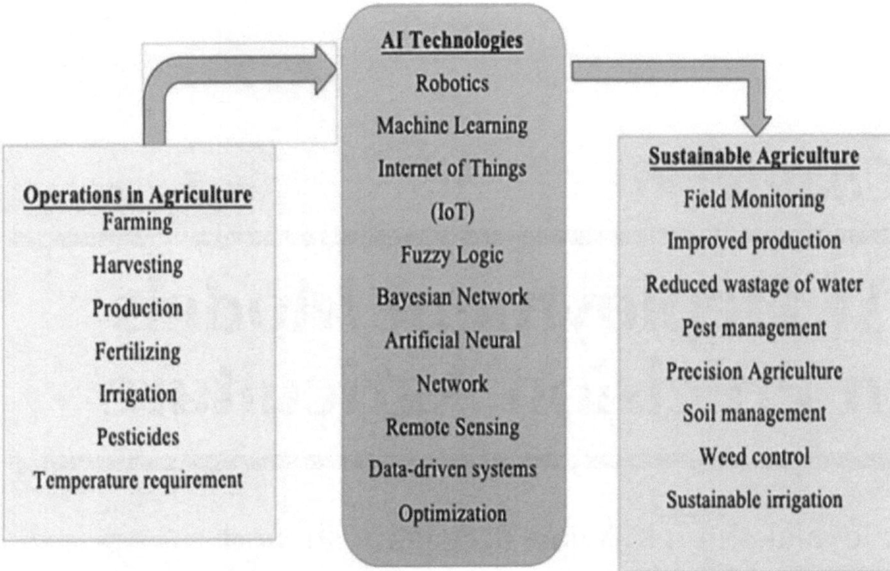

Figure 6.1 Applications of AI technologies and operations[2]

in increasing their yield and return on investment (ROI) by analysing massive volumes of data regarding weather patterns, soil conditions, crop yields, fertiliser use, and other aspects. Furthermore, many agricultural businesses are employing AI technology like machine learning and computer vision to do things like track equipment or keep tabs on the health of cattle. Tools like this can help farmers streamline their processes and work more effectively. The advantages of AI in agriculture are summarised in Table 6.1.

The applications of AI in agriculture are summarised in Table 6.2. Precision farming is made easier with the use of predictive analytics. To further advice on crops that are more resistant to adverse weather, illness, or dangerous pests, AI used data on plant growth in plant breeding. AI systems can examine soil samples' chemical make-up to identify potentially deficient nutrients. Plant infections are another problem that can be solved by AI. Using AI for crop feeding in irrigation may help determine the best times to provide nutrients and when to apply them, as well as help determine the best combination of agronomic items to utilise. AI for harvesting has the potential to increase agricultural yields by determining when harvesting is optimal.[1]

AI helps to optimise the agricultural process and gain more profit. When AI is combined with big data analytics, farmers may obtain advice based on up-to-date, accurate data, leading to higher production and lower overhead Internet of Things (IoT) sensors can gather, evaluate, and store training data on a wide range of factors in real time when integrated with other technologies (e.g., AI drones,

Table 6.1 Advantages of AI in Agriculture

Aspect	*Description*
Data-Based Decisions	AI enables data-driven insights for agriculture, from field analysis to supply chain monitoring. Predictive analytics aids in market analysis, price forecasting, and timing for planting and harvesting. Soil health, weather, and crop management are improved through AI. Farm management software enhances decision-making throughout the cultivation process.
Cost Savings	Precision agriculture with AI optimises resource usage, increasing crop yield while reducing costs. Real-time insights guide irrigation, fertilisation, and pesticide treatment. Innovative practices like vertical farming reduce resource consumption, leading to higher profits and savings.
Automation Impact	Automation, including driverless tractors, smart irrigation, and AI-based robots, addresses labour shortages. AI-driven tools are more efficient and accurate than human labour, revolutionising agriculture.

Table 6.2 Applications of AI in Agriculture

Aspect	*Description*
Risk Management	Predictive analytics powered by AI reduces errors in farming processes, helping farmers make informed decisions to mitigate risks and optimise yields.
Plant Breeding	AI analyses plant growth data to recommend crops that exhibit resilience to extreme weather conditions, diseases, and pests. This assists in more efficient and adaptive plant breeding programs.
Soil and Crop Health	AI algorithms assess soil samples to identify nutrient deficiencies and analyse crop health. AI can also detect and predict crop diseases, aiding in early intervention and prevention measures.
Crop Feeding	AI-driven irrigation systems optimise watering patterns and nutrient application times, leading to improved crop health and resource efficiency. AI suggests the ideal mix of agronomic products for maximum yield and quality.
Harvesting	AI enhances crop yields by providing insights into optimal harvesting times, ensuring that crops are harvested at their peak condition for maximum quality and quantity.

Table 6.3 AI for Agricultural Process Optimisation

Aspect	*Description*
Big Data for Informed Decision-Making	AI coupled with big data analytics provides real-time information for data-driven decision-making in farming, leading to increased productivity and reduced costs.
IoT Sensors for Data Capture and Analysis	IoT sensors, in combination with AI, drones, GIS, and other tools, enable real-time monitoring and data collection on various agricultural metrics, offering accurate and timely insights.
Intelligent Automation and Robotics	AI-powered autonomous tractors and IoT address labour shortages in agriculture, enhancing efficiency. Agricultural robots, known for their precision and endurance, are used for tasks like produce picking, reducing errors, and increasing productivity.

GIS, and other tools). When these tools are used in tandem with AI, farmers have rapid access to reliable data. The common problem of labour shortages is addressed by a combination of AI, autonomous tractors, and the IoT. Automation is also crucial; agricultural robots are already being utilised to perform laborious chores like harvesting fruits and vegetables. Robots have several advantages over human workers on farms, including the capacity to work more hours, more accuracy, and greater resistance to mistakes. The optimisation techniques and technologies are given in Table 6.3.

6.2 Artificial Intelligence in Precision Farming

AI is used in Precision Farming for various purposes. The following section gives a summary of the techniques.

6.2.1 Data Analytics and Predictive Modelling

Predictions of crop yield are crucial to agricultural output planning. In order to learn about and predict crop production in order to fulfil market demand, this research makes use of artificial neural networks (ANN) and multiple linear regression (MLR). The application of predictive analytics in precision farming is revolutionising the agricultural industry by improving yields and product quality. Accurate predictions of the agricultural inputs required to attain maximum yields and product quality are made possible by data collected by sensors and other equipment in the field. Furthermore, using historical and real-time data, predictive

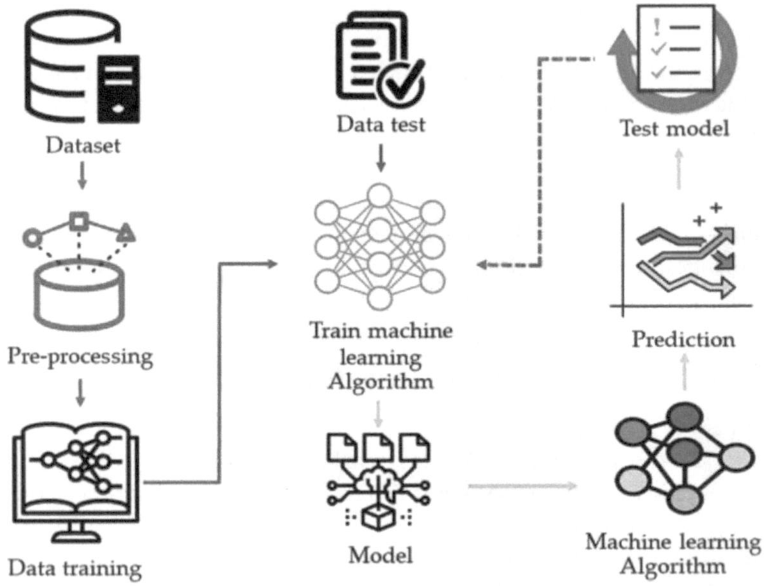

Figure 6.2 Predictive model for agriculture[4]

analytics assists in anticipating climatic conditions, irrigation requirements, and other inputs, which in turn promotes optimal resource management and sustainable agricultural practises. Through improved risk prediction and alternate solution provision, this technology also improves supply chain management, leading to more effective commodity distribution. By optimising supply chains and providing precise estimates of input needs, it helps boost agriculture's bottom line and benefit everyone involved. Figure 6.2 gives a model for Predictive analytics in agriculture.[3]

There are a lot of challenges in predictive analytics in agriculture.[5] Here is a quick rundown of some of the problems with using predictive analytics in farming:

Initial investment: Many farmers may be put off by the need to invest in predictive analytics software. This incorporates the price of software and hardware, in addition to the price of collecting, storing, and analysing data.

Poor data sharing: The reliability and precision of the data used in predictive analytics are crucial. Data collection, storage, and dissemination errors can lead to wrong conclusions and bad choices.

Poor data quality: Predictive analytics fail if the data utilised is of poor quality. This necessitates that the information be reliable, comprehensive, and applicable. Inaccurate outcomes and bad choices might emerge from low-quality data.

Inadequate training: Many farmers lack the education and experience to comprehend and use predictive analytics. As a result, they may struggle to draw useful conclusions from predictive analytics.

In order to face these difficulties, it is crucial to make predictive analytics more accessible to businesses of all sizes by lowering their initial investment through the use of open-source tools and the creation of low-cost data collecting and storage methods.

The quality of data may be increased by creating standardised data formats and protocols and introducing quality assurance checks.[6]

In order for farmers to use predictive analytics effectively, they need education and training in data collection, storage, and use. Training in this area should be developed with the needs of farmers and their enterprises in mind. As a result of these challenges, farmers' decision-making and output could benefit greatly from the application of predictive analytics.

6.2.2 Precision Planting and Irrigation

With the help of precision irrigation, growers can water their crops at just the right time and in just the right amount. When paired with fertilisation, this technique not only hydrates plants but also gives them with nutrients. Businesses in the agricultural sector can benefit financially and environmentally from increased irrigation accuracy since it allows them to use less water and fewer chemicals. Precise moisture delivery is a frugal method that aids in sustainable farming. Figure 6.3 shows smart irrigation.

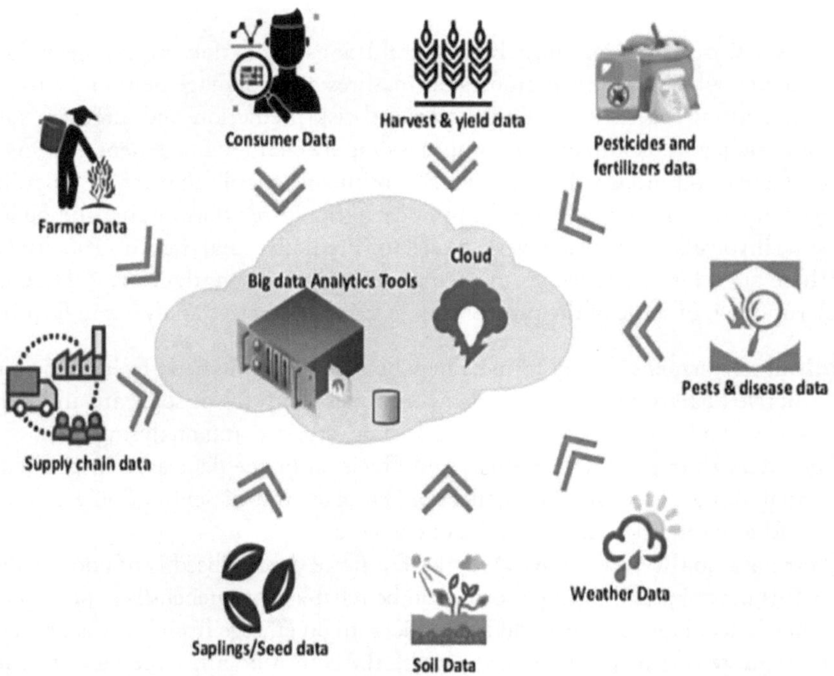

Figure 6.3 Precision irrigation and planting[7]

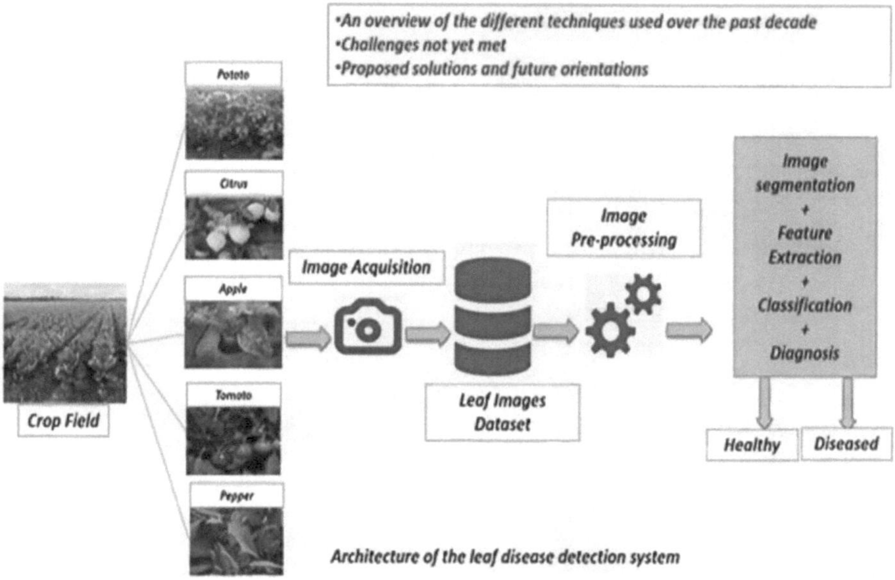

Figure 6.4 Crop disease detection model using AI[9]

6.2.3 Crop Monitoring and Disease Detection

In many regions of the globe, the lack of infrastructure makes it difficult to quickly identify crop diseases,[8] which pose a serious danger to food security. Smartphone-assisted illness detection is now possible thanks to the widespread availability of smartphones and recent advancements in computer vision enabled by deep learning. Figure 6.4 shows the crop monitoring and disease detection model.

6.2.4 Soil Health Assessment

With the help of AI, we can keep tabs on the soil's condition and learn more about its composition and possible problems like erosion and compaction. This information can guide farmers towards actions that will increase soil health and sustain output. The model for soil profile generation is given in Figure 6.5.

6.2.5 Autonomous Farming

Combining precision agriculture methods with technologies like autonomous systems, robots, AI, and machine learning helps farmers gain a deeper, more long-term understanding of the soil they work, plants they tend, and crops they harvest.[11]

Collecting data on the qualities of the soil, sun exposure, moisture, and chemical content, among other important information, greatly aids the planting of crops

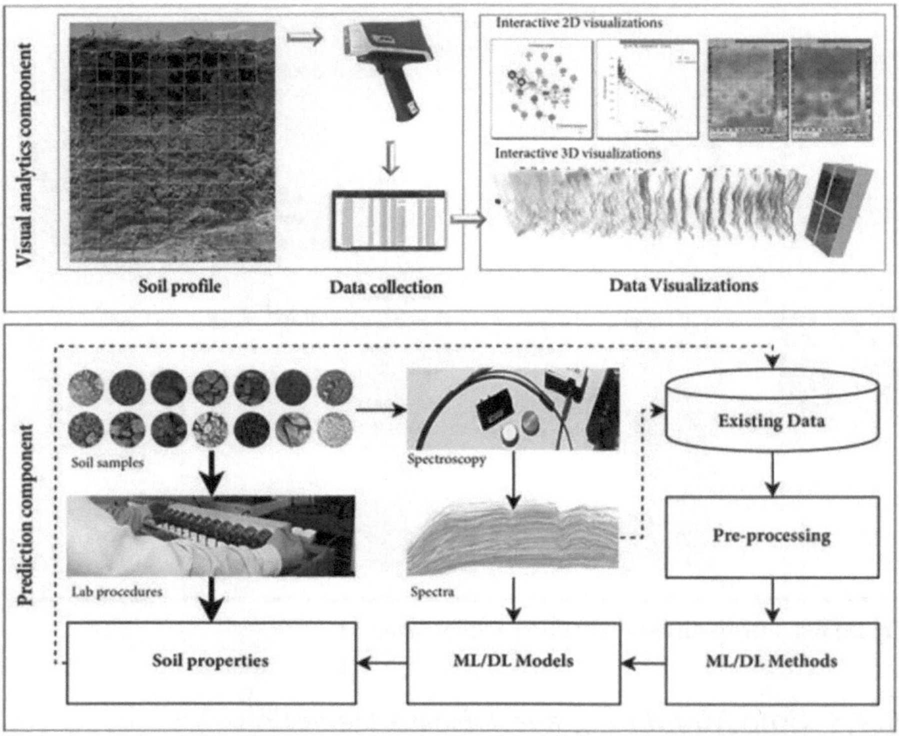

Figure 6.5 Soil profiling model in AI[10]

that are well-suited to the present conditions and yield the biggest yields. Producing larger, higher quality, and more efficient crops may increase earnings per acre. As a result, farmers will be better able to respond to increased demand, diversify their crops, and comply with increasingly stringent regulations.

Multiple sensors are used in autonomous systems and robots to gather information for reliable control. Any mobile autonomous system must have a navigation system at its core. Safe operation relies on precise positioning, and for autonomous tractors, this is especially important for working in the right field or region.

For inspection and precision agricultural applications involving the deployment of unmanned aerial vehicles (UAVs), a navigation system is required to precisely find items of interest using georeferencing (longitude and latitude coordinates). The georeferencing data required by a navigation system is typically provided by GNSS (global navigation satellite system) and associated technologies.

Multiple sensors will be used by autonomous tractors to detect and catalogue the world around them. One example is a camera whose data is analysed in real time by an AI programmed to distinguish between different types of things based on their size and form. Night vision is made possible by infrared cameras, which can also "see" humans and animals by picking up on their body heat.

Numerous sensors, both on the ground and in the air, serve several tasks in precision agriculture. LiDAR (light detection and ranging) sensors are commonly employed in topological and ground surveys to build precise digital representations of the landscape, its features, and its vegetation. Soil moisture, nitrate concentration, pH level, ground and sub-ground temperature, air pressure, relative humidity, and many more can all be measured by sensors in and on the ground.

"Intelligent robotic arms and hands" can now pick and handle delicate produce with ease and accuracy because to developments in robotic control and precise movement technology. It is essential to maintain well-aerated soil that can store moisture for longer, which is why the electrification of power systems is assisting in the miniaturisation of various vehicles; this is because lightweight vehicles compress the soil substantially less than certain older heavy equipment. As autonomous and precision agriculture grows in popularity, investment in R&D for better sensors and more practical mass-produced autonomous agricultural equipment will increase. Figure 6.6 shows the Autonomous Farming process in detail.

Seed may be dropped quickly across wide regions by autonomous UAV drones utilising georeferencing data for precise placement. Similarly, herbicides, insecticides, and

Figure 6.6 Autonomous Farming (Sourced at: https://www.advancednavigation.com/tech-articles/autonomous-farming-a-leap-forward-in-ag-tech/#:~:text=Coupling%20autonomous%20systems%2C%20robots%2C%20artificial,of%20the%20land%2C%20season%20to)

fertilisers may be applied by extremely efficient and targeted spraying and aerial dumping, resulting in minimal waste and focusing treatment precisely where it is required.

6.2.6 Integrated Farming and Livestock Management

Integrated farming is a holistic approach to farm management with the ultimate goal of producing environmentally friendly crops. It's a flexible method that can be used with any kind of agricultural setup, wherever in the world. It requires using data-driven methods of management to optimise every facet of a farm's operations. Depending on the specifics of a farm's location and climate, integrated farming may employ either exclusively contemporary or exclusively traditional methods.

System components include climate, soil make-up, topographical placement, plant and animal genetics, and ecological niche. It includes energy sources, costs, and effects of agricultural policies, as well as management practises, human aspirations and lifestyles, economic possibilities, marketing tactics, and externalities. Everything from the available natural resources and the influence on their usage to wildlife concerns, target and non-target plant and animal species, microorganisms, and so on are all reflected in a system, as are the factors that affect them. Figures 6.7 and 6.8 depict an example of an integrated farming ecosystem. The benefits of an integrated agricultural system are numerous:

- Profitability and output have gone up.
- Sustainability
- Nourishing fare
- Ecological security
- Reusing materials
- Year-Round Income

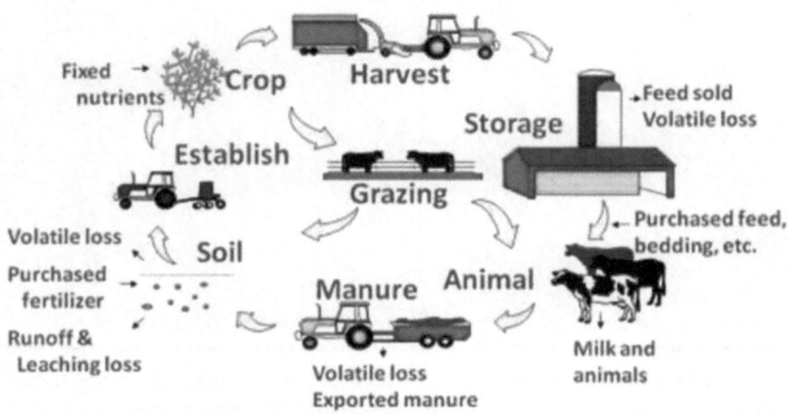

Figure 6.7 Integrated agriculture and processes

Figure 6.8 Integrated farming environment (Sourced at: https://www.agrivi.com/blog/following-integrated-farming-practices-with-agrivi-farm-management-system/)

- Acceptance of Emerging Technologies
- Putting an end to the energy, fuel, and food shortages
- Keeping forests standing
- Efficiency in inputs and outputs.

The goal of integrated farming is to save the environment, sustain the business's profitability, and meet social criteria, all while satisfying the needs of the farmers and their customers. In order to preserve crops while also minimising the negative effects of pesticide use, The most popular strategy now is IPM (Integrated Pest Management). Therefore, IPM is an essential part of Integrated Crop Management, which is a fundamental feature of the all-encompassing Integrated Farming approach to food production.

More and more farmers are adopting integrated farming practices in an effort to lessen their reliance on pesticides, safeguard the health of their ecosystems, and ensure the long-term viability of their farms. For small-scale farmers working with limited means, this solution appears to be the best bet for boosting food production, raising revenue, and improving nutrition without negatively impacting the environment or the agroecosystem. When they use the AGRIVI farm management system, farmers that follow the principles of integrated farming may also monitor the output of their whole farms. It helps them coordinate their workforce, equipment, land, inventories, and funds in the field. Ultimately, it contains an analytics layer with Key Performance Indicator (KPI) dashboards and printed reports that bring everything together and provide a fantastic bird's-eye perspective of farm health.

6.2.7 Supply Chain Optimisation

Agricultural optimisation is essential for saving money, protecting the environment, and producing more food. Accurate fertilisation schedules provide farmers with the ability to anticipate future yields and expenses.[12]

The management of agricultural supply chains might undergo a dramatic transformation as a result of the integration of blockchain and IoT technology. IoT devices, such as sensors, make it possible to gather and analyse data in real time at several locations throughout the supply chain. This improves both quality control and the utilisation of available resources. The blockchain, on the other hand, creates an immutable database of transactions and data connected to the origins of products, as well as their handling and transportation, which guarantees transparency, security, and traceability. Figure 6.9 gives insight about Supply Chain model for precision farming. This combination has a number of opportunities for improvement, including enhanced

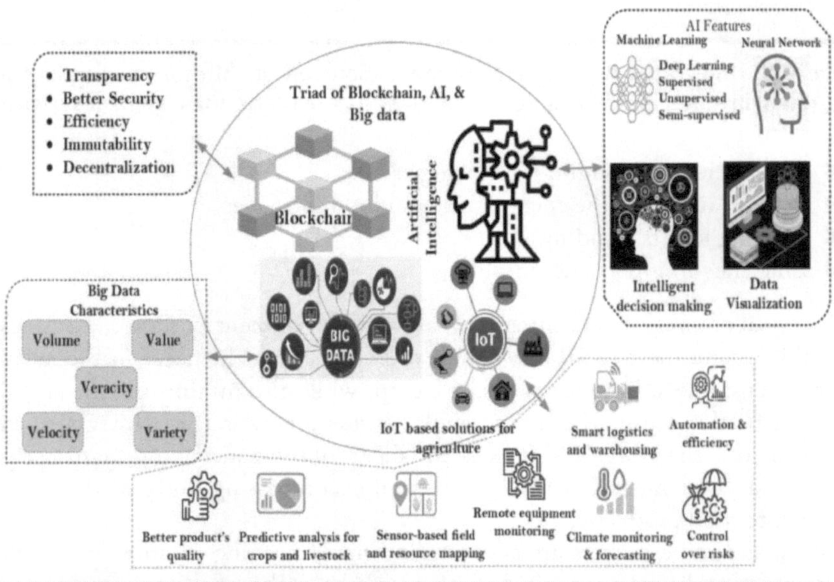

Figure 6.9 Supply chain model for precision farming[14]

traceability, efficiency, quality control, protection against fraud, and the use of smart contracts. However, this does bring forth certain difficulties, such as the expense of implementing the IoT, concerns around online safety, and the requirement for widespread acceptance among supply chain players. In general, the agricultural supply chain stands to benefit from increased openness, increased efficiency, and increased confidence if certain technologies are used.[13]

6.2.8 *Market Price Prediction*

Predicting the price of crops may be done with the help of a number of different techniques, including decision trees, support vector machines, neural networks, deep learning, etc. The supervised machine learning approach known as Decision tree Regressor was utilised in the construction of our model. It is cultivated with a variety of Kharif and Ragi crops, such as paddy, wheat, cotton, and barley, amongst others.[15]

6.2.9 *Weed and Pest Control*

One solution to the problem of monitoring agricultural land and automatically detecting the presence of pests and weeds is to use a machine learning model that processes imagery from satellites and UAVs, which are used for real-time and automated monitoring of agricultural areas by providing data in the visible and radar spectrums. The situation is not hopeless.[16]

6.3 AI Deployment Models for Precision Farming

AI deployment models for precision farming are summarised in Figure 6.10. AI models can be used for various purposes. Over the course of time, technological advancements have rethought farming, and this has had an impact on the agricultural sector in a variety of different ways. Agriculture is the primary source of

Figure 6.10 AI deployment models in agriculture

income for many nations across the world, and the United Nations projects that the world's population will rise from 7.5 billion people now to 9.7 billion by the year 2050.[17] While a result of this projected growth, there will be an increased demand for land, while only an additional 4% of land will be brought under cultivation by that year. This indicates that farmers will be required to produce more with fewer resources. The same poll found that in order to provide food for the additional two billion people, there will need to be a 60 percent increase in food production. However, standard methods are unable to meet the requirements of this enormous demand. Because of this, agricultural businesses and farmers are looking for innovative approaches to boost productivity while also cutting waste. As a direct consequence of this, AI is slowly developing as an integral component of the technical progress of the agriculture business. To meet the demand of an additional two billion people by the year 2050, the world food output must be increased by a factor of fifty, or by 50 percent. Solutions that are powered by AI will not only make it possible for farmers to enhance their efficiency, but they will also improve the quantity and quality of crops and ensure that they are brought to market more quickly.[6]

Every day, farms collect and record a vast quantity of information on the soil's temperature, the amount of water utilised, the weather, and other elements. This data is used in real time with the aid of AI and machine learning models to generate useful insights, such as the best time to plant seeds, crop selections, hybrid seed options to create better yields, and so on.[18]

The use of AI tools in farming, or "precision agriculture," has improved crop yields and crop accuracy. AI aids in plant disease diagnosis, pest control, and dietary enhancement on farms. Sensors powered by AI can detect the presence of weeds, zero in on them, and then decide which herbicide to deploy. Less money will be spent on herbicides if this is implemented. Several tech companies have developed robots that can keep an eye on weeds and spray them with pesticide precisely. Computer vision and AI are utilised by these robots. These machines can cut pesticide application on crops by as much as 80 percent and lower the amount of herbicide needed by 90 percent. In addition, the price of the herbicide might drop by as much as 90%. These innovative AI sprayers have the potential to drastically reduce the amount of pesticides used in the fields, enhancing the quality of agricultural produce while saving farmers and businesses money.[19]

Now, in most situations, it is not the typical farm labourer who is responsible for delivering the food to your kitchen table; rather, it is robotic robots that are capable of undertaking bulk harvesting with greater accuracy and speed. These machines are responsible for the majority of the job.[20] These devices contribute to an increase in the size of the yield and a reduction in the amount of waste caused by crops that are left on the field.[21]

There are a lot of businesses that are focused on making agriculture more productive. Products such as an autonomous strawberry-picking machinel and a vacuum device that can gather ripe apples from trees are already available on the market.

Figure 6.11 Imaging to sense crop maturity

These devices utilise a combination of sensor fusion, machine vision, and AI models to determine the position of the produce that is ready to be harvested and to assist in picking the appropriate fruits.[22]

Following closely behind the defence sector in terms of the number of service robots that have been put into professional usage is the agricultural sector. According to estimates provided by the International Federation of Robotics, as many as 25,000 agricultural robots have been purchased, which is equal to the number of robots utilised by the military. The application of finding ripe fruits using smart imagery is shown in Figure 6.11.

Some key aspects and considerations of any AI deployment:

1. **Data Privacy and Security**: This deployment technique enforces data privacy and security due to the fact that it maintains the AI model and deploys it in its own safeguarded environment, making sensitive data more secure and private. With security being a critical component of AI, this deployment model can be of great use.

2. **Compliance:** Some industries, such as like healthcare and finance, are subject to strict regulatory compliance requirements that affect overall society. On-premises AI deployments can help organisations comply with regulations by enforcing direct supervision, strict policies, and guidelines
3. **Low Latency Requirement**: Certain AI applications, like real-time analytics or industrial automation, require low latency and high-speed data processing.
4. **Customisation and Control:** Organisations that require specific hardware or infrastructure requirements are better met through on-premises deployments. This allows for greater customisation and control over the hardware and software stack. This will help in delivering customised tailor-made digital solutions.
5. **Cost Management:** While cloud services offer scalability and flexibility, they come with ongoing costs that can add up over time. On-premises deployment can provide cost predictability, as the organisation owns and manages the hardware. As there is no third-party medium involved, the costs incurred are significantly lowered.
6. **Resource Management**: Organisations can allocate dedicated resources (e.g., GPUs, TPUs, high-performance servers) for AI workloads as needed without sharing them with other cloud users. This can lead to more efficient resource utilisation.
7. **Network Considerations:** It's essential to have a robust and reliable network infrastructure when deploying AI on-premises, as data transfer and communication between different components can be critical for AI model performance.
8. **Maintenance and Updates:** Organisations are responsible for maintaining, upgrading, and patching both hardware and software components in an on-premises environment. This can require a skilled IT team and careful planning.
9. **Scalability:** Businesses are drawn to on-premise AI deployment because of its many benefits. Probably the most visible benefit is that it gives businesses more control over their data. Businesses may manage who has access to what information by implementing on-premise solutions. They can set their own security requirements to ensure the privacy of their information.
10. **Hybrid Deployments:** Some companies prefer hybrid installations, which combine on-premises and cloud resources for optimal scalability, cost-effectiveness, and data control.

6.4 Conclusion

Models based on AI are required for precision agriculture in order to assist farmers in making more informed choices about crop management. The employment of AI models allows for the prediction of agricultural yields, the identification of pests and diseases, the management of weeds, and the optimisation of the application of fertiliser. These models can assist farmers in making better decisions and increasing their levels of production, both of which can lead to more earnings and greater long-term viability. AI is being used by the University of California, Davis, to construct a model that can anticipate agricultural yields based on soil moisture, meteorological conditions, and the type of crop being grown.

AI is being used by the business CropX to build a system that can recognise and control weeds in agricultural crops.

AI is being used by the business PrecisionHawk to build a system that can map crop fields and detect regions that are vulnerable to being affected by illnesses or pests. Just a few instances of how AI is being applied in precision agriculture are listed above. We should anticipate seeing an even greater variety of creative uses of AI technology in this sector as its underlying technology continues to advance. In conclusion, AI models are essential for precision agriculture because they enable farmers to improve the quality of the judgements they make on crop management. These models may be utilised to make predictions regarding agricultural yields, recognise pests and diseases, maximise the efficiency of irrigation, control weed growth, and improve the usage of fertiliser. Farmers may enhance their decision-making and production by utilising AI models, which can lead to improved revenues and longer-term sustainability.

References

1. Oliveira, Rosana Cavalcante de, and Rogério Diogne de Souza e Silva. 2023. "Artificial Intelligence in Agriculture: Benefits, Challenges, and Trends" *Applied Sciences* 13, no. 13: 7405. https://doi.org/10.3390/app13137405
2. Bhagat PR, Naz F, Magda R, "Artificial Intelligence Solutions Enabling Sustainable Agriculture: A Bibliometric Analysis," PLoS One, 17(6): e0268989, 2022, doi: 10.1371/journal.pone.0268989. PMID: 35679287; PMCID: PMC9182339.
3. Kittichotsatsawat, Y., Tippayawong, N. & Tippayawong, K.Y. Prediction of arabica Coffee Production Using Artificial Neural Network and Multiple Linear Regression Techniques. *Science Report* 12: 14488, 2022. https://doi.org/10.1038/s41598-022-18635-5
4. J. Schöning and M. L. Richter, "AI-Based Crop Rotation for Sustainable Agriculture Worldwide," 2021 IEEE Global Humanitarian Technology Conference (GHTC), Seattle, WA, USA, 2021, pp. 142–146, doi: 10.1109/GHTC53159.2021.9612460.
5. S. K. V, D. B. Koppad, K. Awasthi, P. A. K and V. R, "Application of AI Models in Agriculture," 2022 4th International Conference on Circuits, Control, Communication and Computing (I4C), Bangalore, India, 2022, pp. 387–390, doi: 10.1109/I4C57141.2022.10057718.
6. Lu, Yue, Mingzheng Liu, Changhe Li, Xiaochu Liu, Chengmao Cao, Xinping Li, and Za Kan. 2022. "Precision Fertilization and Irrigation: Progress and Applications" *AgriEngineering* 4(3): 626–655. https://doi.org/10.3390/agriengineering4030041
7. Nagasubramanian, G., Sakthivel, R. K., Patan, R., Sankayya, M., Daneshmand, M., & Gandomi, A. H., "Ensemble Classification and IoT-Based Pattern Recognition for Crop Disease Monitoring System," IEEE Internet of Things Journal, 8(16): 12847–12854, 2021.
8. Orchi, Houda, Mohamed Sadik, and Mohammed Khaldoun. 2022. "On Using Artificial Intelligence and the Internet of Things for Crop Disease Detection: A Contemporary Survey" *Agriculture* 12(1): 9. https://doi.org/10.3390/agriculture12010009

9. Vung Pham, David C. Weindorf, Tommy Dang," Soil Profile Analysis Using Interactive Visualizations, Machine Learning, and Deep Learning", Computers and Electronics in Agriculture, 91, 2021, 106539, ISSN 0168-1699, https://doi.org/10.1016/j.compag.2021.106539.
10. P. Patil, R. Kestur, M. Rao and A. C, "IoT based Data Sensing System for AutoGrow, an Autonomous greenhouse System for Precision Agriculture," 2023 IEEE Applied Sensing Conference (APSCON), Bengaluru, India, 2023, pp. 1–3, doi: 10.1109/APSCON56343.2023.10101100.
11. S. Qazi, B. A. Khawaja and Q. U. Farooq, "IoT-Equipped and AI-Enabled Next Generation Smart Agriculture: A Critical Review, Current Challenges and Future Trends," IEEE Access, 10: 21219–21235, 2022, doi: 10.1109/ACCESS.2022.3152544.
12. S. Arjune and V. Srinivasa Kumar, "Smart Agriculture adoption based on Farmer's Perspective," 2022 International Interdisciplinary Humanitarian Conference for Sustainability (IIHC), Bengaluru, India, 2022, pp. 376–379, doi: 10.1109/IIHC55949.2022.10060306.
13. Bhat, Showkat Ahmad, Nen-Fu Huang, Ishfaq Bashir Sofi, and Muhammad Sultan. 2022. "Agriculture-Food Supply Chain Management Based on Blockchain and IoT: A Narrative on Enterprise Blockchain Interoperability" *Agriculture* 12(1): 40. https://doi.org/10.3390/agriculture12010040
14. C. S. Parr, D. G. Lemay, C. L. Owen, M. J. Woodward-Greene and J. Sun, "Multimodal AI to Improve Agriculture," IT Professional, 23(3): 53–57, 1 May-June 2021, doi: 10.1109/MITP.2020.2986122.
15. S. Torai, S. Chiyoda and K. Ohara, "Application of AI Technology to Smart Agriculture: Detection of Plant Diseases," 2020 59th Annual Conference of the Society of Instrument and Control Engineers of Japan (SICE), Chiang Mai, Thailand, 2020, pp. 1514–1519, doi: 10.23919/SICE48898.2020.9240353.
16. R. Katarya, A. Raturi, A. Mehndiratta and A. Thapper, "Impact of Machine Learning Techniques in Precision Agriculture," 2020 3rd International Conference on Emerging Technologies in Computer Engineering: Machine Learning and Internet of Things (ICETCE), Jaipur, India, 2020, pp. 1–6, doi: 10.1109/ICETCE48199.2020.9091741.
17. P. Ramadoss, V. Ananth, M. Navaneetha and U. Oviya, "E -Xpert Bot -Guidance and Pest Detection for Smart Agriculture using AI," 2023 IEEE 12th International Conference on Communication Systems and Network Technologies (CSNT), Bhopal, India, 2023, pp. 797–802, doi: 10.1109/CSNT57126.2023.10134588.
18. C. Bruno et al., "Embedded Artificial Intelligence Approach for Gas Recognition in Smart Agriculture Applications Using Low Cost MOX Gas Sensors," 2021 Smart Systems Integration (SSI), Grenoble, France, 2021, pp. 1–5, doi: 10.1109/SSI52265.2021.9467029.
19. C. Nicolas, B. Naila and R. -C. Amar, "TinyML Smart Sensor for Energy Saving in Internet of Things Precision Agriculture platform," 2022 Thirteenth International Conference on Ubiquitous and Future Networks (ICUFN), Barcelona, Spain, 2022, pp. 256–259, doi: 10.1109/ICUFN55119.2022.9829675.
20. B. Ragavi, L. Pavithra, P. Sandhiyadevi, G. K. Mohanapriya and S. Harikirubha, "Smart Agriculture with AI Sensor by Using Agrobot," 2020 Fourth International Conference on Computing Methodologies and Communication (ICCMC), Erode, India, 2020, pp. 1–4, doi: 10.1109/ICCMC48092.2020.ICCMC-00078.
21. V. Singh and G. Singh, "Development of AI Enabled Solution for Efficient Feature Enrichment from Multiple Data Sources: An Application in Precision Agriculture," 2023 International Conference on Artificial Intelligence and Applications (ICAIA) Alliance Technology Conference (ATCON-1), Bangalore, India, 2023, pp. 1–4, doi: 10.1109/ICAIA57370.2023.10169451.

Chapter 7

AI Model Deployment: Enhanced Rice Plant Disease Detection and Classification Besides Improved Spatial Pyramid with Enhanced YOLOv7

V. Saranya, Devi T., and N. Deepa

7.1 Introduction

Rice ranks as one of the world's most vital food crops. It is the primary food grain for over 50% of India's residents. India produces 112 million tonnes of rice for domestic consumption and trade with many other countries [1]. Even though agricultural crops are affected by insects, pests, diseases, and weather, annual productivity can vary. Rice plant diseases such as blast, brown spot, sheath blight, bacterial leaf blight, sheath rot, false smut, and grain discoloration are significant challenges. The frequency and magnitude of incidents have fluctuated over time. As a result, crops lose 20%–85% of their yield. These diseases can affect both the quantity and quality of rice. Plant illnesses should be diagnosed early and treated with appropriate composts and fertilizers to address this issue. Artificial intelligence automatic systems can detect diseases earlier and provide findings for crop treatment to avoid low rice yields [2].

DOI: 10.1201/9781003504900-7

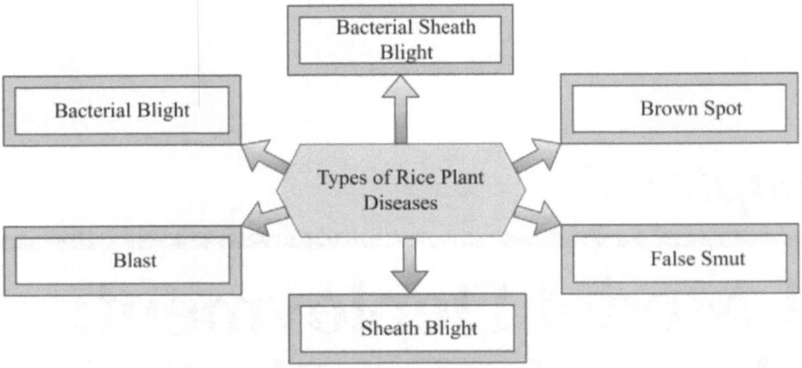

Figure 7.1 Types of Rice Plant Diseases

7.1.1 Rice Plant Diseases

Rice plant illnesses can impact both quality and quantity of rice. Some of the most common diseases include Blast, Brown Spot, Sheath Blight, Bacterial Leaf Blight, and False Smut. Blast is one of the most significant epidemics of rice worldwide, causing considerable damage in regions where agriculture is intensive; disease development is favored by elevated soil nitrogen levels. It affects leaves, neck, and nodes of the plant. The symptom is brown lesions on the leaves. Brown Spot is characterized by number of round brown spots on the affected crops, brown lesions, black discolored roots, and the death of seedlings, leading to reduced grain number and kernel weight. Sheath blight occurs everywhere rice is planted, is the second most important disease after rice blast. Intensive rice production is where it does the most damage by affecting the regional ecosystem, which leads to mislaying biodiversity.[3]. One of the symptoms of sheath blight disease is large oval shapes with gray color center points and red outer lines. Bacterial leaf blight is the most important disease. Its green color turned into yellow stripes on leaf blades, plants may die and the youngest leaves turned yellowish. False Smut strictly affects the grains, causing them to lose weight. It is identified by rice leaves turning orange, at high severity, turning greenish-black. Figure 7.1 displays the types of rice plant diseases.

7.1.2 Artificial Intelligence

Understanding brain function requires a great deal of intricacy. The nervous system functions by electrical impulses that are sent by one neuron and received by another [32]. It processes a wide range of emotions, and these indications are unique from one another. Each of these electrical impulses, which act as the neuron's input, includes some signal transmission to another neuron. This procedure was repeated until the data was delivered to the organ's designated location. This working

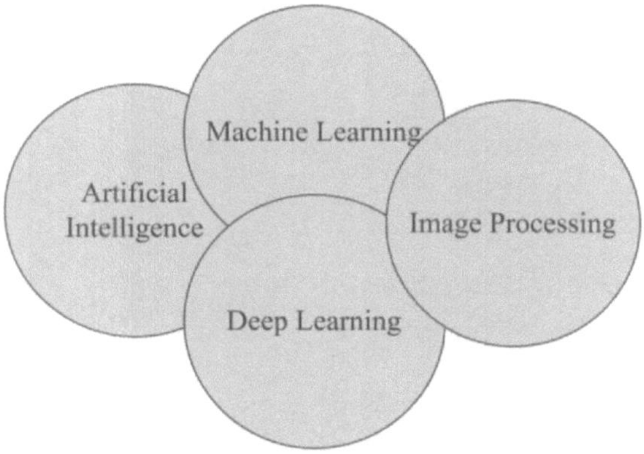

Figure 7.2 Image Processing by AI technology

process of the human brain serves as the cornerstone of artificial intelligence (AI) technology [4]. This is how or through which AI technology processes data. Many datasets utilized by AI have labels, which enable categorizing the data easier. This can analyze and acquire information from a trained dataset [31]. The procedure of learning consists of three stages for determining the result. The technique for training new data focus points entails acquiring the necessary data and selecting the right rules to carry out a certain action. The reasoning step follows and it focuses on selecting the best method or set of algorithms to generate the desired outcome [34]. The final stage, the tuning procedure, continually modifies the set of criteria to identify an outcome that best represents the given input [3]. Figure 7.2 shows the image processing technique by AI technologies.

7.1.2.1 YOLO

It is one of the most emerging technologies of AI, mostly used for object detection. A single CNN may create a large number of boxes with boundaries as well as classes for these boxes [33]. Spatial pyramids with enhanced YOLOv7 can directly execute picture training, assisting in optimizing processing performance. YOLO begins the picture-processing process by loading the neural networks. These networks analyze images and extract information from the input. YOLO's bounding boxes will process the entire image and set the boundaries by frames. This visual data is transmitted by neural networks throughout the procedure of determining features [5]. These networks retrieve the characteristics from the boundary line's boxes.

Each bounding box includes a variety of predictions, which are linked to the mid-point of the boundary box and associated with it; it now begins predicting the confidence score and conditional classes. The training phase of this YOLOv7 trains the images by a single convolutional neural network (CNN) for detecting the

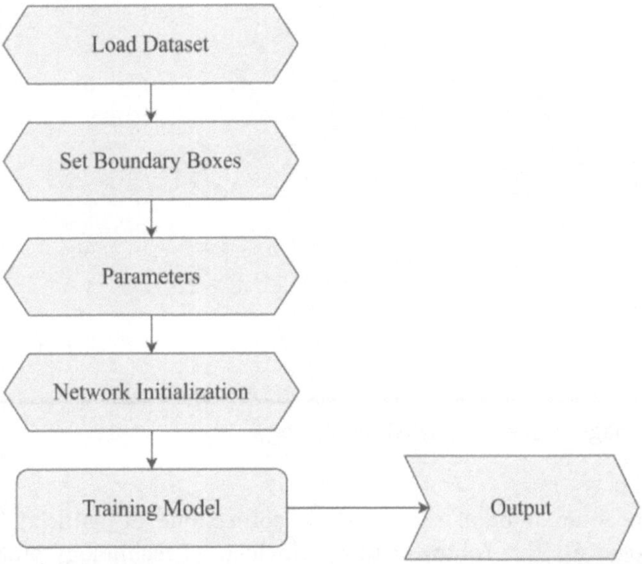

Figure 7.3 Workflow of YOLO

objects. It utilizes the whole image in test time to detect the objects [6]. It detects the objects by boundary box. These boxes can detect a single object at a time on that input image. Figure 7.3 displays the workflow of a general YOLO technique.

7.2 Literature Review

Farming supplies nourishment for all humans, especially in times of rapid demographic expansion. Plant diseases must be predicted at the beginning period in the agriculture sector to provide food to the total population. Yet, it is challenging to forecast illnesses in the starting stages of crop development. The ultimate goal of this study is to educate farmers about cutting-edge technology for reducing illnesses in plant leaves. Because tomato is a readily available fruit, techniques that utilize machine learning and image analysis with a precise algorithm have been found to identify leaf illnesses in tomato plants [7]. In this study, specimens of tomato petals with defects are taken into account. It will be possible for farmers to identify diseases according to initial indications using these pathology samples of tomato leaves. To begin, tomato petal samples are enlarged to 256 × 256 pixels, and then the Histogram Equalized technique is performed to enhance the image quality. K-means clustering is used to split dataspace into Node-link cells. Silhouette tracking is utilized to extract the boundary lines of leaf samples [8].

To extract the informative characteristics of leaf data, several descriptors such as Discrete Wavelet Transform, the analysis of principal components, and Grayscale Co-occurrence Mapping can be used. Finally, the collected characteristics are detected by machine learning techniques such as Support Vector Machine (SVM), CNN, and K-Nearest Neighbor (K-NN). Using tomato disorderly samples, the suggested technique's accuracy is validated and used by 88% in SVM, 97% in K-NN, and 99.6% in CNN [9].

Agricultural production and its related industries offer a living for millions of people in India's remote regions. It is also critical to the development of nations such as India that are developing. Rice, among the key features of Indian agriculture, holds an important part of our nation's and the world's largest food security. Unfortunately, it is plagued by several illnesses that reduce the quantity and quality of rice production.

By lowering the count of parameters connected with the set of networks, this research presents a unique CNN screening for identifying and categorizing usual diseases encountered in rice crops. The suggested set of features was trained on four different types of typical rice plant illnesses. Furthermore, the paper [10] includes 1400 on-field photos of healthier rice leaves source images for detecting disease-free plants. Separate trials were conducted regardless of the better and healthier leaf picture dataset. Many performance metrics are employed to validate the perfect outcome of the suggested model utilizing Adam optimization with Stochastic Gradient Descent with Momentum (SGDM) approaches. The test data obtained from the sample to classify the four varieties of rice plant illnesses reveal that the method using the SGDM approach has the highest accuracy rate of 99.66%, and the technique used for optimizing Adam has the highest accuracy rate of 99.83% also on the validation dataset within the 7th epoch. When the healthier leaf picture dataset was included, the Adam optimizer performed better than the SGDM optimizer, achieving high precision of 99.66% and 97.61% in the 7th epoch, respectively [11].

Ethiopia is Africa's top manufacturer of chickpeas. Fungal diseases, which cause farm produce losses every year, have a significant impact on crop productivity and output in Ethiopia. Chickpea disease, a viral disease, is one of several plant diseases. The more frequent chickpea grain illnesses in Ethiopia that influence crop yield and standard are Ascochyta blight and Fusarium wilt. These disorders must be identified by certified professionals or systems. Although many approaches have been utilized in the past, deep learning (DL) has been proposed as a well-organized strategy for disease prediction and classification because it can dynamically train characteristics derived from the image input [12]. Within that research, a DL strategy for detecting chickpea disease was built using a CNN and Long Short-Term Memory (LSTM) for extracting the features and activation algorithm of SoftMax for classifications. Several picture pre-processing phases such as image size changing, normalizing, and de-noising utilizing a blend of Gaussian filter (GF) and Median filter (MF) were done to build the conceptual approach [13].

To avoid over-fitting, enhancement was performed, and 8391 snapshots were assisted to train and validate the performance of the generated model. Eighty percent of the captured images were utilized for training, 20% were used for evaluation, and 20% of the training model data was utilized for testing. CNN-LSTM, which suggested, worked well in recognizing chickpea illness, including the precision of 92.55%. The proposed CNN-LSTM beats existing approaches, under the report's findings [14].

The agriculture sector is vital in India because of the country's growing demographics and rising food consumption. As a result, the agrarian yield must be increased. Diseases arising from bacteria, fungi, and infections are the main factors that reduce crop yields. This can be avoided and dealt with by employing the identification of plant disease methods. Techniques based on machine learning can be used in the procedure of diagnosing diseases on crops because they depend on data and offer fantastic approaches for detecting plant illnesses. Techniques that utilize machine learning are appropriate for detection and diagnosis because they rely heavily on data and results for an assigned activity [15]. A systematic analysis of the various strategies used in identifying plant illnesses utilizing AI centered machine learning and DL strategies has been conducted throughout this research.

Similarly, DL has grown in importance in the vision-based industry for providing improved performance results for recognizing plant disease. DL improvements were applied to a variety of domains, resulting in outstanding achievement inside this machine learning as well as the computer vision field. A comparative study was conducted regarding machine and DL approaches, as well as their efficiency and usage in different research articles, to demonstrate the superiority of the DL approaches over machine learning approaches. DL has ways to identify plant diseases from collected photos to avoid catastrophic food shortages [16].

The advantages of autonomous learning and segmentation techniques have collected worth of observation in recent days both from professional and educational circles. DL belongs to the category of AI. DL applied to plant disease classification can help avoid problems related to crop disease and productivity losses.

Many recognized machine learning and DL approaches are utilized in combination with an assortment of visualization tools to identify and analyze the indications of plant diseases [17]. This survey addressed the identification of leaf spots by image processing. Image segmentation combined with DL or machine learning models improves leaf diagnostic accuracy. A large dataset can be categorized using picture recognition, and the results are then fed into AI algorithms for disease identification. This study also includes performance indicators from previous research, which provided recommendations for future developments in plant disease identification and prevention strategies [18].

Plant diseases cause losses in yield which have an immediate effect on regional and worldwide food manufacturing systems, leading to financial losses. According to the FAO (Food and Agriculture Organization) and International Plant Protection Convention, pests and plant diseases are responsible for 20%–40% of global food production losses. Plant diseases are estimated to be responsible for 13% of crop yield loss worldwide. These statistics emphasize the importance of identifying plant diseases in order to reduce losses in yield. However, it is critical that you first comprehend the factors that cause illnesses in plants [19]. Plant disease progression is aided by three elements: the host, a conducive environment, and the organism that causes it. The majority of the time, diseases begin exhibiting symptoms and impact a plant from the soil level upward.

After an illness occurs, many plant diseases can spread throughout the entire crop. Crops must therefore be regularly monitored because early disease management may avoid the propagation of the diseases. Many times, after pollination plant illness manifests later in the growing season. Different types of plant diseases affect various plant organs. Pathologists working with plants can most easily identify foliar diseases, or plant diseases that manifest indicators on leaves, by looking at these diseases' unique features. As much as 50% of crop losses are specifically attributable to fungus-related illness [20]. Identification of plant diseases is essential for plant phenotyping and precision agriculture. These two fields depend heavily on data, information, and technology. Since manual visual inspection is required, plant disease diagnosis and detection techniques are costly, consume huge time, reliant on experts, and often inefficient in precision agriculture. Additionally, human bias and fatigue are likely to have an impact on these approaches, resulting in decreased accuracy. Studies have looked into the use of image processing methods on plant images to overcome these inconsistencies in disease identification [21].

Earlier studies used conventional image processing techniques for diagnosing plant diseases. It is a black and white video showing the leaves of potted ferns with blackened foliage.. The plant village dataset was first released, and it has grown in popularity as the most popular database for learning and developing DL-based models for identifying plant diseases and estimating their severity. There are 54,309 images in the entire plant village dataset. The 38 different diseases that affect 14 crops are represented by different images. The majority of the pics were taken in a procedure-based lab setting with a consistent background [22]. To diagnose plant diseases, researchers reviewed 70 research papers, including review articles and studies on machine learning, image processing, and DL. Seven important research questions were compiled and answered. Several publicity accessible datasets and descriptions of common plant diseases were also covered. To identify plant illnesses, various sensors used for imaging and collected data were examined.

Diagnosing plant diseases is crucial for helping farmers make informed management decisions because plant diseases frequently reduce crop yield and product quality. Manual crop inspections are laborious and challenging tasks, often limited to small areas at a time, especially multiple varieties of diseases that have similar symptoms. Farmers can benefit greatly from an automated machine vision screening system for inspection in this regard [23]. Even though numerous algorithms have been developed for diagnosing plant diseases, a straightforward approach that only uses the most basic image data of interest in field situation remains effective.

Black measles, yellow measles, and other grape leaf diseases were identified and categorized in this study using a new technique image processing algorithm and multi-class SVM [24]. For many nations, including Iran, their main factor of income is agriculture.

Detection of plant diseases is crucial because they play a significant role in lowering production levels and quality. Although disease symptoms can appear in various plant parts, it is frequently used to diagnose symptoms that can appear in various plant leaves. An important initial step in minimizing losses brought on by plant infections is an early and accurate diagnosis. An improper diagnosis can result in poor management choices, such as choosing the wrong medicinal application, which could lead to further health losses and decreased productivity. The helpless eye method is a time-consuming, labor-intensive, and subject-to-human-error traditional method of diagnosing diseases, which is not appropriate for large fields [25]. They suggested a structured work for identifying and categorizing RGB images of pepper stem and petal diseases. Their method examined four types of diseases: initial scorch, cottony mound, delayed scorch, and small whiteness.

The color transformation was completed first, and dispersion was done using the K-means cluster algorithm. After feature extraction using the color co-occurrences method, an artificial network classifier was used. A method for sugarcane plant petal disease identification using pictures in RGB has been suggested [26]. Just three sicknesses rust identifies, ring identifies, and yellow identifies have experienced confirmation using the proposed method. Since SVM was employed for classification, feature extraction used a combined method of color and texture features. Four kernel types are used in the SVM classifier. This study examines the precision rate and root mean square error (RMSE) of the forecasting method by utilizing the information on temperatures to predict crop disease using K-Nearest Neighbor (KNN) and Max Voting approaches. The crop illness becomes apparent through analysis. The Kaggle plant image dataset contains sample image sizes of 500 based on temperature conditions. KNN and Max Vote algorithms are utilized for predicting agricultural diseases by considering weather conditions in order to improve accuracy and lower RMSE. Maximum Votes yielded 91% correct results, whereas KNN yielded 88% correct results [26].

The RMSE of the Max Voting technique is 3.97, whereas the KNN method has an RMSE of 4.34. These two measurements differ considerably [20]. The Max Voting approach outperforms the KNN method for plant prediction of diseases based on weather. Agriculture quality is critical to the economy. The high standard of agricultural products influences the economy of a country, and the identification of diseases influences quality. Crop disease detection is a difficult problem for agricultural professionals, requiring scientific methodologies and long-period observation. Plant blight. Analyzing only the leaves of a crop can be challenging if the disease presents identical symptoms. As a consequence, a machine learning system able to diagnose plant illnesses and offer identification of leaf diseases and categorization is lacking [27].

A developing country's population relies on agriculture as a main source of income. Experts in agriculture, identifying plant diseases, require to use the scientific procedures and observation for the treatment of the plant illness. If the medium is exhibited in a similar manner, it may be difficult to identify a plant only based on its leaves. An automated system that can identify the operation, as well as a system that can identify crop diseases and give plant petals disease diagnosis and classification, is needed [28]. The dataset for this study was collected from Kaggle. There are two groups of 500 samples, with 250 taken from each. During the training phase, 90% of the time is spent on training and the remaining 10% on testing [29].

Petal samples also consist of information on temperature, humidity, and color in addition to climate data. A sample dataset exported to Microsoft Excel was employed to validate the suggested and existing models using statistical analysis software. These datasets were trained on a 7th generation Intel i5 processor with 4GB of RAM and a 500 GB hard drive and the evolution of one image with the 24 * 768 pixels was analyzed by the Max voting classifier in sample Group One using the K nearest neighbor (KNN). An innovative method for identifying and categorizing plant leaf illnesses involves using the AdaBoost classifier and comparing its outcomes to the precision of the KNN classifier. In this proposed model, two methods are used: the Max Voting method and the KNN algorithm. The classifier is the most valuable. The RMSE of the Max Voting technique algorithm for forecasting crop diseases using temperature readings is 91%, which is higher than the RMSE of the KNN algorithm, which is 88% [30].

7.3 Proposed Work

Rice is Asia's most frequently farmed crop, but it is prone to a variety of ailments at various phases of development. Rice plant illnesses, and also a major drop in the amount and standard of farming items, can have an impact on food safety and productivity. Plant diseases can completely impede grain harvesting in extreme cases. As a result, automating plant disease identification and diagnosis is in high demand in agriculture. Now this automation is possible through AI techniques. YOLOv7 image object detection algorithm was used for this research. To develop the rice plant disease classification application by AI techniques, aiming to prevent crop reduction and improve the standard of rice grains.

7.3.1 Image Acquisition and Processing

Rice plant images of healthy and diseased plants were downloaded and processed from pen Kaggle datasets. This dataset contains 5447 total pictures of healthy and diseased plants. A total number of 1488 pictures are of healthy rice plants and the remaining 3959 images are of diseased plants affected by false smut, blast, bacterial blight, sheath blight, and brown spot diseases. These images were processed for the rice plant disease classification. All images were resized to 640 × 640 pixels and labeled using image data labeling software. After completion of image labeling, the data was saved in a new file.

7.3.2 Splitting the Training, Testing, and Validating Data

The processed data collection is split into the train, validate, and test images. Seventy percent of images are split into the train set category, 20% of images are stored under the validation category, and the final test set contains 10% of images. Healthy leaves are split into 10% for each disease. The process starts with the training on the images, and processing these images. After the training process, test images are evaluated with the training images to predict and classify the rice plant diseases.

7.3.3 Workflow

One CNN can build a number of bounding boxes and also build a class for these boxes. A spatial pyramid with enhanced YOLOv7 can directly perform training processes on the images, helping to optimize the performance of processing. YOLO starts the image process by loading the neural networks. These networks process the images and extract features from a given input. The bounding boxes built by YOLO will process the whole image and set the boundaries by frames. The neural networks transmit this image data through feature extraction. These networks extract the characteristics from each box of the bounding line. Each bounding box contains a number of predictions; these predictions are connected with the midpoint of the boundary box and it is related to the bound box. Now it starts predicting the confidence score and conditional classes. To find the probability of a class, use the following equation.

$$P(ConClass|A) * P(A) * INOU = P(ConClass) * INOU \tag{7.1}$$

This equation is used to find the confidence score of the confident predictions. Here P denotes the number of predictions; *ConClass* denotes conditional classes; and *INOU* is used for Intersection over union, used to read the area between the assumption and ground truth bounding boxes. It is used to train the images. These predictions are encrypted as follows:

$$Ht * Wt(BB * 5 + ConClass) \tag{7.2}$$

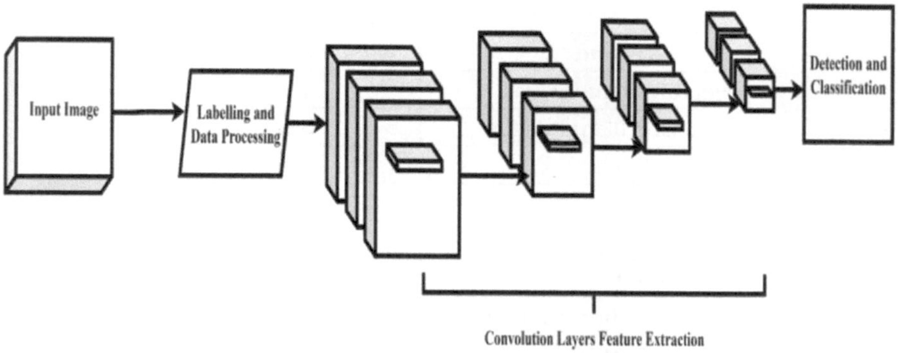

Figure 7.4 Proposed model Workflow

Ht and Wt are the elevation and extent of the grid cell, BB is bounding boxes, 5 is the number of predictions for building a bounding box, and ConClass denotes Conditional Class. *Imter OU* is found by the below equation.

$$Inter\ OU = \frac{Area\ of\ overlapping}{Area\ of\ Union} \tag{7.3}$$

Features are pulled out from the input picture by bounding boxes, and then the photos are moved to the process of detection and classification. Figure 7.4 displays a detailed description of the suggested model working flow.

7.4 Result and Discussion

The spatial pyramid model with enhanced YOLOv7 detects rice plant diseases by the object detection method. It processes the whole image and sets the boundary box for an image to identify the diseases. This boundary box helps to set the boundary with the infected part of a leaf. It helps to quickly identify the variety of diseases in the rice plant. There are 1488 photographs of healthy rice plants and 3959 images of diseased rice plants affected by diseases such as false smut, blast, sheath blight, bacterial blight, and brown spots on petals. These photos are being analyzed for the classification of rice plant sickness. All photos are scaled to 640 × 640 pixels and labeled using image data labeling software. Following the conclusion of picture labeling, the data is saved in a new file. Table 7.1 shows a detailed description of the dataset. The dataset contains six types of unhealthy images and healthy images to determine the rice plant disease prediction and classification.

Seventy percent of photos are saved in the train set category, 20% in the validation category, and 10% in the final test set. Healthy leaves are divided into

Table 7.1 Dataset Description

Disease Name	*Number of Images*
Healthy Plant Images	1488
Diseased Plant Images	3959
Sheath Blight	655
Blast	648
Bacterial Blight	680
False Smut	666
Bacterial Sheath Blight	670
Brown Spot	640

10% for each illness. The initial phase begins with the training photos, which are then processed. Following the training procedure, test photos were compared to the training images in order to predict and classify rice plant illnesses. After the training process, it produces 97.65% accuracy in rice plant disease detection. Figure 7.5 displays the details of the disease detection accuracy range. A spatial pyramid with enhanced YOLOv7 may immediately execute training on

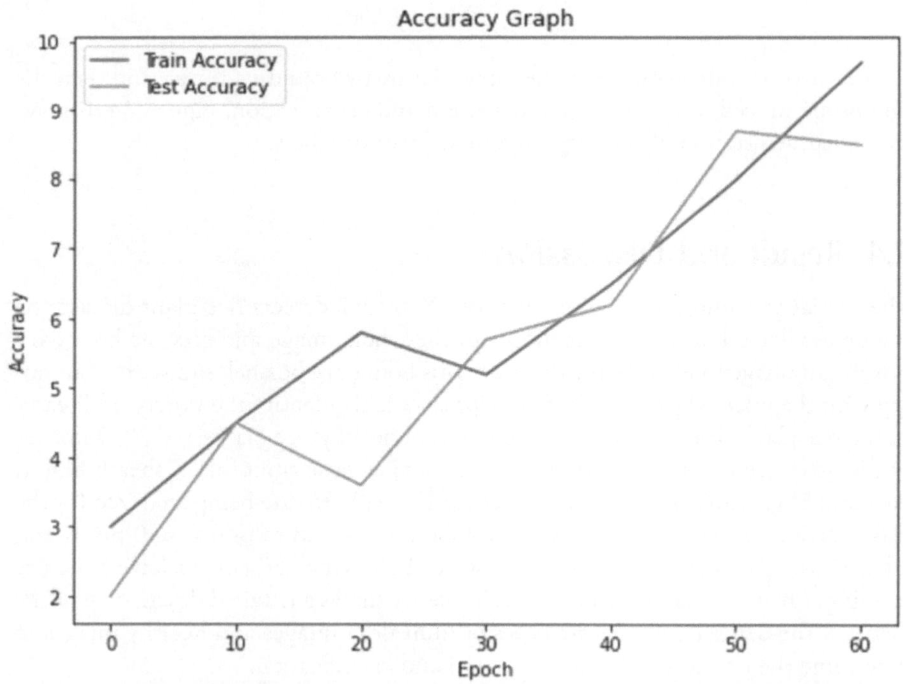

Figure 7.5 Graph of Accuracy Range

images and begin the picture-processing process by loading the neural networks. These networks analyze images and extract information from the input. YOLO's bounding boxes will process the entire image and set the boundaries by frames. This visual data is transmitted by neural networks throughout the feature detection process and retrieves the characteristics from the boundary line's boxes. These inputs are transferred into the output layer for the identification and categorization of rice petal diseases with an accuracy of 97.65% and a loss range of rice plant sickness classification. Figures 7.5 and 7.6 define the accuracy and loss range of the proposed method.

A spatial pyramid with enhanced YOLOv7 produced a 4.57% loss range. This technique generates the best accurate value when adequate rice plant disease prediction is used. YOLOv7 identifies a wide range of disorders at various stages of development. Each stage of the diseased leaves can be identified by the suggested YOLOv7 identification and categorization of each of the six types of rice plant ailments. The main purpose of this strategy was to diagnose the illness earlier in order to protect the plants from dangerous illnesses and treat them sooner. This AI implementation of models is more practical. Users can upload plant photographs in order to identify and cure rice plant ailments for free. Figure 7.5 shows the loss range of the rice leaves

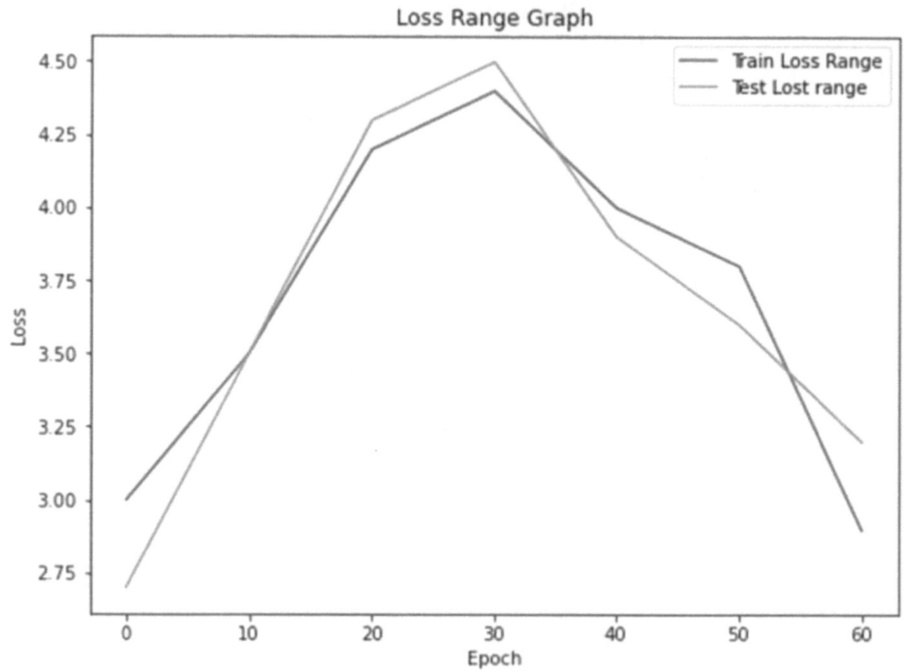

Figure 7.6 Graph of Loss Range

Table 7.2 Comparison Table

Comparison Table of the Proposed and Traditional Models		
Technology Used	*Accuracy Range*	*Loss Range*
DR-IACNN	91.1%	6.76%
Spatial Pyramid with enhanced YOLOv7	97.65%	4.57%

disease identification and classification. This method produces the maximum accurate disease detection of rice crop diseases. Table 7.2 shows the details of the comparison of the DR-IACNN model with the Spatial Pyramid with enhanced YOLOv7.

The table describes the loss range and accuracy range of proposed and existing models to explore the difference between their performance. The prediction and classification of rice plant disease produced by the proposed model demonstrate better accuracy and minimize the loss range for the given input.

This table describes the detailed comparison of the already used method and the proposed model. DR-IACNN framework was used to detect the leaf disease, and it produced 91.1% accuracy with the limitation of dataset insufficiency. Now the proposed Spatial Pyramid with enhanced YOLOv7 method automatically extracts features from the given input images like small spots, plant blasts, and common diseases of the rice plants. This method can identify not only common diseases but also every rice plant disease at all stages.

7.5 Conclusion

With adequate rice plant disease prediction, Spatial Pyramid with enhanced YOLOv7 method yields the most accurate value. It predicts multiple diseases at various phases of development. The proposed YOLOv7 detecting and classifying the six categories of rice plant illnesses can identify each stage of the afflicted leaves. The ultimate goal of this method was to diagnose the sickness at an initial stage in order to safeguard plants from serious infections and treat them sooner. This AI model deployment is more convenient. Users are able to submit plant photos to diagnose rice plant illnesses and cure them without incurring any cost. In the future, this approach can be used for detecting and classifying different types of plant diseases.

References

1. Bashish Al, D., Malik, B. & Bani-Ahmad, S. A framework for detection and classification of plant leaf and stem diseases, in: 2010 International Conference on Signal and Image Processing, 2010, 113–118.

2. Parikh, A., Raval, M. S., Parmar, C., & Chaudhary, S. Disease detection and severity estimation in cotton plant from unconstrained images, in: 2016 IEEE International Conference on Data Science and Advanced Analytics, (DSAA), 2016, 594–601.
3. Ramcharan, A., Baranowski, K., McCloskey, P., Ahmed, B., James, L., & Hughes, D. P. Deep learning for image-based cassava disease detection. Front. Plant Sci., 8, 1852, 2017.
4. Jhatial, M. J., Shaikh, R. A., Shaikh, N. A., Rajper, S., Arain, R. H., Chandio, G. H., ... & Shaikh, K. H. Deep learning-based rice leaf diseases detection using Yolov5. Sukkur IBA J. Comput. Math. Sci., 6(1), 49–61, 2022.
5. Xue, Z., Xu, R., Bai, D., & Lin, H. YOLO-tea: A tea disease detection model improved by YOLOv5. Forests, 14(2), 415, 2023.
6. Harakannanavar, S. S., Rudagi, J. M., Puranikmath, V. I., Siddiqua, A., & Pramodhini, R. Plant leaf disease detection using computer vision and machine learning algorithms. Manag. Glob. Transit., 3(1), 305–310, 2022.
7. Singh, S. P., Pritamdas, K., Devi, K. J., & Devi, S. D. Custom convolutional neural network for detection and classification of rice plant diseases. Procedia Comput. Sci., 218, 2026–2040, 2023.
8. Belay, A. J., Salau, A. O., Ashagrie, M., & Haile, M. B. Development of a chickpea disease detection and classification model using deep learning. Inform. Med. Unlocked, 31, 100970, 2022.
9. Jackulin, C., & Murugavalli, S. A comprehensive review of the detection of plant disease using machine learning and deep learning approaches. Meas.: Sens., 24, 100441, 2022.
10. Sungheetha, A. State of art survey on plant leaf disease detection. JIIP, 4(2), 93–102, 2022.
11. Ahmad, A., Saraswat, D., & El Gamal, A. A survey on using deep learning techniques for plant disease diagnosis and recommendations for development of appropriate tools. Smart. Agr. Technol., 3, 100083, 2022.
12. Javidan, S. M., Banakar, A., Vakilian, K. A., & Ampatzidis, Y. Diagnosis of grape leaf diseases using automatic K-means clustering and machine learning. Smart Agr Technol., 3, 100081, 2023.
13. Nalini, T., & Rama, A. Impact of temperature condition in crop disease analyzing using machine learning algorithm. Meas.: Sens., 24, 100408, 2022.
14. M. Suresha, K.N. Shreekanth, B.V. Thirumalesh, Recognition of diseases in paddy leaves using KNN classifier, in: 2017 2nd International Conference for Convergence in Technology (I2CT), 2017, 663–666.
15. Nagasubramanian, G., Sakthivel, R. K., Patan, R., Sankayya, M., Daneshmand, M., & Gandomi, A. H. Ensemble classification and IoT-based pattern recognition for crop disease monitoring system. IEEE Internet Things J., 8(16), 12847–12854, 2021.
16. J. Abdulridha, Y. Ampatzidis, S.C. Kakarla, P. Roberts, Detection of target spot and bacterial spot diseases in tomato using UAV-based and benchtop-based hyperspectral imaging techniques. Precis. Agric. 21, 955–978, 2020.
17. Abdulridha, J., Ampatzidis, Y., Qureshi, J., & Roberts, P. Laboratory and UAV-based identification and classification of tomato yellow leaf curl, bacterial spot, and target spot diseases in tomato utilizing hyperspectral imaging and machine learning. Remote Sens., 12(17), 2732, 2020.
18. Ojala, T., Pietikinen, M., & Menp, T. Multiresolution gray-scale and rotation invariant texture classification with local binary patterns. Proc. IEEE Trans. Pattern Anal. Mach. Intell., 24, 971–987, 2002.

19. Singh, V. & Misra, A.K. Detection of plant leaf diseases using image segmentation and soft computing techniques. Inf. Process. Agric., 4(1), 41–49, 2017.
20. Al Hiary, H., Bani Ahmad, S., Reyalat, M., Braik, M., & ALRahamneh, Z. Fast and accurate detection and classification of plant diseases. Int. J. Comput. Appl., 17(1), 31–38, 2011.
21. Puspha Annabel, L.S., Annapoorani, T., & Deepalakshmi, P. Machine learning for plant leaf disease detection and classification – a review, in: Proceedings of 2019 International Conference on Communication and Signal Processing (ICCSP), 2019, 538–542.
22. Raut, S., & Fulsunge, A. Plant disease detection in image processing using MATLAB. Int. J. Innov. Res. Sci. Eng. Technol. 6(6) 10373–10381, 2017.
23. Johannes, A., Picon, A., Alvarez-Gila, A., Echazarra, J., Rodriguez-Vaamonde, S., Navajas, A.D., & Ortiz-Barredo, A. Automatic plant disease diagnosis using mobile capture devices, applied on a wheat use case. Comput. Electron. Agric., 138, 200–209, 2017.
24. Kamal, K., Yin, Z., Wu, M., & Wu, Z. Depthwise separable convolution architectures for plant disease classification. Comput. Electron. Agric., 165, 104948, 2019.
25. Kamilaris, A., & Prenafeta-Boldú, F. X. (2018). Deep learning in agriculture: A survey. Computers and electronics in agriculture, 147, 70-90.
26. Karlekar, A. & Seal, A. Soynet: Soybean leaf diseases classification. Comput. Electron. Agric., 172, 105342, 2020.
27. Kim, W.-S., Lee, D.-H, & Kim, Y.-J. Machine vision-based automatic disease symptom detection of onion downy mildew. Comput. Electron. Agric., 168, 105099, 2020.
28. Lee, S.H., Chan, C.S., Mayo, S.J., & Remagnino, P. How deep learning extracts and learns leaf features for plant classification. Pattern Recognit., 71, 1–13, 2017.
29. Lee, S.H., Goëau, H., Bonnet, P., & Joly, A. New perspectives on plant disease characterization based on deep learning. Comput. Electron. Agric., 170, 105220, 2020.
30. Devi, T., Alice, K., & Deepa, N. Plant Disease Detection Using Enhanced EfficientNet Architecture in Comparison with DenseNet to Analyze the Severity in Leaves with Performance Measures, in: 2022 International Conference on Data Science, Agents & Artificial Intelligence (ICDSAAI) (Vol. 1), 2022, December, 1–5. IEEE.
31. Devi, T., Deepa, N., Chokkalingam, S. P., Gayathri, N., & Kumar, S. R. (2021). An Artificial Intelligence-Based Smart Task Responder: Android Robot for Human Instruction Using LSTM Technique. In: Dubey, A. K., Kumar, A. Rakesh Kumar, S., Gayathri, N., & Das, P. (eds.), AI and IoT-Based Intelligent Automation in Robotics, 149–163. Wiley.
32. Devi. T., Deepa, N., Kumar, S. R., Ganesan, R., & Gayathri, N. (2021). Towards Analyzing Skill Transfer to Robots Based on Semantically Represented Activities of Humans. In: Dubey, A. K., Kumar, A. Rakesh Kumar, S., Gayathri, N., & Das, P. (eds.), AI and IoT-Based Intelligent Automation in Robotics, 75–86. Wiley.
33. Deepa, N. E-TLCNN classification using densenet on various features of hypertensive retinopathy (HR) for predicting the accuracy, in: 2021 5th International Conference on Intelligent Computing and Control Systems (ICICCS), 2021, May, 1648–1652. IEEE.
34. Deepa, S. P. C. N., & Chokkalingam, S. P. A deep learning method on medical image dataset predicting early dementia in patients Alzheimer's disease using convolution neural network (CNN). Int. J. Recent Technol. Eng., 8, 604–609, 2019.

Chapter 8

Sustainable Farming Through AI-Enabled Precision Agriculture

D. Prabha, Siva Subramanian R., M.G. Dinesh, and Girija P.

8.1 Introduction

With the incorporation of Artificial Intelligence (AI) technology, precision agriculture has witnessed a tremendous transformation that has drastically changed contemporary agricultural practices. In addition to improving farming's sustainability and efficiency, this dynamic fusion of cutting-edge technology and agricultural practices has created opportunities for previously unheard-of levels of output (Smith MJ 2018). An age when data-driven insights and precisely directed actions enable farmers to optimize resource allocation, reduce risks, and increase yields has arrived as a result of AI and agriculture coming together. Traditional agricultural techniques often depended on generalized strategies that treated whole fields similarly, resulting in inconsistent yields and ineffective resource use. However, with the development of AI, this paradigm has changed to one that is more subtle and adaptable. Farmers can view their fields in high detail because of AI's skills in analyzing enormous volumes of data from many sources, including satellite images, sensors implanted in the ground, and weather predictions (Eli-Chukwu NC 2019). The foundation of AI-driven precision agriculture is this improved visibility. Here we will explore the extensive ramifications of AI integration in Precision Agriculture in following sections. AI-driven techniques provide solutions that are not only timely but also

DOI: 10.1201/9781003504900-8

grounded in data-driven accuracy, from real-time crop monitoring and disease diagnosis to predictive modeling of yields and optimum resource allocation. The revolutionary potential of AI in Precision Agriculture rests not only in its capability to simplify present procedures but also in its ability to support ecologically friendly and sustainable farming. The capacity to monitor crops in real time is one of the biggest innovations made possible by AI. AI algorithms analyze the collected data using remote sensing technology, such as satellites and drones, to provide useful insights. These insights include a wide range of essential elements, including soil moisture content, nutrient availability, and growth trends (Bannerjee G et al. 2018). Armed with this knowledge, farmers may take focused actions to reduce the danger of undernourished or waterlogged crops, such as precision irrigation and fertilization. Additionally, AI is adept at spotting illnesses and pests before they become problems. AI systems can quickly spot irregularities in crop health by analyzing images and data. This makes it possible for farmers to stop potential outbreaks before they spread, minimizing crop loss and obviating the need for pesticide application without regard for the intended target. The result is a safer and healthier product for customers as well as greater yields and less environmental impact. Another aspect of AI called predictive analytics is essential for modernizing agricultural tactics. With the use of both historical and current data, machine learning (ML) algorithms are able to predict yields quite accurately. Beyond the farm gate, this forecasting supports supply chain management and market planning. Farmers that have the benefit of foresight may allocate and distribute crops wisely, maximizing profits and minimizing waste. Sustainable agriculture's foundational principle is the effective distribution of resources, and AI adds a fresh perspective to this idea. Intelligent algorithms recommend the best times for planting, irrigation, and harvesting based on factors including soil composition, weather predictions, and consumer demand (Sharma R 2021). Farmers may improve output and reduce resource consumption by adjusting their behaviors in response to these findings, which is an important step toward agricultural sustainability. It is crucial to understand the wider ramifications of this development as will explore the world of AI-driven precision agriculture in this study. Beyond specific sectors, AI makes it possible to develop data-driven agricultural models that may guide policy decisions and solve issues with global food security. AI may help to shape solutions that address problems like climate change, land degradation, and population expansion by combining findings from many sources. Beyond simple convenience, the blending of technology and agriculture has the power to alter the very foundation of our food systems.

In summary, the integration of AI methods with Precision Agriculture heralds a paradigm change in contemporary agricultural practices. Farmers become active data-driven decision-makers according to the symbiotic link between technology and agriculture, which gives them previously unheard-of authority and transforms them from passive land stewards. In order to fully understand the many advantages these technologies provide to agriculture, this chapter sets out on a trip across the worlds of AI-driven crop monitoring, disease diagnosis, yield prediction, and resource optimization. As we enter the age of AI in agriculture, it is crucial to move cautiously and make the most of these developments for everyone's well-being as well as the sustainability and resilience of our planet as a whole.

8.1.1 Objectives

This chapter aims to thoroughly examine how AI methods are incorporated into precision agriculture and how they have a revolutionary effect on contemporary agricultural practices. The following goals are the focus of the survey: (1) Examine the different AI applications used in Precision Agriculture, such as real-time crop monitoring, disease detection, yield prediction, and resource optimization. (2) Recognize and list the advantages AI provides to Precision Agriculture, including increased crop yields, less resource waste, better decision-making, and environmental sustainability. (3) Explore challenges: Identify the obstacles and restrictions to using AI in agricultural practices, taking into account technological, financial, and moral considerations. (4) Evaluate Future Implications: Examine the long-term effects of AI-driven Precision Agriculture, including how it can help with issues like environmental protection, food security, and the changing position of farmers in a technologically advanced world.

8.1.2 Motivation

The introduction of AI technology is causing quick and significant changes in the agriculture industry, which is the driving force for completing this study. The purpose of this survey is driven by a number of following factors: Knowledge Deficit—A thorough study that explores the subtleties of AI's integration in precision agriculture is scarce, despite the fact that there is growing awareness of its effects on agriculture. By offering a thorough examination of the topic, this study seeks to close this knowledge gap. Second, timeliness AI approaches are at the forefront of the fast change that is taking place in the area of precision agriculture. It's important to document the present condition of the industry and any prospective future developments as AI continues to change agricultural practices. Practical Relevance—Farmers, agricultural researchers, policymakers, and stakeholders need knowledge about the advantages and difficulties of using AI in agriculture. The purpose of this survey is to provide useful information that assists in strategic planning and decision-making in the agriculture industry. Efficiency and sustainability AI-driven Precision Agriculture offers a viable answer in a society struggling with environmental issues and the need for sustainable resource management. It is crucial to comprehend its potential to increase productivity and decrease the environmental impact. Interest in a variety of fields—The survey caters to a large audience that includes academics, practitioners, and from both domains and strives to close the gap between agriculture and AI. It aims to promote multidisciplinary cooperation and information sharing. A look toward the future—The study promotes a forward-looking mindset by examining current trends and their potential repercussions.

This survey aims to illuminate the complex ways in which AI is incorporated into precision agriculture. Its motivation stems from the need for a thorough understanding of this revolutionary fusion of technology and agriculture, with an emphasis on the advantages, difficulties, and prospects for the future.

8.2 Overview of Artificial Intelligence

One of the most ground-breaking developments in modern technology is the introduction of AI. The creation of computer systems that are capable of carrying out activities that traditionally need human intellect is what AI is all about at its core. These jobs cover a wide range, requiring the ability to learn from data and adapt to new information often (Khandelwal PM and Chavhan H 2019). They range from simple arithmetic to complex decision-making procedures. The overview of AI is given in Figure 8.1.

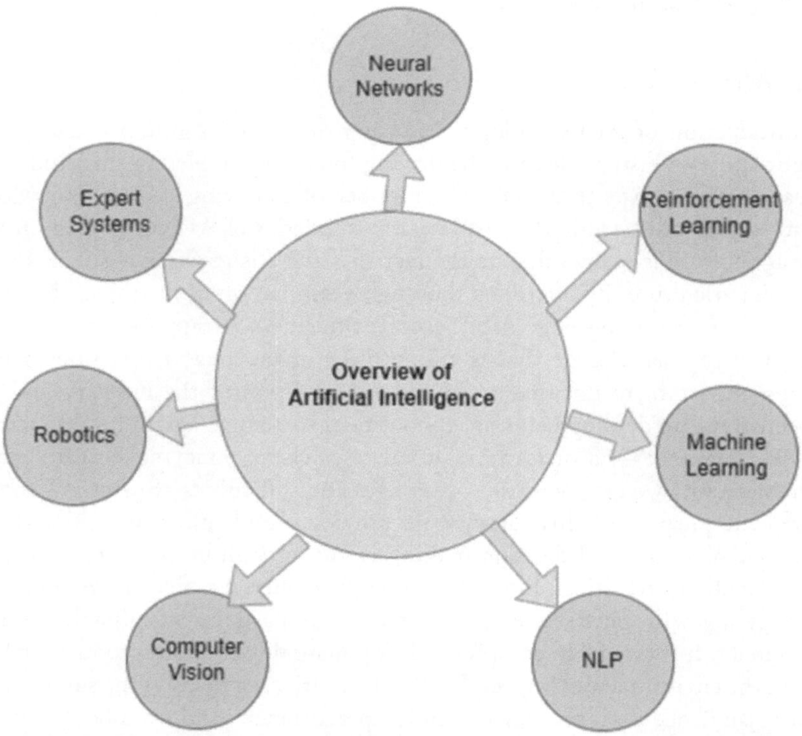

Figure 8.1 Overiew of Artifical Intelligence in Agriculture

8.2.1 The Development of AI

The idea of robots imitating human intellect initially appeared in the middle of the 20th century, which is when AI first emerged. The Dartmouth Workshop in 1956 was when the phrase "artificial intelligence" was first used, thus establishing the discipline. Symbolic reasoning was the focus of early AI research, and programs created to manipulate symbols and logic for problem-solving were used. The development of AI, however, proceeded at a slower rate than first anticipated, resulting in what is now referred to as the "AI winter" between the late 1970s and early 1980s. As the limits of the current technology became clear, funding for research decreased. AI didn't make a comeback until the 1990s, fueled by improvements in computing power and cutting-edge algorithms.

8.2.2 AI Domain

AI spans a variety of fields, each tackling a unique aspect of intellect akin to that of a person. Among the notable AI domains are as follows:

8.2.2.1 Machine Learning

Making algorithms for ML requires teaching computers to learn from data. These algorithms are adept in applications like image identification, natural language processing, and recommendation systems because they continuously improve their performance on particular tasks as they meet more data (Sharma A et al. 2020).

8.2.2.2 NLP

The goal of natural language processing (NLP) is to make it possible for computers to comprehend, translate, and create human language. Language translation, sentiment analysis, chatbots, and voice assistants are all supported in this area.

8.2.2.3 Computer Vision

Computer vision enables machines to receive and analyze visual data from the outside environment in a manner similar to how humans see visual information. It powers programs that control autonomous cars, object detection, and face recognition.

8.2.2.4 Robotics

AI and physical machinery are combined in robotics to produce autonomous, intelligent systems that can carry out tasks in the real world. This includes anything from drones and humanoid robots to industrial robots.

8.2.2.5 Expert Systems

In some fields, expert systems mimic human decision-making processes. In order to provide well-informed advice or solutions, they depend on norms and information bases.

8.2.2.6 Neural Networks

A subset of ML is neural networks, which are modeled after the way the human brain is organized. They are made up of networked information processing and transmission nodes (neurons). Deep learning, a branch of neural networks, excels at difficult tasks like voice and picture recognition.

8.2.2.7 Reinforcement Learning

Reinforcement learning is the process of teaching an algorithm to do a series of actions that maximize cumulative rewards. This method is useful in situations when an agent interacts with the environment to discover the best tactics.

The deployment of AI across businesses, from healthcare and finance to entertainment and agriculture, is the result of the merging of AI with several disciplines. The ability of AI to analyze data, spot trends, and arrive at well-informed judgments has sparked fundamental shifts in how farmers manage their crops and resources. Let us explore how agricultural practices and resource management have changed as a result of AI's inclusion into Precision Agriculture as our survey evolves. The roles that AI plays in crop monitoring, disease identification, yield prediction, and resource optimization will be explored in the following sections, emphasizing the real effects that AI-driven initiatives are having on the agricultural environment.

8.3 Applications of Machine Learning Algorithms

In the field of Precision Agriculture, ML algorithms have become potent tools that are revolutionizing how farmers manage crops, identify conditions, and forecast harvests. In-depth discussions of some of the most important ML techniques, such as decision trees, support vector machines, and random forests, are explored in this section, along with an emphasis on how they are crucial for yield prediction, disease detection, and crop monitoring. The overview of ML in Precision Agriculture if represented in Figure 8.2.

8.3.1 Knowledge of ML Algorithms

The many methods used by ML algorithms allow computers to learn from data and enhance their performance on certain tasks. These algorithms may be divided into many categories, such as reinforcement learning, unsupervised learning, and

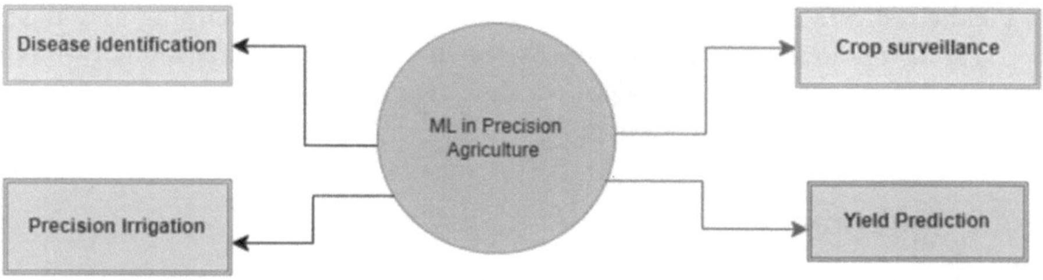

Figure 8.2 ML in Precision Agriculture

supervised learning. Here, we concentrate on three widely used categories of algorithms:

Decision Trees: Decision trees are hierarchical frameworks that facilitate decision-making by outlining probable outcomes for each conceivable course of action. They work especially well for classification work where the objective is to classify data according to its attributes.

Support Vector Machines (SVM): SVM is a supervised learning technique that performs well in both regression and classification applications. It operates by locating a hyperplane that maximizes the margin between data points of distinct classifications.

Random Forests: Multiple decision trees are combined in random forests, an ensemble learning technique, to increase resilience and accuracy(Nagasubramanian et al. 2021). During training, they produce a large number of decision trees, and they combine their predictions to produce a final outcome (Rajendran T et al. 2023).

8.3.2 Precision Agriculture Uses

Innovative applications that improve agricultural practices have resulted from the incorporation of ML algorithms into Precision Agriculture:

Crop Surveillance: To monitor crops in real time, ML algorithms analyze data from sources including satellite images, drones, and ground sensors. They are able to recognize elements including plant health, growth patterns, and signs of stress. Based on these variables, crop conditions are often classified using decision trees and random forests (Carvajal-Yepes M et al. 2019).

Disease Identification: To stop the spread of infections and minimize agricultural loss, early disease diagnosis is essential. SVM and other ML algorithms analyze data, such as plant images, to find abnormalities and signs of illness. These algorithms may be trained to recognize healthy from unhealthy plants, enabling farmers to take quick action.

Yield Prediction: Crop yield predictions assist farmers in planning and allocating resources wisely. To properly predict future yields, ML algorithms, in particular regression models, use past data on elements like weather, soil parameters, and crop attributes.

8.3.3 Advantages

There are various benefits of using ML algorithms in precision agriculture.

Data-Driven Insights: Farming decision-makers may now use real-time data from ML algorithms to swiftly analyze massive volumes of data.

Accuracy: Algorithms have the ability to spot small patterns and correlations in data that may go undetected by people, leading to more precise forecasts and treatments.

Efficiency: Automated procedures powered by ML algorithms save time and resources by streamlining workflows and cutting down on waste.

8.3.4 Obstacles

Data Quality: ML algorithms rely on trustworthy, high-quality data. Biased or inaccurate data might provide incorrect forecasts and judgments.

Interpretability: Understanding the logic behind certain algorithms' predictions may be tricky, such as complicated neural networks, which make it challenging to comprehend their results.

Model Robustness: It is essential to make sure ML models generalize successfully to fresh, unexplored data. A possible problem is overfitting, which occurs when a model performs well on training data but badly on fresh data.

8.4 Deep Learning Applications and Algorithms

Precision Agriculture is being revolutionized in many different ways because of Deep Learning, a subset of ML. This section explores how deep learning methods, such as convolutional neural networks (CNNs) and recurrent neural networks (RNNs), are used in the fields of yield estimate, plant phenotyping, and picture analysis (Kamilaris A, Prenafeta-Boldú FX 2018).

8.4.1 Knowledge of Deep Learning Algorithms

Artificial neural networks with numerous layers—known as "deep architectures"—are used in deep learning techniques to extract complex properties from data. Deep learning networks may be divided into following two categories:

8.4.1.1 Convolutional Neural Networks

CNNs are designed to examine visual information like pictures and movies. To automatically extract hierarchical information from pictures, they use specialized layers, such as convolutional and pooling layers.

8.4.1.2 Recurrent Neural Networks

RNNs work well with sequential data, such as text or time-series data. They are skilled at activities requiring context and temporal relationships because they have internal memory to interpret sequences.

8.4.2 Precision Agriculture Uses

Precision agriculture has benefited from revolutionary uses made possible by deep learning algorithms:

Image Analysis: In agriculture, CNNs are often used for image-based activities. They are able to recognize and categorize a wide range of things in images, including crops, weeds, pests, and illnesses. This supports accurate diagnosis and early illness intervention.

Plant Phenotyping: Deep learning makes it easier to collect precise data about plant traits from images. CNNs that can recognize tiny differences in plant characteristics improve the accuracy and efficacy of phenotyping, which entails assessing plant attributes.

Yield Prediction: Understanding the numerous elements that impact growth is important for predicting crop yields. To reliably predict yields, deep learning algorithms, like as RNNs, can analyze past data on the climate, soil, and plant development trends (Chandra AL et al. 2020).

8.4.3 Advantages

Deep learning benefits Precision Agriculture in a number of ways:

Attributes Extraction: CNNs eliminate the need for human feature engineering by automatically extracting pertinent features from pictures.

Complex Relationships: RNNs are efficient for jobs requiring sequences and time-series data because they capture temporal relationships.

High Accuracy: Deep learning models often outperform conventional approaches in complicated problems where they may reach extraordinary accuracy.

8.4.4 Obstacles

Data Requirements: In agricultural contexts, it may be difficult or costly to get the large amounts of labelled data that deep learning models need for training.

Computing Resources: Training deep learning models requires a lot of computing time and power, which may be a problem in contexts with limited resources.

Interpretability: It might be difficult to comprehend the rationale behind deep learning model choices, particularly in complicated models.

Deep learning algorithms have changed the way precision agriculture is practiced because of their ability to extract complex patterns and representations. Deep learning provides farmers with knowledge that improves decision-making, resource management, and crop health via applications in image analysis, plant phenotyping, and yield calculation. As deep learning technology develops, its application to precision agriculture has the potential to spur more innovation and environmentally friendly farming methods.

8.5 AI Model Generation Platforms

The availability of strong platforms and tools designed to speed up the development process has considerably aided in the development of AI models. This section gives a brief introduction to popular platforms that make it possible to create AI models, such as TensorFlow, PyTorch, and specific AutoML tools designed specifically for agricultural applications.

8.5.1 TensorFlow

Google created TensorFlow, an open-source ML framework that is now widely used. It gives users access to a whole ecosystem for creating and refining different kinds of AI models, including neural networks. Both low-level APIs for increased flexibility and high-level APIs for convenience of use are available with TensorFlow. Because of its adaptability, developers may work on a range of projects, including natural language processing and picture identification.

8.5.2 PyTorch

Another well-liked open-source deep learning framework is PyTorch, which is renowned for its dynamic computation graph and user-friendly layout. PyTorch, which was created by Facebook's AI Research lab (FAIR), is favored for being simple to use and working with intricate and dynamic neural network topologies. Its "define-by-run" concept makes it especially suitable for research-oriented jobs since it helps researchers and developers to build models more naturally.

8.5.3 AutoML Instruments for Agriculture

AutoML (automated machine learning) systems have arisen to streamline the creation of AI models by automating a number of tasks, including model selection and optimization as well as data pretreatment. These technologies are specially designed for the agricultural industry, using data and experience from that industry to speed up model generation. They often combine data pipelines, hyperparameter optimization, and model validation, which increases the usability of AI for ML novices.

Due to their capacity to quickly provide useful models for certain agricultural activities, AutoML technologies have proved beneficial in precision agriculture. Agriculture applications include:

Detection of Crop Disease: With excellent accuracy, AutoML algorithms can analyze photos of crops and identify disease signs to support early intervention.
Yield Estimation: These technologies anticipate crop yields based on previous soil and climatic data, providing farmers with information for resource allocation and planning.
Precision Irrigation: AutoML models optimize irrigation methods by using meteorological and soil moisture data, reducing water waste.
Detection of Weed: AutoML technologies can identify crops and weeds from photographs of fields, allowing for more precise weed control.

Platforms for creating AI models are crucial in driving the implementation of AI in precision agriculture. They enable both professionals and novices to use AI, producing better, more precise answers that promote innovation in agricultural techniques. A big step has been taken toward realizing the full potential of AI in contemporary agriculture with the development of platforms like TensorFlow, PyTorch, and specialized AutoML tools designed for the agricultural sector (Lohith R et al. 2022).

8.6 Exposing AI Models as Microservices Driven by API

The conversion of these models into API-driven microservices has tremendously aided the integration of AI models into agricultural systems and applications. This section discusses the importance of presenting AI models as microservices, how it facilitates seamless integration, and how it improves Precision Agriculture's capabilities.

8.6.1 Microservices Architecture

In a microservices architecture, large, complicated programs are divided into smaller, autonomous services that interact with one another using APIs. Each microservice specializes in a particular activity or feature, fostering modularity, scalability, and simpler maintenance. This strategy allows for the independent development, deployment, and updating of many AI models, which is ideally suited to the varied demands of precision agriculture.

8.6.2 Making AI Models Visible as Microservices

Encapsulating the trained model within a service that can be accessed using APIs is the first step toward exposing AI models as microservices. This enables smooth interaction between the AI model and other applications like farm management software or IoT gadgets. The microservices strategy has the following advantages when used in precision agriculture:

Modularity: The ability to deploy AI models as standalone microservices makes it simpler to update, swap out, or scale specific models without affecting the performance of the overall system.

Scalability: Microservices design supports horizontal scalability, allowing for the deployment of extra instances of the same microservices to accommodate rising demand.

Interoperability: Through APIs, many parts of an agricultural system may interact with one another, allowing diverse services to operate in unison.

Flexibility: Microservices may be created using a variety of programming languages or frameworks, allowing the appropriate tool to be used for the job.

8.6.3 Integrating Agricultural Systems

Exposing AI models as microservices makes it easier to integrate them into different agricultural applications and systems:

Farm Management Applications: AI-powered insights may be easily incorporated into the software that farmers use to make decisions.

IoT Hardware: AI models may be installed on edge devices like drones or sensors, allowing for on-the-spot analysis and decision-making in real time.

Market Place Integration: AI models may be used in agricultural markets to provide farmers with specialized suggestions and services.

Data Aggregation: Microservices have the ability to combine data from several sources, providing thorough insights for improved decision-making.

Finally, by making AI models accessible as API-driven microservices, Precision Agriculture is able to fully use the potential of AI. The modular, adaptive design of agricultural systems is in line with this strategy, allowing for smooth integration and improved capabilities. Precision Agriculture is better positioned to make use of AI's promise and drive innovation and efficiency in contemporary agricultural practices by using a microservices architecture.

8.7 AI Models Deployment

Efficiency, real-time processing, and data security are important considerations for using AI models in Precision Agriculture settings. This section looks at several deployment methods for AI models, including edge computing, cloud deployment, and hybrid approaches, to make sure they work well with agricultural systems and are integrated into them.

8.7.1 Edge Computing

Edge computing reduces the need to send huge volumes of data to a centralized location by processing data closer to the source. Edge deployment in Precision Agriculture has the following benefits:

Decision-Making in Real-Time: AI models installed on edge devices, such as sensors or drones, may take prompt actions in response to current data without depending on distant servers.

Reduced Latency: Because of its ability to lower latency, edge computing is perfect for time-sensitive applications like disease diagnosis and irrigation control.

Bandwidth effectiveness: Edge deployment lessens the requirement for high-bandwidth transmission, which may be expensive or unreliable in rural locations, by processing data locally.

8.7.2 Cloud Deployment

Hosting AI models on cloud servers enables scalability and accessibility and is known as cloud deployment. Benefits of cloud deployment for precision agriculture include:

Scalability: AI models can expand according to demand thanks to cloud infrastructure, which can handle shifting workloads during busy times of the year or special occasions.

Centralized Management: Cloud deployment offers centralized control over AI models, simplifying management of upgrades and upkeep.

Resource Effectiveness: For resource-intensive applications like large-scale data processing or complicated model training, the cloud's computational capability is suitable (Choudhary SK et al. 2016).

8.7.3 Hybrid Methodologies

Edge computing and cloud services are combined in a hybrid deployment, which makes use of each strategy's advantages. Hybrid methods in precision agriculture provide flexibility:

Data Processing Distribution: Hybrid techniques divide up work across edge and cloud resources according to compute needs and data volume.
Real-Time Analysis and Aggregate: While aggregated data is forwarded to the cloud for additional in-depth analysis or long-term storage, edge devices are capable of doing real-time analysis.
Cost-Saving Measures: By using edge resources for rapid processing and cloud resources for less time-sensitive workloads, hybrid solutions may aid in cost management.

8.7.4 Consideration

Data Privacy and Security: Whether handled at the edge or in the cloud, make sure that sensitive agricultural data is kept safe.
Network Connectivity: Consider the accessibility and dependability of network connections, particularly in isolated agricultural settings.
Resource Constraints: The sophistication of models that may be deployed may be hampered by the edge devices' constrained computational capabilities.
Latency Requirements: To determine whether edge deployment or cloud deployment is more appropriate, determine the allowable latency for certain applications.
Scalability Needs: Take into account the deployment's size and if the ability to scale up or down fast is necessary.

8.7.5 Impact on Precision Agriculture

The implementation of AI models has a significant influence on precision agriculture:

Real-Time Response: Edge computing improves disease detection and irrigation control by enabling quick reactions to changing circumstances.
Scalability: AI solutions may be deployed on the cloud to provide flexibility and efficiency under a range of workloads.
Resource Optimization: Hybrid techniques optimize resource utilization by balancing processes depending on resource availability and demand.

The effectiveness, reactivity, and scalability of AI solutions in Precision Agriculture are determined by the deployment of AI models, which is a crucial phase. Stakeholders can guarantee that AI models easily integrate into agricultural systems, delivering real-time information and spurring innovation in contemporary farming techniques, by deliberately deciding between edge computing, cloud deployment, and hybrid methods.

8.8 Data Ingestion and Integration Techniques

The successful integration of many data sources is essential for the success of AI applications in Precision Agriculture. This section explores the techniques for combining data from satellite photography, Internet of Things (IoT) devices, remote sensing, and other sources. In order to ensure smooth processing and analysis, it also investigates methods for feeding these data into AI pipelines.

8.8.1 Integrating Data

Several data sources must be integrated in order to provide a complete picture of agricultural systems. The sources listed below are often incorporated:

Remote Sensing: High-resolution insights into crop health, growth patterns, and soil conditions are provided by data from drones, satellites, and aerial images.

IoT Hardware: Sensors placed in fields gather information on environmental variables such as humidity, temperature, and soil wetness.

Satellite Imagery: Satellite data provides comprehensive coverage and aids in monitoring and tracking changes over time in bigger agricultural regions.

Weather Data: Forecasts for yield, disease, and irrigation schedules are all impacted by current weather conditions.

8.8.2 Data Ingestion Methodologies

Data ingestion is the process of gathering, cleaning, and storing data before it is fed into pipelines for AI. Data ingestion methods include:

Extract, Transform, and Load (ETL) Pipelines: These pipelines collect data from a variety of sources, clean and prepare it, and then load it into databases or data lakes.

Lakes of Data: Data lakes and other centralized repositories hold a variety of raw, processable data. This strategy encourages adaptability and scalability.

API Integration: APIs provide real-time data updates by enabling direct integration of data from external sources into AI pipelines.

Streaming data: Streaming data approaches handle data as it comes in, allowing for quick responses to changing circumstances and real-time insights.

8.8.3 Challenges and Factors to Take into Account

There are issues with data input and integration that need to be resolved:

Data Quality: To avoid inaccurate analysis and judgments, it is essential to ensure data accuracy, consistency, and relevance.
Data Privacy: To protect sensitive information, integrating data from diverse sources requires cautious treatment.
Data Volume: Effective techniques for storing, processing, and retrieving data are required to manage huge amounts of data.
Interoperability: Interoperability solutions are required because different data sources may utilize different formats and protocols.
Latencies: In order to allow quick reactions to changing circumstances, minimizing data latency is essential for real-time systems.

8.8.4 The Effect on AI Applications

AI applications in Precision Agriculture are strengthened by efficient data integration and ingestion:

Accurate Analysis: Integrating many data sources improves the precision of predictions and suggestions made by AI models.
Timely Decisions: The ability to react quickly to altering environmental circumstances is made possible by real-time data input.
Holistic Insights: Combining satellite images, IoT data, and remote sensing enables a thorough comprehension of agricultural systems.
Optimized Resource Management: "Sustainable practices" are made possible by precise data integration, which helps to optimize resource allocation.

Successful AI solutions in Precision Agriculture are built on effective data integration and ingestion techniques. Data-driven insights that support informed choices, better resource management, and improved agricultural outcomes are made available to farmers and stakeholders when data from different sources is properly gathered, cleaned, and preprocessed, as well as when it is smoothly integrated into AI pipelines.

8.9 Data Analytics Techniques

The different datasets used in Precision Agriculture may be used to glean insightful information thanks in large part to data analytics methods. This section explores several data analytics techniques used for pattern identification, anomaly detection,

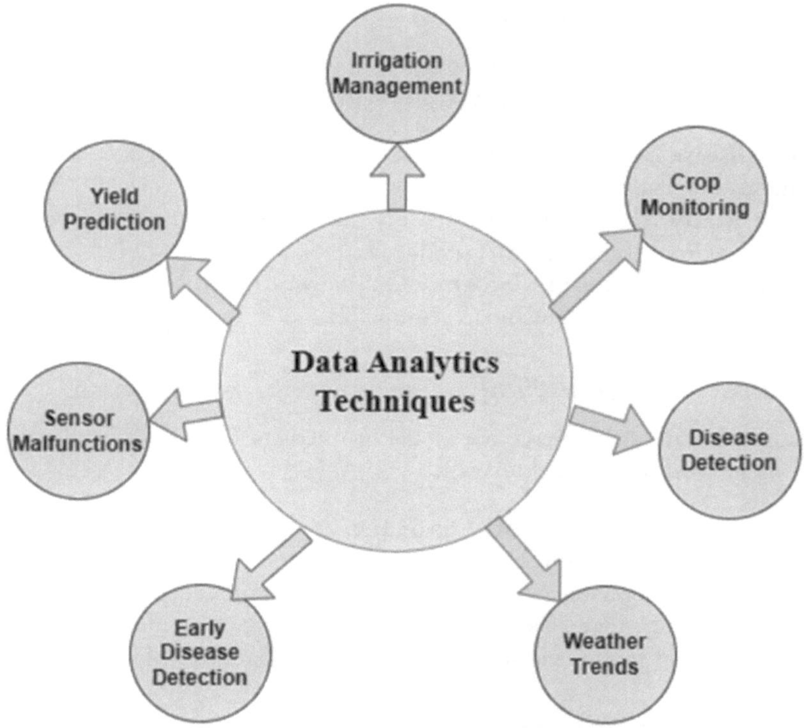

Figure 8.3 Data Analytics Techniques.

and predictive modeling, all of which improve agricultural practices and allow for more informed decision-making. The overview of data analytics techniques is given in Figure 8.3.

8.9.1 Pattern Recognition

Finding repeating patterns in data entails finding linkages, structures, and trends. Pattern recognition in Precision Agriculture can:

Crop Monitoring: Use data from remote sensing and IoT devices to analyze growth patterns, stress indicators, and health trends.

Disease Detection: Identify abnormalities and disease signs in plant photographs to enable early intervention.

Weather Trends: Recognize persistent weather trends that affect crop growth and disease outbreaks (Raj MP et al. 2015).

8.9.2 Anomaly Recognition

The goal of anomaly detection is to locate data points that drastically depart from the norm. Using anomaly detection in precision agriculture, you can:

Early Disease Detection: Look for out-of-the-ordinary changes in crop health indicators that might indicate disease outbreaks.
Sensor Malfunctions: Find IoT sensors that aren't working properly so that false environmental data isn't produced.
Weird Weather Patterns: Recognize unusual weather patterns that can affect crop development (Catalano C et al. 2022).

8.9.3 Predictive Modeling

Utilizing previous data, predictive modeling predicts future results. Predictive modeling in Precision Agriculture can:

Yield Prediction: To anticipate crop output, use climatic information, soil characteristics, and previous yields.
Irrigation Management: Predict the best irrigation times depending on the soil moisture content and the upcoming weather. Forecasting market demand for certain crops might help with planting choices (Sangeeta SG 2020).

8.9.4 Data Analytics Methods

Precision agriculture uses a number of data analytics approaches, including:

Machine Learning: Techniques for supervised, unsupervised, and reinforcement learning examine data to identify patterns, generate predictions, and categorize them.
Time-Series Analysis: It is possible to spot trends, seasonal patterns, and probable relationships by analyzing data that has been gathered over time.
Cluster Analysis: Similar data points may be grouped together to uncover underlying patterns and classifications (Sagar BM and Cauvery NK 2018).
Regression Analysis: A method for predictive modeling that finds links between dependent and independent variables.
Ensemble techniques: The accuracy and robustness of predictions are improved by combining different models.

8.9.5 Impact on Precision Agriculture

Precision Agriculture is considerably improved by data analytics techniques:

Informed Decisions: Analytics methods provide useful insights for the best planning, resource allocation, and decision-making.

Risk Mitigation: Resource wastage and crop loss are reduced by early disease identification and anomaly notifications.

Resource Optimization: Predictive models support the effective use of resources such as fertilizers, water, and others.

Sustainability: Data-driven agricultural techniques help to promote environmental sustainability.

Precision Agriculture relies on data analytics techniques to effectively analyze data gathered from various sources. These methods provide farmers and stakeholders with actionable insights that boost productivity, efficiency, and sustainability in contemporary agricultural practices by identifying trends, spotting anomalies, and forecasting future consequences.

8.10 Agriculture and Demystifying Precision

The revolutionary ideas of smart agriculture and precision agriculture make use of technology to change agricultural methods. This section gives a thorough explanation of these ideas, their goals, and any possible advantages they may have for farmers and the environment.

8.10.1 Precision Agriculture

Precision farming, often known as precision agriculture, is the process of using technology to optimize and customize agricultural practices depending on particular field circumstances. The following are some of precision agriculture's goals:

Decision-Making: With precision agriculture, choices about irrigation, fertilization, and pest control are informed by data gathered from a variety of sources, including sensors, drones, satellites, and historical records.

Resources Should Be Optimized. Precision Agriculture reduces resource waste by customizing interventions to the precise requirements of each section of a farm. This entails maximizing the use of pesticides, fertilizers, and water.

Yield Maximization: Precision Agriculture tries to increase crop yields by pinpointing and resolving certain issues that restrict growth, including nutrient shortages or disease outbreaks.

Reduced Environmental Impact: Precision Agriculture promotes ecologically friendly farming by accurately directing inputs and reducing wasteful practices.

8.10.2 Intelligent Agriculture

Precision Agriculture is enhanced by smart agriculture, sometimes referred to as digital agriculture or agriculture 4.0, which incorporates cutting-edge technology like the IoT, data analytics, and AI. Along with more general aims, Smart Agriculture's goals include those of Precision Agriculture (Kim WS et al. 2020).

Automation and Connectivity: IoT devices are used in "smart agriculture" to gather real-time data from farms, allowing automated procedures and smooth device connection.

Predictive Analytics: Smart Agriculture anticipates future trends, such as weather patterns or disease outbreaks, using data analytics and AI, enabling farmers to proactively handle issues.

Integration of the Ecosystem: By examining elements outside of the field itself, such as market demand, supply chain dynamics, and sustainability, smart agriculture takes the greater ecosystem into account.

Accessibility: By allowing farmers to exchange knowledge and learn from one another, smart agriculture fosters a group-based approach to problem-solving.

8.10.3 Environmental and Farmer Advantages

Both smart agriculture and precision agriculture have several advantages:

Increased Efficiency: Farmers optimize resource utilization and lower operating costs by accurately focusing interventions.

Higher Yields: Improved Crop Health and Higher Yields result from tailored treatments to specific regions of a field.

Environmental Sustainability: More sustainable agricultural techniques are aided by less resource waste and improved procedures.

Risk Mitigation: Using predictive analytics, farmers may foresee problems and lessen the effects of unfavorable circumstances.

Decision-Making Driven by Data: Both ideas provide farmers with data-driven information to help them make wise choices.

Economic Viability: Profitability is increased thanks to increased effectiveness, greater yields, and lower expenses.

Significant changes in the agricultural sector are being brought about through precision agriculture and smart agriculture. These ideas provide farmers the ability to improve practices, boost yields, and lessen their effect on the environment by using technology, data, and sophisticated analytics. In the end, they support agriculture that is more productive and sustainable, tackling the difficulties of feeding a population that is expanding while protecting the resources of our planet.

8.11 Future Challenges and Directions

The fusion of AI with precision agriculture has enormous potential to influence how farming is done in the future. This voyage is not without its difficulties, however. This section examines possible future developments in this area and talks about pressing issues that need to be resolved if AI in agriculture is to be used to its full potential.

8.11.1 Future Trends Include

Advanced sensing technologies—more precise and real-time data may be gathered as sensing technologies advance, allowing for greater insights into crop health, soil conditions, and environmental influences.

Edge Computing: By processing data closer to the data source, edge computing enables real-time decision-making without substantially depending on cloud infrastructure.

AI-Enabled Robotics: AI-powered robots have the potential to revolutionize labor-intensive operations like planting, harvesting, and weeding while also cutting expenses.

Hyper-Personalization: AI might make it possible to develop highly individualized strategies for each plant or crop, optimizing treatments and resources at the cellular level.

Multimodal Data Fusion: A more complete knowledge of agricultural systems could be possible by combining data from many sources, such as satellite photography, IoT devices, and drones.

8.11.2 Challenges

Data Security and Privacy: Data security and privacy are issues that are brought up by data collection. It is essential to preserve sensitive agricultural data.

Interoperability: It's possible that different platforms, technologies, and data sources won't work together perfectly. It is essential to establish standards for data exchange and interoperability.

Data Quality: The quality of AI models depends on the data they are trained on. Data quality and accuracy assurance is an ongoing concern.

Accessibility and Affordability: For small-scale farmers or those living in areas with low resources, access to cutting-edge AI technology may be restricted.

Education and Training: To properly use the advantages of AI, stakeholders, and farmers must comprehend its principles and technology.

Bias and Fairness: Biases found in training data may be carried over into AI models. It's crucial to guarantee justice and equality in AI results.

8.11.3 Future Direction

Several tactics may be used to overcome obstacles and clear the way for a prosperous future:

Data Governance: Adopt stringent data governance procedures to guarantee the security, privacy, and quality of your data.

Standardization: Create and embrace industry-wide standards for interoperability, data formats, and APIs.

Education and Training: Provide farmers and stakeholders with training programs that will enable them to comprehend and successfully employ AI technology.

Ethical AI: Create AI models that are built with a solid ethical foundation, eliminating prejudice, and guaranteeing justice. Encourage cooperation among farmers, researchers, technologists, and policymakers to jointly solve problems.

Policies and Rules: Governments and regulatory organizations can help foster the deployment of AI in agriculture while protecting data privacy and justice.

Future of Precision Agriculture and AI—With the use of cutting-edge technology, data-driven insights, and sustainable agricultural methods, agriculture offers immense potential. To realize this promise, it is necessary to overcome issues with data protection, interoperability, and accessibility. The agricultural sector can overcome these obstacles and steer toward a future in which AI improves farming for the better by adopting ethical AI practices, encouraging cooperation, and putting intelligent rules in place.

8.12 Conclusion

In conclusion, a new age of innovation and sustainability in farming has been ushered in by the symbiotic interaction between AI and precision agriculture. The landscape of agriculture has changed significantly as a result of the introduction of AI technology, moving away from conventional methods and towards one where data rules. This research has shown the critical role AI plays in all facets of precision agriculture, from the original conceptualization of problems to the ultimate deployment of solutions that are morally responsible. The combination of AI's skills, which include ML and deep learning, has sparked important innovations. AI-driven insights have evolved into crucial tools for farmers, helping them anticipate yields and quickly spot crop diseases. These precisely created AI models have successfully integrated themselves into agricultural systems by harnessing the potential of microservices and data integration. Decision-makers may now get actionable insights from detailed patterns and trends because of data analytics and visualization, allowing them to make wise decisions in an agricultural world that is always changing. The AI revolution has made the fundamental aims of precision and smart agriculture, which include wise resource management, increased yields, and sustainable practices, more realistic. Looking forward, the development of AI will have an impact on precision agriculture. Untapped potential is suggested by promising futures like cutting-edge sensing, edge computing, and AI-driven robotic help. However, these options also present issues with data security, interoperability, and the need for education. Despite these obstacles, the direction is clear: The incorporation of AI into agriculture will transform farming into a dynamic, data-driven, and environmentally sensitive activity. The full scope of AI's revolutionary impact goes beyond simple technological development. Instead, it represents a paradigm change in how we think about agriculture—one in which every data point, anomaly, and prediction holds the key to improving food security, streamlining resource use, and preserving our planet's priceless resources. The synergy of technology and agriculture is ready to assure a fed global population and peaceful cohabitation with our environment as AI continues to reveal the way ahead. In fact, the union of AI with precision agriculture is more than simply a revolution; it is a fundamental development that has the potential to completely reimagine farming in the future.

References

Bannerjee G, Sarkar U, Das S, Ghosh I, "Artificial intelligence in agriculture: A literature survey", *International Journal of Scientific Research in Computer Science Applications and Management Studies*. 2018 May; 7(3):1–6.

Carvajal-Yepes M, Cardwell K, Nelson A, Garrett KA, Giovani B, Saunders DG, Kamoun S, Legg JP, Verdier V, Lessel J, Neher RA, "A global surveillance system for crop diseases", *Science*. 2019 Jun 28; 364(6447):1237–9. DOI: 10.1126/science.aaw1572

Catalano C, Paiano L, Calabrese F, Cataldo M, Mancarella L, Tommasi F, "Anomaly detection in smart agriculture systems", *Computers in Industry*. 2022 Dec 1; 143:103750. https://doi.org/10.1016/j.compind.2022.103750

Chandra AL, Desai SV, Guo W, Balasubramanian VN, "Computer vision with deep learning for plant phenotyping in agriculture: A survey," *arXiv preprint arXiv:* 2006.11391. 2020 Jun 18. https://doi.org/10.48550/arXiv.2006.11391

Choudhary SK, Jadoun RS, Mandoriya HL, "Role of cloud computing technology in agriculture fields", *Computing*. 2016; 7(3):1–7.

Eli-Chukwu NC, "Applications of artificial intelligence in agriculture: A review", *Engineering, Technology & Applied Science Research*. 2019 Aug 1; 9(4). DOI: 10.48084/etasr.2756

Kamilaris A, Prenafeta-Boldú FX, "Deep learning in agriculture: A survey", *Computers and Electronics in Agriculture*. 2018 Apr 1; 147:70–90. https://doi.org/10.1016/j.compag.2018.02.016

Khandelwal PM, Chavhan H, "Artificial intelligence in agriculture: An emerging era of research", Research Gate Publication. 2019 Sep; 1(1):56–60.

Kim WS, Lee WS, Kim YJ, "A review of the applications of the internet of things (IoT) for agricultural automation", *Journal of Bio Systems Engineering*. 2020 Dec; 45:385–400. https://doi.org/10.1007/s42853-020-00078-3

Lohith R, Cholachgudda KE, Biradar RC, "PyTorch Implementation and Assessment of Pre-Trained Convolutional Neural Networks for Tomato Leaf Disease Classification," In *2022 IEEE Region 10 Symposium (TENSYMP)*. 2022 Jul 1 (pp. 1–6). IEEE. DOI: 10.1109/TENSYMP54529.2022.9864390

Nagasubramanian G, Sakthivel RK, Patan R, Sankayya M, Daneshmand M, Gandomi AH, "Ensemble classification and IoT-based pattern recognition for crop disease monitoring system", *IEEE Internet of Things Journal*. 2021; 8(16):12847–12854.

Rajendran T, Ramana TV, Nalini M, Subramanian RS, Devi DC, Anuradha M, "Paddy Plant Leaf Disease Classification based on Artificial Intelligence", In *2023 4th International Conference on Electronics and Sustainable Communication Systems* (ICESC). 2023 Jul 6 (pp. 1002–1009). IEEE. DOI: 10.1109/ICESC57686.2023.10193536

Raj MP, Swaminarayan PR, Saini JR, Parmar DK, "Applications of pattern recognition algorithms in agriculture: A review", *International Journal of Advanced Networking and Applications*. 2015; 6(5):2495–2502.

Sagar BM, Cauvery NK, "Agriculture data analytics in crop yield estimation: A critical review. Indonesian", *Journal of Electrical Engineering and Computer Science*. 2018 Dec;12(3):1087–1093. http://doi.org/10.11591/ijeecs.v12.i3.pp1087-1093

Sangeeta SG, "Design and implementation of crop yield prediction model in agriculture", *International Journal of Scientific & Technology Research*. 2020; 8(1):544–549.

Sharma R, "Artificial intelligence in agriculture: a review", In *2021 5th International Conference on Intelligent Computing and Control Systems (ICICCS)*. 2021 May 6 (pp. 937–942). IEEE. DOI: 10.1109/ICICCS51141.2021.9432187

Sharma A, Jain A, Gupta P, Chowdary V, "Machine learning applications for precision agriculture: A comprehensive review," *IEEE Access*. 2020 Dec 31; 9:4843–73. DOI: 10.1109/ACCESS.2020.3048415

Smith MJ, "Getting value from artificial intelligence in agriculture", *Animal Production Science*. 2018 Nov 21; 60(1):46–54. https://doi.org/10.1071/AN18522

Chapter 9

The Significance of Data Integration and Ingestion in Enhancing Agricultural Operations: A Comprehensive Exploration

V. Ebenezer, M. Robinson Joel, M. Vargheese, N. Karthik, K. Arul Jeyaraj, and G. Gandhi Jabakumar

9.1 Introduction

Data integration is the procedure of merging and converting the various sources of raw data in agriculture into a single, coherent dataset for use in decision-making and analysis. Data ingestion is the first step in the collection of raw data from various sources in agriculture. Utilising the potential of information-driven understanding in contemporary agriculture requires both of these processes. Integration [1] is necessary for two reasons: Firstly, an integrated view may be developed from a collection of already-existing information systems to make it easier to access and reuse information from a single centralised access source. Secondly, given a specific information requirement, information from many complementary information systems is merged to provide a more thorough foundation to meet the requirement.

DOI: 10.1201/9781003504900-9

The process of merging data from multiple places into just one, cohesive perspective is known as data integration. ETL refers to Extract, Transform, and Load. Integration involves procedures like cleaning. Data warehousing [2] is a technology that facilitates data analysis applications for industries including retail, banking, telecommunications, and bioinformatics. For instance, a network of retail stores like Walmart is interested in combining information from its sales and inventory databases from many shops in various places, and its marketing from various departments. The field of communications [3] can also benefit from data warehousing. Programmes for services that gather user data and then identify use trends, identify fraudulent activity, and improve overall performance. What qualities does an API (Application Programming Interfaces) [4] have? Accessibility gateways to an application have been collectively referred to as "APIs". But as time went on, the contemporary API developed several distinctive qualities that have genuinely changed the technological landscape. First off, contemporary APIs follow particular standards (usually HTTP (HyperText Transfer Protocol) and REST (REpresentational State Transfer) [5]), which make them developer-friendly, self-descriptive, simple to use, and widely recognised. The process of abstracting data from a variety of information sources, including relational databases, information assets exposed through internet services, XML archives, and others, in order to enable access to them regardless of their tangible storage or varied structure, is known as statistical virtualisation [6]. This covers cloud computing, SOA designs, post-acquisition system integration, visualisations, mashups, sites, disclosure, and information warehousing [7]. RDBMS [8] stands for Relational Data Base Management System and ERP [9] stands for Enterprise resource planning ERP system, which supports robotics and procedures in finance, personnel, production, logistical services, purchasing, and other areas, aiding in managing your complete firm. Customer relationship management (CRM) [10] is a collection of data-driven, integrated software tools that manage, track, and store information about the existing and future clients of your business.

Master data management (MDM) [11] is a procedure to generate a consistent collection of data about customers, goods, suppliers, and other business entities across various IT systems. MDM evolved from earlier distinct techniques, particularly customer data integration (CDI) [12] and product information management (PIM) [13], which were aimed at combining data for particular entities. CDI and PIM are still functioning subcategories; however, MDM combined them into one category with a wider focus. Systems for processing transactions are necessary for corporate operations, and BI and analytics are increasingly used to power advertising campaigns, client engagement initiatives, supply chain management, and other business operations. However, a lot of businesses lack a unified consumer view. The need for real-time data integration [14] has grown as businesses try to make data-driven choices and increase operational effectiveness in today's business environment. Businesses must combine data from numerous sources in real-time due to the increasing amount of data if they want to swiftly get insights and make wise decisions.

The practice of correcting or deleting inaccurate, damaged, improperly formatted, duplicate, or missing data from a dataset is known as data cleaning [15]. There are several ways for information to be replicated or incorrectly labelled when merging different data sources. Even though results and algorithms appear to be right, they are unreliable if the data is inaccurate. Remove duplicate or pointless observations [16] as well as undesirable observations from your dataset. The majority of duplicate observations will occur during data gathering. Duplicate data can be produced when you merge datasets from several sources, scrape data, or get information from consumers or other departments. One of the most important factors to take into account in this procedure is de-duplication. Data correctness, consistency, and dependability throughout an organisation are the main objectives of MDM [17].

Creating a uniform format and structure for data coming from multiple source systems so that it may be successfully merged into a single data repository. Putting safeguards in place to preserve delicate master information and ensure adherence to data privacy laws like GDPR (General Data Protection Regulation) [18], HIPAA (Health Insurance Portability and Accountability Act) [19], and others. Maintaining the information that gives master data items context and definitions. Data must be sent from the source to the destination in real-time or almost real-time. This data can be sent through a number of technologies, including change data capture (CDC) [20], event-driven systems, message queues, and streaming platforms. Data is moved from source systems to a data warehouse or other destination systems using the ETL (Extract, Transform, Load) [21] process. Schema transformation and mapping are essential steps in this procedure. Cloud-based integration [22] frequently necessitates information modification and mapping since various systems may employ various data types and structures.

Data ingestion is the procedure of gathering, importing, and ingesting data into a storage or computer system from numerous sources, frequently in a systematic and organised way. A variety of sources, including databases, files, APIs, streaming platforms, sensors, and more, can provide this data. Real-time or streaming data [23] intake is the act of acquiring and loading data without arranging it into the correct area as soon as it is produced. Since it needs to be monitored and used in circumstances when timing is crucial, it is expensive. In the hybrid procedure known as lambda data ingestion, the latter two are both utilised. Demographic information [24], local trends, and geographical analysis may all be gleaned from the dataset by include location data such as geocodes, postal locations, and coordinates. Businesses may get insights into client preferences, behaviour, and interactions by connecting social media accounts to customer information. When integrating external data sources, it is crucial to protect data privacy and security, especially if they include personally identifiable information (PII) [25]. NoSQL [26] tables, data warehouses, flat files, APIs, storage in the cloud, and other types of databases might be included. This entails searching data sources for certain information. In this regard, SQL (Structured Query Language) is frequently used in databases to retrieve data based on specific criteria. Before being used, data frequently has to be processed, cleansed, or aggregated. By utilising agile methods for data, a DataOps approach [27] to pipelines for data automates every step as possible and eliminates the "how" of implementation. The "what" of the information as well as meeting business criteria may be the areas of focus for data engineers.

9.2 Data Integration and Ingestion in Agriculture

The process of merging and analysing numerous sets of agricultural information from different places in order to obtain insightful knowledge, better decision-making, and increase overall productivity in the agricultural sector is referred to as data integration in agriculture. The data generated by modern agriculture comes from a variety of sources, including sensors, satellite imaging, weather forecasts, machine telemetry, crop health evaluations, and more. Integrating various data sources can help decision-makers become more knowledgeable and data-driven, improving agricultural practices' sustainability, effectiveness, and yield.

Collecting information from many sources, including Internet of Things (IoT) sensors, drones, communications satellites, weather observatories, and machinery. This data includes details regarding the state of the soil, weather patterns, crop development, the frequency of pest and disease outbreaks, equipment operation, and more. Preserving gathered data on platforms or databases that permit simple access, retrieving, and interpretation is essential. Data authenticity, safety, and accessibility for authorised users are all guaranteed by appropriate data storage. Putting the original information into a standardised format that can be analysed after cleaning, normalising, and transformation. To ensure the correctness of the following studies, this stage entails deleting mistakes, inconsistencies, and duplication. Creating a single dataset by combining data from many sources and formats. Based on shared identities, dates and times, and geolocation information, individual data points are matched in this manner. Multiple analytical methods, including machine learning [28–32], statistical modelling, and geographical analysis, are used to extract valuable insights from combined data. In agricultural operations, this may be used to spot developments, patterns, connections, and abnormalities. Making informed decisions about planting, irrigation, fertilisation, pest management, and other elements of farming using the knowledge acquired from data analysis can result in higher agricultural yields and optimal resource utilisation. Precision farming is built on data integration, which enables farmers to target their efforts to certain fields depending on variability revealed in the combined data. As a result, there is less waste and more efficiency since resources may be used precisely.

Data integration may cover the full agricultural supply chain and go beyond the farm gate. Data integration from manufacturing to distribution can result in

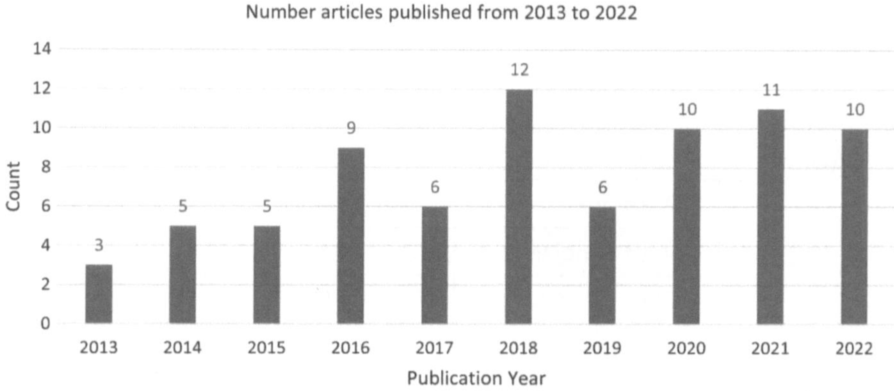

Figure 9.1 Publications published between 2013 and 2022 in quantity.

improved supply chain management, less waste, better transportation, and fresher items reaching customers. Farmers can forecast future hazards, such as outbreaks of illness, harsh weather occurrences, and market swings, by analysing historical and current data. This makes it possible to take proactive steps to lessen undesirable effects. Farmers, academics, and other agricultural sector stakeholders may collaborate and exchange knowledge using integrated agricultural data to encourage innovation and advancements. Figure 9.1 shows the different publications in year between 2013 and 2022, which shows the growing interest of people in agriculture field.

For the initial search, Figure 9.2 and Table 9.1 explain the publications and their domains from the journal, including Google Scholar and other domains. Compared

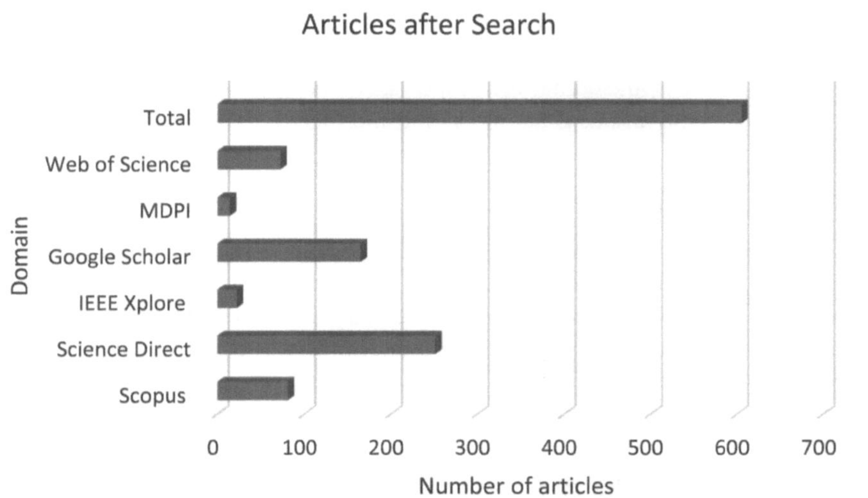

Figure 9.2 Articles available after search.

Table 9.1 Articles Available After Search

Databases	*Articles after Search*
Scopus	83
Science Direct	252
IEEE Xplore	24
Google Scholar	170
MDPI	16
Web of Science	70
Total	615

to Science Direct data sources, Scopus data sources have a significantly less number of publications available. Around 605 articles are available overall in the digital database. Very few publications are accessible to scholars as open sources via platforms like MDPI. Predicting daily rainfall boosts agricultural output and ensures a reliable supply of food and water to maintain healthy populations. Several studies have been done to forecast rainfall using data analysis and machine learning approaches on environmental datasets from various nations. The agriculture, which is the foundation of the nation's economy, is impacted by the country's uneven rainfall distribution.

The procedure of gathering, retrieving, and importing various forms of data pertaining to agricultural operations into a centralised system or database is referred to as agriculture data ingestion. Numerous sources, including sensors, satellites, equipment, handwritten notes, meteorological observatories, and more, can provide this data. Once gathered, this information may be applied in the agriculture sector for a variety of purposes, including research, decision-making, optimisation, and more. Figure 9.3 shows the various Data Integration and Ingestion Elements in Agriculture.

9.2.1 Data Collection

Making wise decisions, streamlining procedures, and raising output all depend on gathering and analysing data in agriculture. Here is a step-by-step manual for gathering data for agricultural analysis: Outline your data collecting goals and objectives in detail. Decide what precise data you need to collect and what enquiries you want analysis to address. Select the categories of data you must gather. This might include information on the weather, the soil, crop growth, pest and disease information, machinery performance information, and more. Choose the most effective data collection techniques. These may involve on-site measurements, surveys, interviews, sensor networks, remote sensing (using satellites and drones [33]), and on-field measurements.

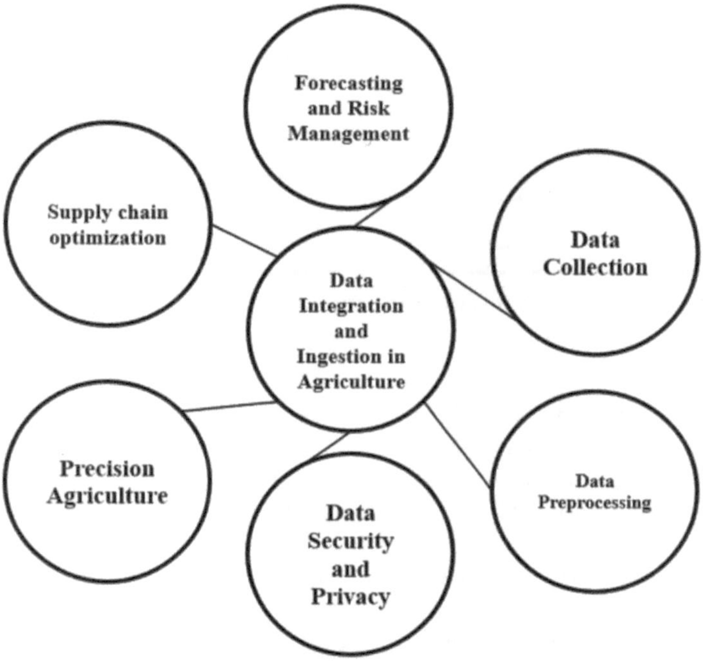

Figure 9.3 Data integration and ingestion elements in agriculture.

In the field of agriculture, data collection refers to gathering knowledge and observations on many elements of agricultural practices, crops, animals, and environmental conditions. This data is gathered to improve farming operations, increase production, and promote sustainability. Modern agricultural data collection frequently combines conventional techniques with cutting-edge technology. Here are some popular techniques and tools for gathering agricultural data: Data are frequently gathered manually through field research and observation by farmers and agronomists. Assessing agricultural growth phases, the presence of pests and diseases, the state of the ground, and seasonal variations are some examples of this. These observations offer important information on the condition of the vegetables and the ecosystem at large.

Remote sensing is the process of gathering information about agricultural land across great distances utilising aerial vehicles, satellites, and other airborne platforms. Information regarding crop health, levels of nutrients, water content, and other topics may be obtained by remote sensing. Drone surveys and satellite photos may swiftly cover enormous regions and give a more comprehensive view of agricultural landscapes. Farm-based weather stations gather information on weather variables such as humidity, wind gusts, temperature, precipitation, and others. Making informed decisions about water supply, planting, and harvesting requires the use of this information. To gather real-time information on the soil's moisture level, temperature,

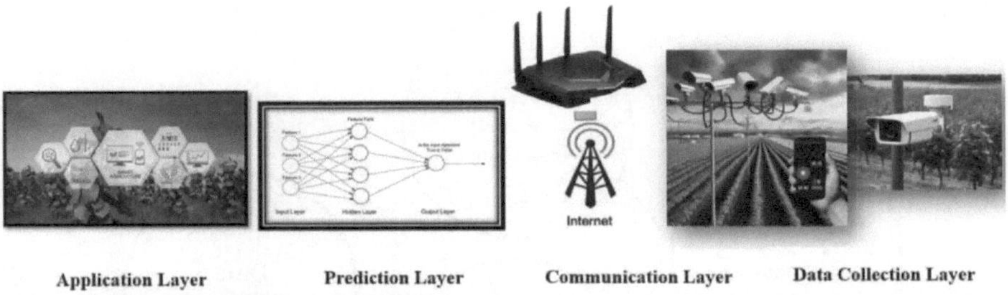

Figure 9.4 Internet of Things (IoT) for agriculture: data collecting and forecasting [34].

amount of nutrients, and other characteristics, several types of equipment are placed in the field. Farmers can optimise irrigation and fertiliser application with the aid of these sensors, which offer precise and ongoing monitoring.

As shown in Figure 9.4, Geographic information systems (GIS) [35] and GPS technologies enable farmers to track the whereabouts of their equipment, keep an eye on crop development, and make digital maps of their farms. This geographical data enables focused operations, which aid in precision agriculture. IoT devices collect data and send it over the internet, such as linked machinery and smart irrigation systems. These tools may automate processes and give farmers up-to-the-minute information on the state of their businesses. Farmers may receive and distribute data on market pricing, forecasts for the weather, and farming methods thanks to satellite communication, even in distant locations without dependable internet connectivity. Farmers may collect data, inspect vegetables, track stocks, and access markets using a variety of mobile applications and software platforms.

9.2.2 Data Preprocessing in Agriculture

As in every other industry that uses data analysis, data pretreatment is essential in agriculture. In order to make raw agricultural data appropriate for analysis and modelling, preprocessing information in agriculture aims to clean, convert, and organise the data as shown in Figure 9.5. Here are some crucial actions and ideas for

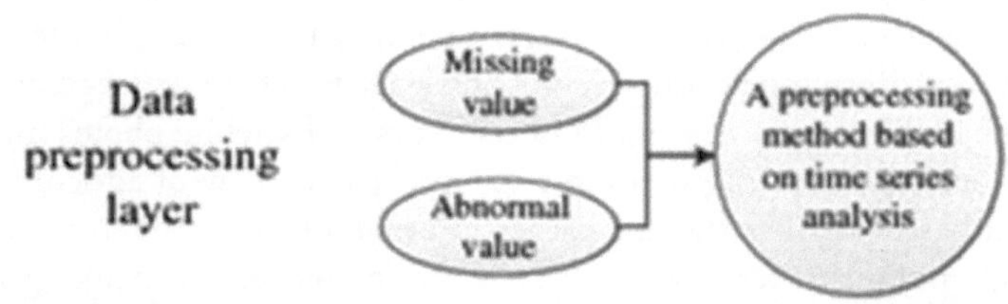

Figure 9.5 Data preprocessing in agriculture.

agricultural data preparation. Getting pertinent information from multiple sources is the first stage. This might include information on the weather, soil, agricultural yields, pests and diseases, satellite imaging, and more. When collecting data, it's crucial to ensure its accuracy and quality. Managing Missing Values: Sensor failures, human error, and other factors are frequently to blame for missing values in agricultural data. Scales for various characteristics might vary. It is ensured that each characteristic is evaluated similarly by normalising or scaling the data. Develop novel functions that can deliver more insightful data. For instance, you may determine characteristics associated with weather patterns or compute indices of plants from satellite photos. Combining data from several sources, including meteorological data and agricultural production data, is frequently required for agricultural analysis. Make sure the data is properly merged. Some aspects might not be pertinent to your analysis. To enhance model performance and cut down on complexity, identify and choose the most crucial elements. Data about agriculture may have several dimensions. Dimensionality can be decreased by using methods like Principal Component Analysis (PCA) or strategies for selecting features. Use methods like one-hot encoding or label coding to transform categorical variables into numerical representations.

Handle time dependencies correctly if your data contains time series information (such as information on crop growth over time). Think about methods like exponential smoothing or lag characteristics. Consider using spatial interpolating methods to construct continuous geographical surfaces and patch in the values that are absent for spatial data (such as soil qualities throughout a field). To effectively evaluate the performance of your models, divide your preprocessed data into the following three stages: training, validation, and testing sets. Unbalanced classes, like healthy vs. diseased plants, can be seen in agriculture. To rectify the imbalance between classes, employ methods like excessive sampling, undersampling, or creating synthetic data. To keep your preprocessed data accurate and indicative of the agricultural operations, periodically check and evaluate its quality.

9.2.3 Data Security and Privacy in Agriculture

Like every other industry that depends on technology and data, the agriculture sector must take the privacy and security of information seriously. It's crucial to ensure that private data is safeguarded and individuals' privacy is upheld as agriculture adopts more digital technology and data-driven practices. Ensure that only relevant data is acquired and data gathering is conducted using trustworthy techniques. Use effective encryption methods to safeguard data while it is being stored and sent. To avoid unauthorised access, use encrypted cloud storage options or servers on your premises with appropriate access restrictions. Data access should be restricted according to roles and responsibilities. Not all employees in the company require reference to all data. Consider anonymising or eliminating the identification of the data before distributing or analysing it to get rid of sensitive information. As a result, data may be used without endangering people's privacy. Learn about pertinent data protection laws, such as the Health Insurance Portability and Accountability Act (HIPAA) in the US or the General Data Protection Regulation (GDPR) in Europe. To prevent legal problems, make sure your data practices are in compliance with these laws. When sending data over networks, use secure communication methods (like HTTPS). Send confidential information using secure means rather than regular email. Conduct routine security audits to find system vulnerabilities. Update your hardware, firmware, and software with the most recent security fixes.

As shown in Figure 9.6, due to the growing use of advances in technology and solutions based on data in the agriculture industry, safety and confidentiality of data are important issues to take into account. A large quantity of data is produced by agricultural activities, including data on crop yields, weather patterns, animal health, and supply chains. To guarantee the accuracy, discretion, and compliance of agricultural activities, this data must be protected. Here are some important things to think about when it comes to the privacy and security of data in farming. A variety of tools are used in agricultural operations to collect data, including sensors, drones, satellite imaging, and machinery. Sensitive information regarding animals, soil, and crops may be included in this data. To guarantee that data is not altered during transmission, secure data collection techniques should be used, such as encryption and authentication.

Data encryption ensures that even if unauthorised people access the data, they won't be able to decode it without the right decryption key. Data transmitted and stored should be encrypted to avoid interception and unauthorised access. Put in place stringent access restrictions to guarantee that only people with the proper authorisation may access private agricultural data. To specify various degrees of access for various stakeholders depending on their positions and duties, utilise role-based access control (RBAC). Consider masking or eliminating any personally identifiable information in the acquired data in order to safeguard individual privacy.

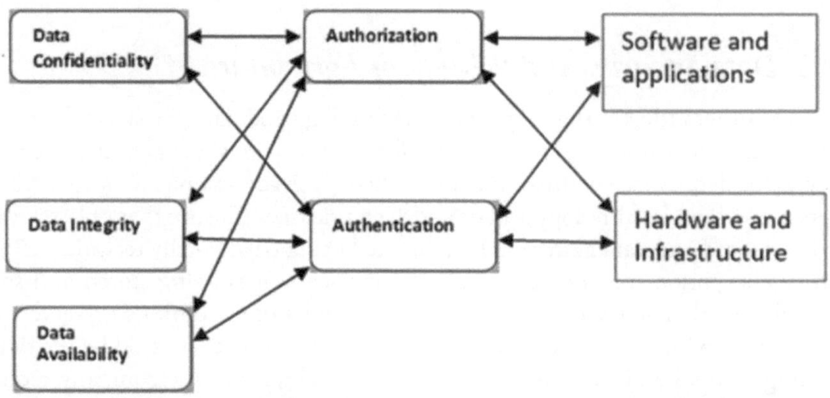

Figure 9.6 Data security and privacy in agriculture.

Data that has been aggregated and anonymised can nevertheless offer insightful information without endangering the privacy of specific persons.

9.2.4 Precision Agriculture

Precision farming, sometimes known as precision agriculture, is a modern approach to farming that makes use of data, technology, and a variety of strategies to maximise the effectiveness, efficiency, and long-term viability of agricultural practices. Instead of using standard farming practices over a whole field, precision agriculture aims to cater farming practices to the unique demands of specific plants or regions of a field. This strategy helps farmers to manage resources more wisely and make better decisions, which eventually results in increased agricultural yields, lower costs, and reduced environmental effects. Data about the field is gathered from a variety of data sources, including soil characteristics, climate patterns, crop performance, and insect populations. Sensors, drones, orbiting satellites, and other remote sensing technologies are used to gather this data.

As shown in Figure 9.7, farmers can map and thoroughly analyse their farms thanks to GIS technology [35]. Farmers may spot geographical variances and take targeted actions by superimposing several layers of information, such as the kind of soil, terrain, and yield historical. High-resolution photographs of fields are provided by satellite photography and drone technology, enabling farmers to track crop health, identify stresses, and evaluate the effectiveness of different treatments. VRT (Variable Rate Technology) [36] entails tailoring the usage of inputs like water, herbicides, and fertilisers to the particular requirements of various fields. By doing this, waste is prevented and resources are utilised effectively. For a crop to grow uniformly [35], the proper depth and spacing of the seed planting are essential. With the use of precision

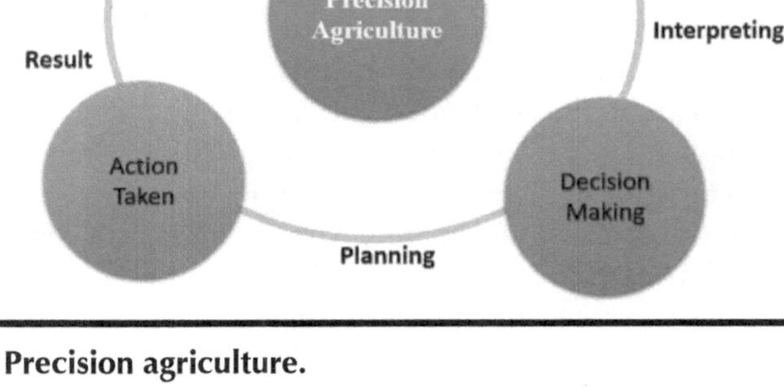

Figure 9.7 Precision agriculture.

planting equipment, it is possible to accomplish ideal planting of seeds and spacing, which improves germination and the growth of the crop as a whole.

By fusing geographical data and technology, geographic information systems (GIS) [35] enable farmers make informed decisions regarding resource allocation, crop productivity, and other aspects of contemporary agriculture. Using GIS in agriculture, researchers may gather, examine, and visualise spatial and geographic data to learn more about many facets of farming operations. With GIS software, farmers may make accurate maps of their agricultural land that include details on soil types, terrain, drainage systems, and other important factors. These maps provide a visual representation of the terrain, assisting farmers in understanding the variation in their fields.

By enabling farmers to analyse geographical data and make focused decisions, GIS facilitates precision agriculture practices. For instance, depending on the distinct requirements of various locations within a field, producers might develop prescription maps that direct the variable rate delivery of inputs such as fertilisers, herbicides, and irrigation water. Based on elements like the soil's condition, slope, sunshine exposure, and drainage features, GIS may be utilised to pinpoint optimal places for cultivation while planning fresh crop harvests or expansions. To track the growth and health of crops, satellite images and drone data may be incorporated into GIS. This enables early detection of problems like insect infestations, illnesses, or nutritional deficits by farmers, allowing for prompt responses.

By examining information on weather patterns, plant water requirements, and soil moisture levels, GIS can aid in the optimisation of irrigation practices. As a result, water is administered more effectively, resulting in less waste and more crop output. To produce yield maps, harvested yield data may be georeferenced and combined with GIS. These maps offer information on yield variability within a field, assisting farmers in determining the variables that affect output outcomes. GIS may be used to track the movement of pollutants, manage erodibility-prone regions, and adopt conservation practices, among other environmental monitoring and management tasks related to farming. GIS can help determine how susceptible crops are weather-related dangers such as frost, extreme weather, or deluge. Farmers may prepare for probable effects and reduce losses using this knowledge.

9.2.5 Supply Chain Optimisation in Agriculture

Application of ideas, technologies, and practices to improve the effectiveness, long-term viability, and the overall outcome of the agricultural supply chain is known as supply chain optimisation. The production and harvesting phases are followed by manufacturing, shipping, and consumption in the agricultural supply chain. This supply chain's optimisation [37] can result in lower prices, better resource utilisation, higher-quality products, and more environmental sustainability. Farmers and producers may more efficiently organise their farming and shipping operations with accurate customer demand forecasting. This lessens the possibility of either overproducing or underproducing while reducing waste and boosting overall effectiveness. To prevent stockouts or surplus inventory, agricultural businesses must manage their inventories effectively.

Crop planning and production may be improved by utilising data-driven insights, precision agricultural methods, and cutting-edge technology like sensors and drones. This entails selecting the best crops for a certain location, selecting the best planting dates, and modifying cultivation techniques in response to the environment and the state of the soil. Reducing delays, minimising product deterioration, and cutting transportation costs all depend on efficient transportation and distribution. Some tactics include route optimisation, real-time monitoring, and making use of temperature-controlled containers. Better communication and coordination of information can result from working with other supply chain participants, such as suppliers, distributors, and retailers. This may lead to better demand management and a more fluid product flow. Advanced data analytics systems can offer information about past patterns, consumer behaviour, and supply chain efficiency. Artificial intelligence (AI)-powered systems that enable decision-making can help with pricing, distribution, and production-related decisions. Implementing traceability systems makes it possible to follow items from the farm to the table, promoting accountability and transparency. Product quality and customer confidence may be preserved with the use of quality control procedures, such as routine inspections and certifications. Understanding trade laws, tariffs, and compliance standards is crucial for a successful supply chain operation for agricultural goods that travel across international boundaries.

Preserving the quality and safety of consumable crops, such as fruits and vegetables, dairy goods, and meat, by regulating the temperature from manufacturing through transportation to consumption, is the main goal of the agricultural supply chain's crucial cold chain management system, as shown in Figure 9.8. In order

Figure 9.8 Cold chain management in agriculture.

to protect food from spoiling, decrease post-harvest losses, increase shelf life, and guarantee that consumers receive safe, high-quality food items, proper refrigeration chain management is essential. It's crucial to keep the appropriate temperature across the whole supply chain. Temperature regulation throughout harvesting, processing, packing, storage, shipping, and retailing are all included in this. It's important to match the right environment with the specific kind of product being sold since different crops and goods have different requirements for temperature.

Cold chain management relies heavily on refrigeration technology, including cold storage amenities, refrigerated vehicles, and refrigerated containers. These establishments and vehicles are made to maintain items at the right temperature in order to avoid spoiling. Agricultural products must be properly packaged to safeguard them from temperature changes and physical harm. The cold chain is maintained by using packaging methods including vacuum-sealed bags, insulated containers, and refrigerated pallets. It is crucial to continuously check the humidity and temperature levels. Temperature data is recorded and transmitted across the supply chain using data loggers and sensors. If temperatures stray from the specified range, this real-time monitoring enables prompt remedial action.

By guaranteeing the efficient and safe transportation of perishable agricultural goods from fields to markets, manufacturing facilities, and consumers, refrigeration transport plays a critical role in the agriculture business. To preserve their quality and freshness, perishable agricultural items including fruits and vegetables, dairy, and meat must be transported at a specified temperature and humidity level. Systems for refrigeration aid in delaying the process of ripening and degradation processes, so increasing the duration of storage of these goods. Transport temperature requirements for various agricultural goods vary. For instance, certain kinds of produce should be stored just above freezing, but frozen meat has to be stored at temperatures below zero. To meet these needs, refrigerated vehicles such as trucks and containers come with temperature controls that may be adjusted.

9.2.6 Forecasting Management in Agriculture

Agriculture relies heavily on prediction and handling risks to assist farmers make wise decisions and reduce possible losses caused by unforeseen changes in the weather, market circumstances, and other variables. In forecasting, future events are predicted using analysis and data from the past. Farmers use accurate weather forecasts to schedule putting down roots, irrigation, and harvesting tasks. Crop selection and climate adaptation are made easier with the help of long-term climate projections. Crop yield forecasting aids farmers in resource allocation planning, income estimation, and educated marketing and distribution choices. Farmers may take precautions like applying the right pesticides or implementing crop rotation techniques by anticipating epidemics of diseases and pests. Farmers are helped in choosing which crops to plant and at what times to sell for the most profitability by predicting market conditions and needs.

Figure 9.9 Pest and disease forecasting in agriculture.

A proactive strategy for controlling and reducing the effects of pests and illnesses on crops and animals is known as pest and disease forecasting, as shown in Figure 9.9. In order to forecast the likelihood and severity of outbreaks of diseases and pests, information, technology, and scientific understanding are used. Farmers that employ efficient illness and pest forecasting may safeguard their crops and livestock in a timely and economical manner while using less pesticides and other treatments. Pest and disease forecasting must include ongoing field and livestock monitoring and surveillance. To find early indications of vermin or disease outbreaks, this may entail using detectors, traps, satellite-based equipment, and on-site inspections.

9.2.7 Risk Management in Agriculture

Agribusiness operations are inherently hazardous because of variables including weather whims, market swings, and biological uncertainty. Effective risk management techniques can reduce prospective losses and improve resilience in general. By distributing the risk among several crops, planting a range of crops can lessen the effects of just one crop failure. Insurance for agriculture provides protection [38] from weather-related occurrences, failure of crops, and other damages. It offers monetary security during difficult times. To reduce the risk of price volatility, farmers can utilise vegetable futures and forward-looking contracts to secure a lock in prices for what they grow before harvest. Agritourism and value-added

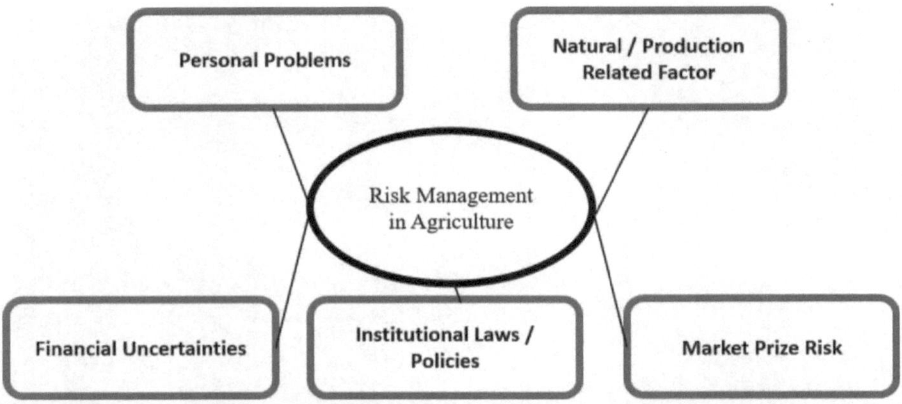

Figure 9.10 Risk management in agriculture.

processing are two examples of activities that can provide additional cash. Precision resource management is made possible by technology such as GPS, detectors, and unmanned aircraft, which optimises inputs and reduces waste.

As shown in Figure 9.10, For farmers and other agricultural sector stakeholders risk management in agriculture is essential to minimise possible losses and preserve the long-term viability of their businesses. Inherent risks to agriculture include those associated to the weather, market instability, pest and disease outbreaks, and legislative changes. Farmers may overcome these difficulties and make informed decisions with the use of efficient risk management solutions. A crucial instrument for controlling the hazards involved in agricultural production is crop insurance. It offers farmers financial security in the event of crop losses brought on by unfavourable weather, pests, or illnesses. To promote adoption, governments frequently support and subsidise crop insurance programmes. Risk can be dispersed by diversifying animals and crops. If a single crop or animal species is impacted by unfavourable conditions, diversifying the crops or livestock raised can assist reduce losses. Rotating crops helps to lower the risk of disease and insect infestation. Droughts, floods, and severe temperatures are just a few examples of weather-related threats that may have a big influence on agriculture. To address these risks, farmers can use a variety of techniques, such as irrigation, weather forecasts, and crop types that are climatically adaptable.

9.2.8 Collaboration in Agriculture

In agriculture, collaboration is the process of stakeholders, groups, and individuals coming together to address issues, share resources, and work towards common objectives. This cooperative strategy acknowledges that many agricultural problems are intricate and varied, necessitating the knowledge, effort, and involvement of several partners to develop practical solutions. At numerous sizes, encompassing the local area regional, national, and perhaps worldwide ones, collaboration is possible. Here are a few elements and instances of agricultural cooperation. To exchange information, best practises, and innovations in agriculture, extension agents, researchers, and agricultural professionals work together. This might enhance agricultural practises, insect control methods, and general output.

University, research, and industry partners frequently work together on agricultural research projects. Collaboration in research can result in the creation of sustainable technology, better agricultural methods, and new crop types. Collaboration among participants in the agricultural supply chain promotes a more seamless transfer of goods from producers to consumers. To improve logistics and distribution, this may entail collaboration between agriculturalists, processors, who process suppliers, and retailers. Services for agricultural extension offer farmers helpful knowledge and assistance. Experts collaborate with local farmers in collaborative extension programmes to solve particular issues and provide solutions catered to their need. Governments, private businesses, and non-governmental organisations (NGOs) frequently work together through public-private partnerships (PPPs) to address agricultural concerns. PPPs can provide funding for initiatives aimed at improving infrastructure, conducting research, and disseminating technology.

Promoting sustainable agriculture practices requires cooperation. To lessen the influence of agriculture on the environment, cooperation between farmers, environmental groups, and regulatory entities may be necessary. Through training programmes, workshops, and information sharing, collaboration may improve the capacities and skills of farmers. This gives farmers the ability to embrace contemporary methods and enhance their standard of living. Collaboration is crucial for organising relief operations, allocating resources, and assisting impacted farmers in recovering after catastrophes or natural disasters. Value-added processing facilities may be established through cooperative efforts. For instance, a group of farmers may work together to construct a cooperative processing facility to enhance the value of their crops before they are sold. The creation and deployment of agricultural innovations can be accelerated via cooperation.

In order to develop cutting-edge solutions like precision agricultural instruments or smart farming systems, this may include sharing data, knowledge, and resources. The development of agricultural policy frequently calls for input from a variety of parties, including farmers, business representatives, researchers, and governmental organisations. Collaboration in the policy-making process ensures that laws are sensible and well-informed. Effective dialogue, mutual trust, mutually agreed upon goals, and a willingness to cooperate for the common benefit are requirements for successful collaboration in agriculture. Better results for customers, farmers, and the environment at large can result from it, as can increased agricultural output and sustainable practises.

9.3 Conclusion

This chapter explains Data Integration and Data Ingestion in detail as well as their pros and cons in agriculture. In order for farmers to leverage the power of data to make informed decisions, enhance yields, lower costs, and assure sustainability in the face of diverse problems including climate change and population expansion, data integration and ingestion are crucial. In modern agriculture, data integration and data acquisition play crucial roles in helping farmers and agricultural groups make well-informed choices, improve operations, and boost total productivity. These steps involve compiling, integrating, and harmonising data from many sources across the agricultural ecosystem. Data about agriculture is gathered using a variety of instruments and techniques, such as sensors, unmanned aircraft, satellites that are weather stations, machinery, and human inputs. Information from these many sources is combined into a single system or database through data integration. Different data sources may utilise different formats, measures, or data structures. Standardising and integrating this data is typically a component of data integration operations to ensure interoperability and consistency.

References

1. Ziegler P, Dittrich KR. Data Integration — Problems, Approaches, and Perspectives. In: Krogstie, J., Opdahl, A.L., Brinkkemper, S. (eds) Conceptual Modelling in Information Systems Engineering. Springer, Berlin, Heidelberg; 2007. https://doi.org/10.1007/978-3-540-72677-7_3.
2. Singhal A. An Overview of Data Warehouse, OLAP and Data Mining Technology. In: Data Warehousing and Data Mining Techniques for Cyber Security. Advances in Information Security, vol 31. Springer, Boston, MA; 2007. https://doi.org/10.1007/978-0-387-47653-7_1.
3. Molina Hector Garcia. Distributed and Parallel Computing Issues in Data Warehousing. In: Proceedings of ACM Conference on Distributed Computing, Puerto Vallarta, Mexico, 1999. https://doi.org/10.1145/277697.277701.
4. Hari Krishna SM, Sharma Rinki. Survey on application programming interfaces in software defined networks and network function virtualization. Global Trans Proc. 2021;2(2):199–204. https://doi.org/10.1016/j.gltp.2021.08.018.
5. Severance C. Roy T. Fielding: Understanding the rest style. Computer. 2015;48(6):7–9
6. Bologa Ana, Botha Iuliana. "A Perspective on the Benefits of Data Virtualization Technology", Informatica Economica 2011;15(4): 110–118.
7. Eckerson, Wayne. The Evolution of Data Federation, 2009. https://tdwi.org/blogs/tdwi-blog/2010/06/the-evolution-of-data-federation.aspx
8. Dilling TJ. Artificial intelligence research: The utility and design of a relational database system. Adv Radiat Oncol. 2020 Jul 13;5(6):1280–1285. doi: 10.1016/j.adro.2020.06.027.
9. Matende Samwel, Ogao Patrick. Enterprise Resource planning (ERP) system implementation: A case for user participation. Procedia Technol. 2013;9:518–526. https://doi.org/10.1016/j.protcy.2013.12.058.

10. Guerola-Navarro V, Gil-Gomez H, Oltra-Badenes R. et al. Customer relationship management and its impact on entrepreneurial marketing: A literature review. Int Entrep Manag J. 2022;20:507–547. https://doi.org/10.1007/s11365-022-00800-x.
11. Otto Boris, Reichert Andreas. Organizing master data management: findings from an expert survey. In: Proceedings of the 2010 ACM Symposium on Applied Computing (SAC '10). Association for Computing Machinery, New York, NY, USA, 2010. 106–110. https://doi.org/10.1145/1774088.1774111.
12. Barker D. Customer data integration: Reaching more consumers with certainty. J Database Mark Cust Strategy Manag. 2011;18:214–219. https://doi.org/10.1057/dbm.2011.19.
13. Lin Heng-Chang, Chen Yu-Cheng, Lai Hsin-Hsi. Integrating product information management (PIM) with internet-mediated transactions (IMTs). J Convergence Inf Technol. 2010;5(10):99–108.
14. Fikri N, Rida M, Abghour N. et al. An adaptive and real-time based architecture for financial data integration. J Big Data. 2019;6:97. https://doi.org/10.1186/s40537-019-0260-x.
15. Huang F. Data Cleansing. In: Schintler, L., McNeely, C. (eds) Encyclopedia of Big Data. Springer, Cham; 2019. https://doi.org/10.1007/978-3-319-32001-4_300-1.
16. Kwon Y, Lemieux M, McTavish J, Wathen N. Identifying and removing duplicate records from systematic review searches. J Med Libr Assoc. 2015 Oct;103(4):184–188. doi: 10.3163/1536-5050.103.4.004.
17. Martins J, Mamede HS, Correia J. Risk compliance and master data management in banking - A novel BCBS 239 compliance action-plan proposal. Heliyon. 2022 Jun 3;8(6):e09627. doi: 10.1016/j.heliyon.2022.e09627.
18. Marikyan Davit, Papagiannidis Savvas, Ranjan Rajiv, Rana Omer. General data protection regulation: an individual's perspective. In: Proceedings of the 14th IEEE/ACM International Conference on Utility and Cloud Computing Companion (UCC '21). Association for Computing Machinery, New York, NY, USA, 2022. Article 8, 1–6. https://doi.org/10.1145/3492323.3495620.
19. Theodos K, Sittig S. Health information privacy laws in the digital age: HIPAA Doesn't apply. Perspect Health Inf Manag. 2020 Dec 7;18(Winter):1l.
20. Patel VL, Kushniruk AW, Yang S, Yale JF. Impact of a computer-based patient record system on data collection, knowledge organization, and reasoning. J Am Med Inform Assoc. 2000 Nov-Dec;7(6):569–585. doi: 10.1136/jamia.2000.0070569.
21. Astriani Winda. Trisminingsih Rina. Extraction, transformation, and loading (ETL) module for hotspot spatial data warehouse using geokettle. Procedia Environ Sci. 2016;33:626–634. https://doi.org/10.1016/j.proenv.2016.03.117.
22. Patil Charushila, Chaware Anita. Integration of Internet of Things, Cloud Computing: Review. IOP Conf Ser: Mater Sci Eng. 2021;1022:012099. https://dx.doi.org/10.1088/1757-899X/1022/1/012099.
23. Almeida A, Brás S, Sargento S. et al. Time series big data: A survey on data stream frameworks, analysis and algorithms. J Big Data. 2023;10:83. https://doi.org/10.1186/s40537-023-00760-1.
24. Salgado A, Li W, Alhasoun F. et al. Street context of various demographic groups in their daily mobility. Appl Netw Sci. 2021;6:43. https://doi.org/10.1007/s41109-021-00382-7.
25. Song Z, Ma H, Sun S. et al. Rainbow: Reliable personally identifiable information retrieval across multi-cloud. Cybersecurity. 2023;6:19. https://doi.org/10.1186/s42400-023-00146-z.

26. Grolinger K, Higashino WA, Tiwari A. et al. Data management in cloud environments: NoSQL and NewSQL data stores. J Cloud Comp. 2013;2:22. https://doi.org/10.1186/2192-113X-2-22.
27. Ataei P, Staegemann D. Application of microservices patterns to big data systems. J Big Data. 2023;10:56. https://doi.org/10.1186/s40537-023-00733-4.
28. Ebenezer, V., Daniel Arsha Anna, Lillipushpam P. Getzi Jeba, Edwin E. Bijolin, Thanka M. Roshni, Devassy Rosebel. Classification and Segmentation of Leaf Images based on Deep Learning for Peanut Plant Disease Detection. In: 2023 4th International Conference on Electronics and Sustainable Communication Systems (ICESC), Coimbatore, India, 2023, 1414–1420.
29. Edwin, Bijolin, Veemaraj Ebenezer, Parthiban Pradeepa, Devarajan Joseph Pushparaj, Mariadhas Vargheese, Arumuganainar Ahila, Reddy Maheshwar. Smart agriculture monitoring system for outdoor and hydroponic environments. Indones J Electr Eng Comput Sci. 2022;25(3):1679–1687.
30. Thanka, M. Roshni, Edwin E. Bijolin, Duela J. Shiny, Ebenezer V. Automated diagnosis of skin disease multiclass image classification using deep convolution neural network. J Green Eng. 2020;10(10):7646–7660.
31. Thanka M. Roshni, Edwin E. Bijolin, Ebenezer V., Sagayam K. Martin, Reddy B. Jayakeshav, Günerhan Hatıra, Emadifar Homan. A hybrid approach for melanoma classification using ensemble machine learning techniques with deep transfer learning. Comput Methods Prog Biomed Update. 2023;3:100–103.
32. Joel MR, Ebenezer V, Jeyaraj KA, Navaneethakrishnan M, Arunadevi R, Jenifa DR. Decoding and Analysing Consumer Feedback for Companies and Goods using Machine Learning. In: 2023 Third International Conference on Artificial Intelligence and Smart Energy (ICAIS), Coimbatore, India, 2023, 1085–1091.
33. Budiharto W, Irwansyah E, Suroso JS. et al. Mapping and 3D modelling using quadrotor drone and GIS software. J Big Data. 2021;8:48. https://doi.org/10.1186/s40537-021-00436-8.
34. Jin, Xue-Bo, Yang Nian-Xiang, Wang Xiao-Yi, Bai Yu-Ting, Su Ting-Li, Kong Jian-Lei. Hybrid deep learning predictor for smart agriculture sensing based on empirical mode decomposition and gated recurrent unit group model. Sensors. 2020;20(5)1334. https://doi.org/10.3390/s20051334.
35. Teixeira S. Qualitative geographic information systems (GIS): An untapped research approach for social work. Qualit Social Work. 2018;17(1):9–23. https://doi.org/10.1177/1473325016655203.
36. Masi, Margherita, Di Pasquale Jorgelina, Vecchio Yari, Capitanio Fabian. Precision farming: Barriers of variable rate technology adoption in Italy. Land. 2023;12(5):1084. https://doi.org/10.3390/land12051084.
37. Chowdary VT, Robinson Joel M, Ebenezer V, Edwin B, Thanka R, Jeyaraj A. 2022. A Novel Approach for Effective Crop Production using Machine Learning. In: 2022 International Conference on Electronics and Renewable Systems (ICEARS), Tuticorin, India, 2022, 1143–1147. doi: 10.1109/ICEARS53579.2022.9751798.
38. Guo Yuhan, Yin Yue. Analysis on optimization of agricultural products supply chain based on dynamic system. Discrete Dynamics Nat Society. 2022;2022:1656636. https://doi.org/10.1155/2022/1656636.

Chapter 10

Revolutionising Crop Yield Prediction in Agriculture: Harnessing Machine Learning and Environment Factors for Effective Solutions in the Face of Climate Change

M. Robinson Joel, V. Ebenezer,
S. Stewart Kirubakaran, E. Bijolin Edwin,
M. Roshni Thanka, and E. John Alex

10.1 Introduction

This chapter focuses mostly on works relating to the agricultural industrial sector and the public forum. Researchers are interested in doing research to determine the best outcome and provide a method to determine the best crop for a given soil. The soil and other factors, such as the environment, have a major impact on crops. We investigate these crop forecasting publications in order to examine how much attention academics have given to the welfare and development of farmers.

DOI: 10.1201/9781003504900-10

Farmers make their production plans based on their knowledge gained from their prior production outcomes, the local temperature, and encouraging remarks from their elders. They also learn about the worth of the land and the crop's production costs from the young farmer and their other farmers. The primary tenet of the Indian economy depends on agricultural and crop output.

According to the study challenge and research objectives, a machine learning (ML) framework[38] can be either descriptive or prognostic. Predictive models are used to make forecasts about the future, whereas descriptive models are utilised to learn from the data collected and clarify exactly what has occurred. When attempting to create a high-performance prediction approach, ML research faces a variety of difficulties. To tackle the challenge beforehand, it is essential to choose the appropriate methods. Furthermore, the algorithms and underlying platforms must be able to handle a great deal of data.

We conducted a systematic literature review (SLR) to gain an overview of the work that has been conducted regarding the application of ML [33–37] in agricultural production prediction. An SLR identifies possible research gaps in a specific field of study and provides guidance for practitioners and academics who desire to do novel research on that field of investigation.

Crop yield forecasting is a critical duty for timely decision-making at both the federal and geographic levels. An effective crop production forecast model can help farmers decide whether to cultivate, as well as when and where to plant. Various techniques can be used to forecast agricultural production. In this literature review research, the use of ML to forecast production from agriculture is currently examined.

The publication's status as a survey or typical review piece is one of the many exclusion criteria utilising the gathered data from the articles. In reality, the excluded works—which are described in this section—are related. A review study was conducted by Chlingaryan [1] and Sukkarieh on the application of ML to evaluate nitrogen status. Ramu [2] and his collaborator use ML methods that take into account soil and other parameters to assess the crop yield studies put out by other researchers.

In his evaluations, Dharani describes a method for automatically forecasting production and keeping track of the crop. Research has been done on deep learning [3] and its hybrid techniques, which include artificial neural networks, deep neural networks, and recurrent neural networks. A review study on the general use of data mining for decision-making in the agriculture sector was published by Gandhi [4] and Armstrong.

10.2 Machine Learning on Agriculture

ML technology has transformed many aspects of farming, crop management, and food production and has considerably improved the agriculture sector. For the nation to be prosperous, agriculture is crucial. Numerous issues have arisen for the science of agriculture sector as a result of climate change. ML is a great deal of

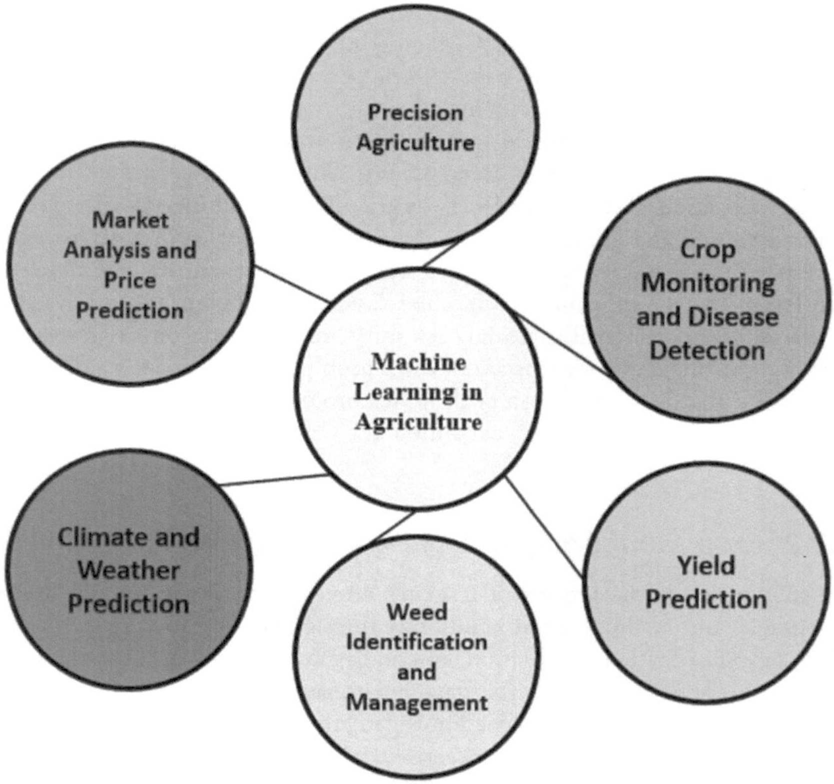

Figure 10.1 Overall elements available in machine learning.

effective technique to address issues effectively and efficiently. Crop yield prediction is calculating a crop's anticipated output using existing information in order a range of variables, such as the local environment, the quality of the soil, the level of moisture, and environmental conditions. Different elements of ML concepts in agriculture are shown in Figure 10.1.

10.2.1 Precision Agriculture

Precision farming, also known as precision agriculture, is a method of using cutting-edge tools, data, and technology to improve the sustainability and efficiency of agricultural output. Precision farming aims to increase yields while minimising resource waste and the environmental effect of farming operations. International Positioning System Farmers can precisely map and track their fields, equipment, and other assets thanks to GPS technology. For the exact application of inputs, this information is essential. Farmers can use Geographical Information Systems (GIS) software to build comprehensive maps of their fields that display changes in the kinds of soil, the availability of nutrients, and other factors. Making informed decisions about irrigation, fertilisation, and planting is easier with the help of this knowledge.

Based on the unique requirements of various locations, VRT (Variable Rate Technology) [5] enables farmers to administer inputs like fertilisers, insecticides, and water at various rates across a field. This reduces waste and improves resource utilisation. Tractors and other agricultural machinery with sophisticated automation and guiding systems can carry out duties with great accuracy. As a result, there is less overlap, less soil is compacted, and efficiency is increased. Countless pieces of information on their land are gathered and analysed by farmers. When analysed using analytics software, this data offers conclusions regarding the efficiency of crops and aids in prompt decision-making. Different sensors are utilised to gauge crop health, meteorological characteristics, and soil conditions. These sensors offer real-time data that may be utilised to modify farming practices right away.

10.2.2 Crop Monitoring and Disease Detection

Crop monitoring is the process of tracking numerous indicators linked to crop development and environmental conditions throughout time. Making educated decisions about irrigation, fertilisation, pest management, and other crucial farming practices is easier with the use of this data by farmers and agronomists. Crop development, wellness, and stress information is gathered through satellite photography and drones with recording devices or sensors with multiple spectral bands. Remote sensing makes it possible to evaluate broad regions rapidly and might reveal areas that require attention. Current information on moisture in the soil, temperature, relative humidity, and various other factors affecting the environment is gathered using on-field sensor networks. By using this information, irrigation and the application of nutrients are optimised, avoiding resource waste or misuse. Farmers can remotely monitor conditions and make timely choices thanks to IoT (Internet of Things) equipment like weather observatories and moisture in the soil sensors that wirelessly send data. To find patterns and trends, the acquired data is analysed using a variety of algorithms and methodologies. Farmers may modify their practices to meet the unique requirements of their crops with the help of this study.

For the purpose of limiting agricultural losses and halting the spread of illnesses, early disease identification is essential. Technology is essential for quickly diagnosing illnesses. Subtle alterations to plant colour and structure that signify the existence of parasites can be found using cameras with excellent resolution and sensors with multiple spectral bands. Then, in order to pinpoint individual diseases, these photos are examined using ML techniques. The electromagnetic energy that plants emit or reflect is measured using spectrometers. Even before physical symptoms show, shifts in wavelength patterns might signal stress or the presence of illness. Pathogens in plant samples may be precisely identified using DNA-based techniques such as Polymerase Chain Reaction [6] (PCR) and others. These methods are particularly helpful for locating certain microorganisms that cause illness. Large datasets of spectral and picture data can be used to train ML systems.

10.2.3 Yield Prediction

The technique of calculating the anticipated result or manufacturing of a particular crop or agricultural good is known as yield prediction. This forecast may be based on a number of elements, including historical data, current weather, the soil's condition, and other pertinent variables. In contemporary agriculture, the ability to anticipate yield [32] has grown in importance as it enables farmers to make educated choices regarding growing, reaping, the use of resources, and the management of risks. Amass pertinent information, including past yield records, meteorological data (temperature, moisture, precipitation, etc.), data on the soil (nutrient levels of difficulty, pH, etc.), and any additional pertinent factors that may have an impact on crop output. Clean up and preprocess the gathered data to get rid of any mistakes, discrepancies, or outliers. Techniques like feature engineering, data normalisation, and missing value imputation may be used for this.

10.2.4 Weed Identification and Management

Identification and control of weeds are crucial components of the agriculture sector, landscaping, and management of resources. Unwanted plants known as weeds can compete for resources like nourishment, water, and sunshine with agricultural products or attractive plants. Both recognising distinct weed species and using the right tactics to remove or control them are necessary for effective weed management. Look at the leaves size, shape, colour, and arrangement. The features of the leaf can reveal information about the weed's identification. Keep an eye out for the weed's blossoms and seeds. These characteristics are frequently recognisable and can help with identification. Examine the weed's stem structure (such as the branching patterns) and general growth pattern (upright, creeping). Take into account the weed's habitat, environmental conditions, and nature of the soil because these variables might affect its identification.

Utilise methods that inhibit weed development, such as mulch application, cover crop use, and appropriate crop spacing between crops. Use physical means to get rid of or control weeds, such as hoeing, cultivating, trimming, and flame weeders. Keep in mind that weed control involves a combination of tactics that are adapted to your particular circumstance. The kind of plant life present, the extent of the contamination, and the management resources available will all influence the approach selected. Additionally, it's crucial to think about how various control systems may affect the environment and choose which ones are both efficient and sustainable. Consulting local gardening groups, agricultural extension agencies, or experts in the field can help with this difficult weed problem and offer insightful advice.

10.2.5 Climate and Weather Prediction

Short-term atmospheric conditions, usually lasting a few hours to a few days, are the main emphasis of weather forecasting. To analyse the present atmospheric conditions and produce forecasts, meteorologists employ a variety of data sources, including meteorological satellites that are meteorological radars, observational observatories, along with computer models. On the basis of intricate equations that explain the fundamental laws of the atmosphere, computer models, sometimes referred to as mathematical weather forecasting models, mimic the behaviour of the atmosphere. Meteorologists can forecast the future course of the weather by feeding recent data into these models. Numerous uses for weather forecasts exist, such as daily planning, disaster preparedness, aerospace, the agricultural sector, and more.

Weather forecasts, however, lose accuracy as the period of prediction gets longer owing to the turbulent structure of the environment. Contrarily, climate prediction focuses on long-term patterns and trends in atmospheric conditions, typically extending decades to centuries. A larger variety of variables, such as the temperature of the ocean, quantity of greenhouse gases, exposure to sunlight, and more, are considered by climate models. These models represent the interactions of the atmosphere, seas, land, and ice, among other parts of the Earth system. In order to estimate future changes and better understand how various causes affect the temperature of the planet over longer time scales, climate predictions are made.

10.2.6 Market Analysis and Price Prediction

Due to the many variables that affect agricultural markets, such as the environment, dynamics of supply and demand, governmental regulations, and international commerce, market evaluation and price forecasting can be difficult in this industry. Here is one method for approaching market research and price forecasting in the agricultural sector. Recognise the present agricultural product supply (crops, cattle, etc.) and contrast it with the demand from the public, processors, and exporters. This study may be impacted by elements including population growth, dietary patterns, and export demand. Keep an eye on weather forecasts and trends since they have a direct impact on crop production. Extreme weather disasters like droughts, floods, and others can significantly impact supply and prices.

10.2.7 Articles Related to Agriculture Using Machine Learning

Planning the review is the first step. Research-related questions arise and are established at this important curious stage. A protocol or a model is then created for analysis and search. Finally, the protocol or the method is evaluated to see whether the method is workable. The publishing venues or the domain or its location, beginning search terms and ideas, and publication selection of journal criteria are established in addition to the research questions.

The systematic review will be guided by the research questions below:

Question 1: Which ML methods have been applied to agricultural yield prediction in the literature?
Question 2: What techniques for remote sensing are used alongside ML methods to estimate crop yield?
Question 3: What agricultural indices and ecological factors are utilised to forecast crop yield?

The first step is to organise the assessment. At this step, research questions are identified, a protocol is made, and then the protocol is assessed to determine whether the approach is practical. In order to further the goals of the study, publication locations, initial search keywords, and article selection requirements are created. All of this information is defined, and the procedure is then revised once again to see how well it correctly reflects the assessment's approach. Figure 10.2 shows the internal phases of the Set up Analysis stage.

The publication has nothing to do with ML mixed with yield prediction in the agriculture industry. The second factor is whether the published work is written in English or another language. Then, to determine whether they are relevant to the area, published articles are retrieved from another database. The complete text of several journals is unavailable for free download or viewing. Before 2013, we should have released our article.

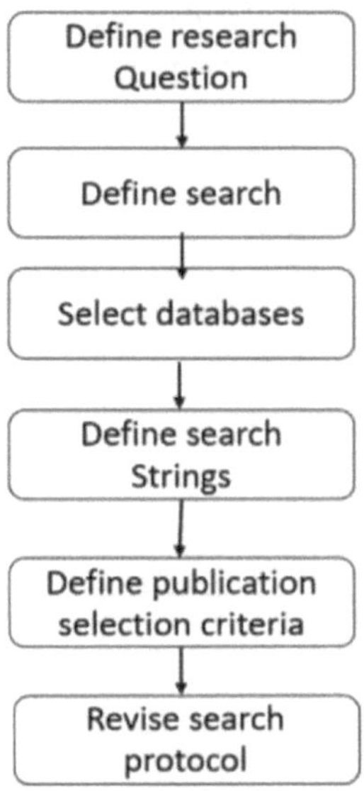

Figure 10.2 Information about the plan assessment step.

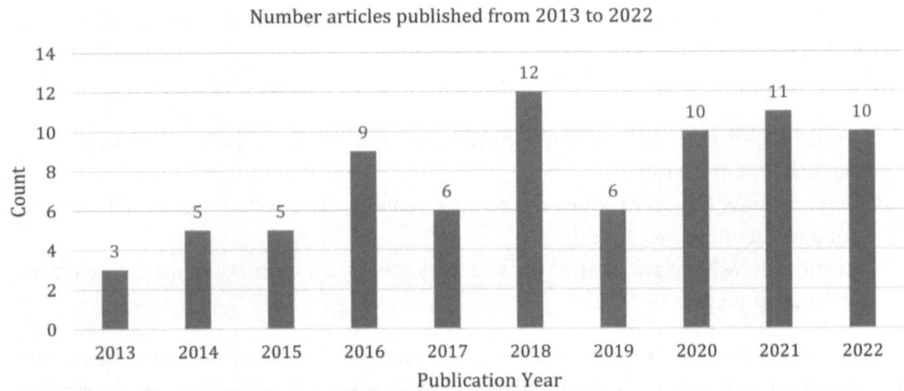

Figure 10.3 Publications published between 2013 and 2022 in quantity.

The amount of research conducted during the course of the most recent ten-year period, from 2013 to 2022, is discussed and shown in precise detail in Figure 10.3. The information demonstrates the researchers' keen interest in identifying issues and assisting farmers with their challenges to improve the farmer's lifestyle and make a profit. A large number of articles have been analysis during the last ten years compared with the previous decade.

Table 10.1 shows the number of publications available, written by several researchers in different fields. These are gathered using the search criteria listed above and the questions. These sources are reviewed and prepped after choosing such data in order to choose the precise data sources needed for the investigation. This decision was made in response to the queries posed by the individuals gathered through internet resources. The basic concept analysis selection was repeated from a variety of sources, and extra unneeded information was gathered.

Table 10.1 Number of articles published for the past ten years

Databases	*Articles after Search*	*Articles after Removing Duplicate*
Scopus	81	6
Science Direct	251	13
IEEE Xplore	22	8
Google Scholar	165	15
MDPI	14	4
Web of Science	72	5
Total	605	51

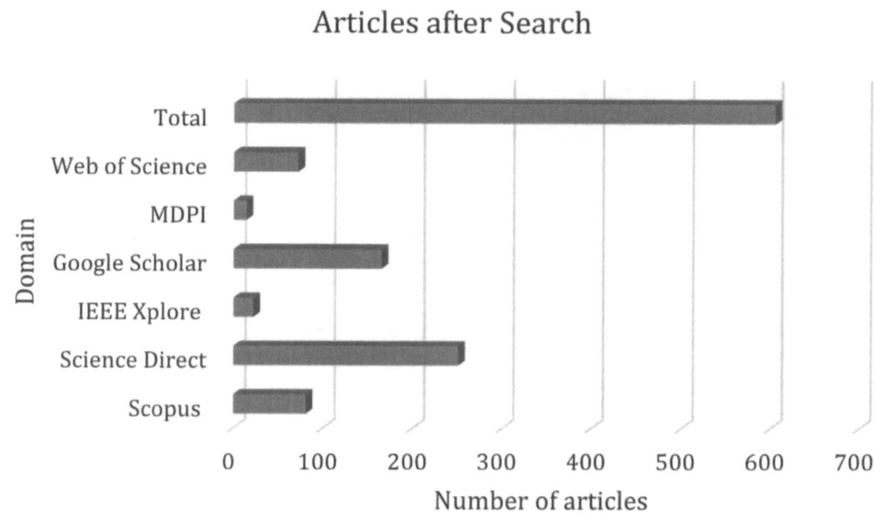

Figure 10.4 Articles after search.

Figure 10.4 explains the articles and their domain of the journal like Google Scholar and other domains for the first round of search. Available articles in Scopus data are considerably fewer compared with Science Direct data sources. The total number of articles available is around 605 on the digital database. Very few in the number of articles are available as open sources for researchers through sources like MDPI. Figure 10.5 shows the count after removing duplicates and repeated

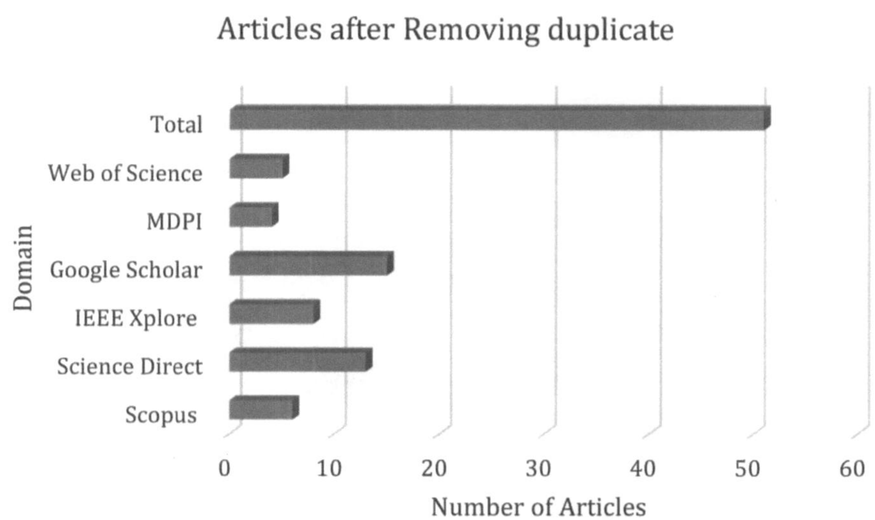

Figure 10.5 Articles after removing duplicates.

Table 10.2 Frequencies of Machine learning algorithm

Machine Learning Algorithms	*Number of Times Used*
Neural Networks	27
Linear Regression	14
Random Forest	12
Support Vector Machine SVM	10
Gradient Boosting Tree	4

publications in the main domain like Scopus and another database. Google Scholar has more publications compared to other domain databases. With this, we can easily find out the required analysis on the digital database.

Table 10.2 mentions different algorithms analysed by researchers. The neural network performs well compared to many other algorithms and it helps farmers to gain knowledge about cropping to yield the best result. Next to the Neural networks algorithm, the algorithm Linear Regression was used to get good results which in turn helps farmers. If the farmers' health and wealth increase, the nation's wealth will automatically increase.

Figure 10.6 shows the pie chart of the algorithms employed in the research work done by good searchers. This will help the agriculture field at the maximum level to the higher level. Forty percent of the work concentrates on the neural networks, which helps the searchers to achieve the maximum results.

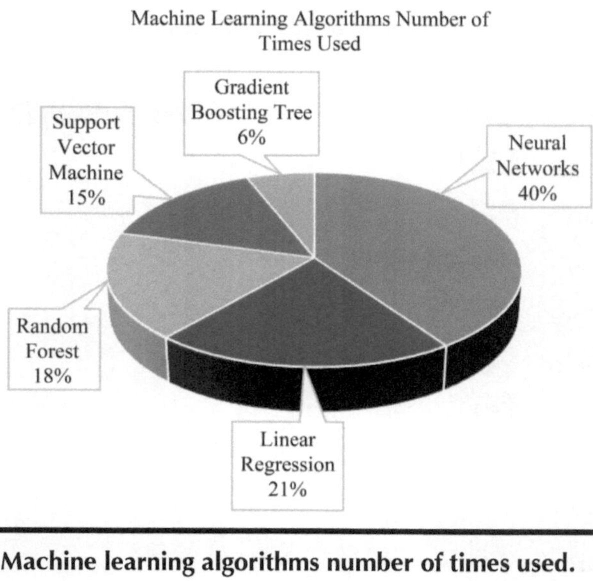

Figure 10.6 Machine learning algorithms number of times used.

Evaluation parameters are termed as Simple average ensemble which is termed as SAE. Similarly, the term Reference change values are noted as RVC. Similarly, the term Reference change values noted as Multi factored evaluation termed MFE. Similarly, the term Reference change values noted as Reduced simple average ensemble termed as RSAE. Similarly, the term Reference change values are noted as Mean absolute percentage error termed MAPE. Similarly, the term Reference change values noted as Mean square error termed MSE. Similarly, the term Reference change values noted as Mean absolute error termed MAE. Similarly, the term Reference change values noted as R-squared termed R^2. Similarly, the term Reference change values noted as Root mean square error termed as RMSE. Keyword parameters are noted as Matthew's correlation coefficient termed MCC. Similarly, Keyword parameters are noted as Lin's concordance correlation coefficient termed as LCCC. To address research question 3, evaluation standards were chosen. The frequency of usage for each evaluation parameter is shown in Table 10.3. RMSE is the research criterion most frequently employed, as seen in Table 10.3.

The number of available publications, written by several researchers in different fields, is shown in Table 10.4. These are gathered using the filtering criteria listed above in addition to the questions. These avenues are reviewed and prepped after choosing such data in order to choose the precise data sources needed for the investigation. This decision was made in response to the queries posed by the individuals gathered through internet resources. The basic concept evaluation selection was

Table 10.3 Types of evaluation parameter

Evaluation Parameter (Keyword)	*Count Researchers Used*
SAE	1
RVC	1
MCC	1
LCCC	2
MFE	2
RSAE	4
MAPE	5
MSE	6
MAE	9
R2	22
RMSE	32

Table 10.4 Types of Feature related to agriculture

Feature	*Count (Times)*
Weather	25
Type of soil	19
The quantity of rain	16
details regarding grains	15
soil charts	15
The moisture content	13
pH-value	12
Renewable energy sources	11
The falling of rain	10
Pictures	9
Manufacturing region	8
The process of reproduction	7
Nitrogen (N)	7
Ability to exchange cations	7
Watering	6
P- Potassium Material	5
Breeze velocity	5

performed again from a variety of sources, and extra unneeded information was gathered. The parameters such as soil texture, soil type, pH value, cation exchange capacity, and production area, which make up the feature category "soil information" are discussed in Table 10.4. The use of earth maps and the data contained on the illustrations varied depending on the publication.

Soils formed from rocks are very different from the nature of the rocks. Soil fertility can be determined through soil testing. Plants get nitrogen, phosphorus, potassium, lime, magnesium, sulphur, iron, manganese, boron, copper, zinc, chlorine, molybdenum, etc. as nutrients from the soil. Soil deficiencies include soil acidity and salinity. Their nature can be fertilised by nature and change the fertility of the soil. Application of chemical fertilisers also degrades good quality soil. Soil organisms, nitrogen-fixing bacteria, earthworms, and soil composition are some of the features. These details are collected and evaluated using many algorithms, as explained in Table 10.5.

Table 10.5 Types of parameters for soil prediction

Parameters	*Count*
knowledge of the soil	56
The sun's intelligence	40
The moisture content	39
Vitamins and minerals	31
Other	25
Details regarding crops	17
Farm supervision	14

Figure 10.7 clearly shows the features extracted from a number of articles available on the web by many researchers. Many researchers concentrate on soil information mainly which helps the farmers about the growth of the plant and yield they can get from their investment. Many farmers fail to get exact results by doing agriculture and other sources. In Table 10.6, the Neural Network is denoted as NN, and Multiple Linear Regression is denoted as MLR. Similarly, the Gradient boosting tree and linear regression are denoted as GBT and LR, respectively. Random forest, multivariate polynomial regression, K-nearest network, and Support vector machine are denoted as RF, MPR, KNN, and SVM, respectively.

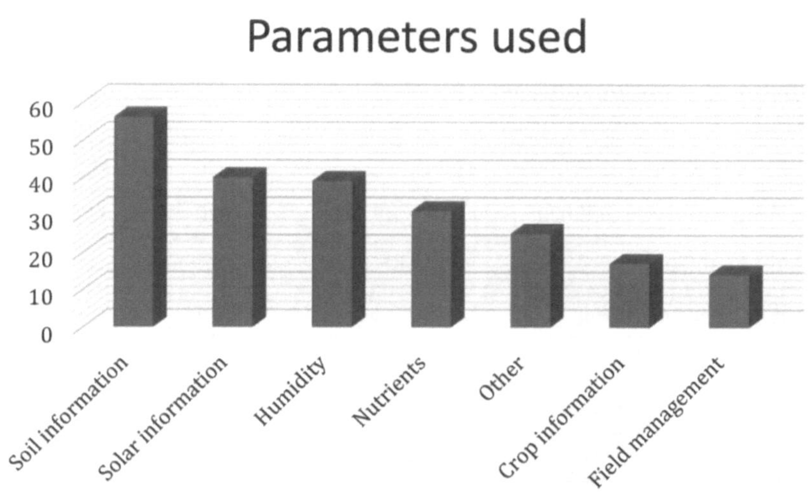

Figure 10.7 Parameters used in articles.

Table 10.6 Literature survey of machine learning algorithms in Agriculture

Authors[7–32]	*Algorithms*	*Year*	*Retrieved From*
Ali et al. (2017)	NN, MLR	2017	Google Scholar
Kouadio et al. (2018)	NN	2018	Science Direct
Goldstein et al. (2018)	GBT, LR	2018	Springer Link
Zhong et al. (2018)	RF, LR	2018	Scopus
Crane-Droesch (2018)	NN	2018	Scopus
Villanueva et al. (2018)	NN	2018	Scopus
Girish et al. (2018)	SVM, LR, KNN	2018	Google Scholar
Khanal et al. (2018)	SVM, RF, and NN	2018	Google Scholar
Taherei Ghazvinei et al. (2018)	NN	2018	Google Scholar
Ahmad et al. (2018)	SVM, RF, DT	2018	Springer Link
Shah et al. (2018)	SVM, RF, MPR	2018	Springer Link
Monga (2018)	NN	2018	Springer Link
Wang et al. (2018)	NN	2019	Google Scholar
Xu et al. (2019)	RF, SVM	2019	Science Direct
Filippi et al. (2019b)	RF	2019	Scopus
Rao & Manasa (2019)	NN	2019	Google Scholar
Ranjan & Parida (2019)	LR	2019	Springer Link
Charoen-Ung & Mittrapiyanuruk (2019)	RF	2019	Springer Link
Patil, A et al.(2020)	NN	2020	Google Scholar
Van Klompenburg,et al.(2020)	SVM, RF, DT	2020	Science Direct
Ranjani J et al.(2021)	SVM,RF, MPR	2021	Scopus
Ansarifar, Jet al.(2021)	NN	2021	Google Scholar
Miklós Neményi et al.(2022)	NN	2022	Springer Link
Annabella Zelenák et al.(2022)	RF SVM	2022	Springer Link
Uzair Ahmad et al.(2022)	RF SVM	2022	Springer Link
Vamsi Tej Chowdary et al (2022)	LR	2022	Scopus

As a result, the second component of our research focused on determining the extent to which deep learning algorithms have been used to estimate agricultural productivity. We created a new search strategy centred on deep learning and carried out a fresh search in the same computerised databases that were previously utilised during the SLR research in order to widen our analysis and include newer uses of deep learning algorithms in yield prediction.

10.3 Conclusion

The use of ML in agriculture has an opportunity to enhance sustainability, boost yields, cut down on resource waste, and promote global food security. To fully realise the advantages of these applications, issues including data accessibility, model interpretability, and technological accessibility still need to be resolved. This chapter shows that, according to the nature of the investigation and the accessibility of data, the selected articles employ a number of features. Each article examines yield prediction using ML; however, the characteristics vary. The extent, physical site, and crop of the research also vary. The dataset's accessibility and the study's objectives influence the attributes used. Studies have also shown that models with additional characteristics do not necessarily perform as well in predicting yield. To decide which model works the best, models with more and fewer features should be compared. Numerous studies have used a range of algorithms. The results show that it is tough to identify the best model, but they also clearly show that certain models based on ML are used more frequently than others. The gradient boosting tree, as well as neural networks, also the linear regression, and finally random forest are the most often used models. The majority of the research examined multiple models using ML to see which method produced the most accurate predictions.

References

1. Chlingaryan, A., Sukkarieh, S., Whelan, B. 2018. Machine learning approaches for crop yield prediction and nitrogen status estimation in precision agriculture: A review. Comput. Electron. Agric. 151, 61–69. https://doi.org/10.1016/j.compag.2018.05.012.
2. Ramu, K., Priyadarsini, K. 2021. A Review on Crop Yield prediction Using Machine Learning Methods. In: 2021 2nd International Conference on Smart Electronics and Communication (ICOSEC), Trichy, India, 1239–1245. doi: 10.1109/ICOSEC51865.2021.9591764.
3. Dharani, M.K. et al. 2021. Review on Crop Prediction Using Deep Learning Techniques. In: Journal of Physics: Conference Series, Volume 1767, International E- Conference on Data Analytics, Intelligent Systems and Information Security & ICDIIS 2020, Pollachi, India, 1767, 012026. doi: 10.1088/1742-6596/1767/1/012026.

4. Gandhi, N., Armstrong, L. 2016. Applying data mining techniques to predict yield of rice in humid subtropical climatic zone of India. In: Proceedings of the 10th INDIACom; 2016 3rd International Conference on Computing for Sustainable Global Development (INDIACom), New Delhi, India, 1901–1906. https://ieeexplore.ieee.org/abstract/document/7724597/.
5. Masi, M., Di Pasquale, J., Vecchio, Y., Capitanio, F. 2023. Precision farming: Barriers of variable rate technology adoption in Italy. Land. 12(5):1084. https://doi.org/10.3390/land12051084.
6. Garibyan, L., Avashia, N. 2013. Polymerase chain reaction. J Invest Dermatol. Mar;133(3):1–4. doi: 10.1038/jid.2013.1.
7. Ali, I., Cawkwell, F., Dwyer, E., Green, S. 2017. Modeling managed grassland biomass estimation by using multitemporal remote sensing data—a machine learning approach. IEEE J. Sel. Top. Appl. Earth Obs. Remote Sens. 10(7):3254–3264. https://doi.org/10.1109/JSTARS.2016.2561618.
8. Kouadio, L., Deo, R.C., Byrareddy, V., Adamowski, J.F., Mushtaq, S., Phuong Nguyen, V. 2018. Artificial intelligence approach for the prediction of Robusta coffee yield using soil fertility properties. Comput. Electron. Agric. 155:324–338. https://doi.org/10.1016/J.COMPAG.2018.10.014.
9. Goldstein, A., Fink, L., Meitin, A., Bohadana, S., Lutenberg, O., Ravid, G. 2018. Applying machine learning on sensor data for irrigation recommendations: Revealing the agronomist's tacit knowledge. Precis. Agric. 19(3):421–444. https://doi.org/10.1007/s11119-017-9527-4.
10. Zhong, H., Li, Xiaocheng, Lobell, D., Ermon, S., Brandeau, M.L. 2018. Hierarchical modeling of seed variety yields and decision making for future planting plans. Environ. Syst. Decis. 38:458–470.
11. Crane-Droesch, A. 2018. Machine learning methods for crop yield prediction and climate change impact assessment in agriculture. Environ. Res. Lett. 13(11):114003. https://doi.org/10.1088/1748-9326/aae159.
12. Villanueva, M.B., Louella, M., Salenga, M., 2018. Bitter melon crop yield prediction using machine learning algorithm. IJACSA. 9(3). http://dx.doi.org/10.14569/IJACSA.2018.090301.
13. Girish, L., Gangadhar, S., Bharath, T., Balaji, K. 2018. Crop Yield and Rainfall Prediction in Tumakuru District using Machine Learning. Ijream.Org. Retrieved from https://www.ijream.org/papers/NCTFRD2018015.pdf.
14. Khanal, S., Fulton, J., Klopfenstein, A., Douridas, N., Shearer, S. 2018. Integration of high resolution remotely sensed data and machine learning techniques for spatial prediction of soil properties and corn yield. Comput. Electron. Agric. 153:213–225. https://doi.org/10.1016/J.COMPAG.2018.07.016.
15. Taherei-Ghazvinei, P., Hassanpour-Darvishi, H., Mosavi, A., Yusof, K.W., Alizamir, M., Shamshirband, S., Chau, K. 2018. Sugarcane growth prediction based on meteorological parameters using extreme learning machine and artificial neural network. Appl. Comput. Fluid Mech. 12(1):738–749. https://doi.org/10.1080/19942060.2018.1526119.
16. Ahmad, I., Saeed, U., Fahad, M., Ullah, A., Habib-ur-Rahman, M., Ahmad, A., Judge, J. 2018. Yield forecasting of spring maize using remote sensing and crop modeling in Faisalabad-Punjab Pakistan. J. Indian Soc. Remote Sens. 46(10):1701–1711. https://doi.org/10.1007/s12524-018-0825-8.
17. Shah, A., Dubey, A., Hemnani, V., Gala, D., Kalbande, D.R. 2018. Smart Farming System: Crop Yield Prediction Using Regression Techniques. Springer, Singapore, 49–56. https://doi.org/10.1007/978-981-10-8339-6_6.

18. Monga, T. 2018. Estimating vineyard grape yield from images, In: Bagheri, E., Cheung, J. (eds) Advances in Artificial Intelligence. Canadian AI 2018. Lecture Notes in Computer Science, vol 10832. Springer, Cham. 339–343. https://doi.org/10.1007/978-3-319-89656-4_37.
19. Wang, A., Tran, C., Desai, N., Lobell, D. 2018. Deep transfer learning for crop yield prediction with remote sensing data. COMPASS '18: Proceedings of the 1st ACM SIGCAS Conference on Computing and Sustainable Societies. https://dl.acm.org/citation.cfm?id=3212707.
20. Xu, X., Gao, P., Zhu, X., Guo, W., Ding, J., Li, C., Wu, X. 2019. Design of an integrated climatic assessment indicator (ICAI) for wheat production: A case study in Jiangsu Province. China. Ecol. Ind. 101:943–953. https://doi.org/10.1016/j.ecolind.2019.01.059.
21. Filippi, P., Jones, E.J., Wimalathunge, N.S., Somarathna, P.D.S.N., Pozza, L.E., Ugbaje, S.U., Bishop, T.F.A. 2019. An approach to forecast grain crop yield using multi- layered, multi-farm data sets and machine learning. Precis. Agric. https://doi.org/10.1007/s11119-018-09628-4.
22. Rao, T. V. N., & Manasa, S. (2019). Artificial neural networks for soil quality and crop yield prediction using machine learning. International Journal on Future Revolution in Computer Science & Communication Engineering, 5(1), 57-60.
23. Ranjan, A.K., Parida, B.R. 2019. Paddy acreage mapping and yield prediction using sentinel-based optical and SAR data in Sahibganj district, Jharkhand (India). Spat. Inf. Res. 27:399–410. https://doi.org/10.1007/s41324-019-00246-4.
24. Charoen-Ung, P., Mittrapiyanuruk, P. 2019. Sugarcane yield grade prediction using random forest with forward feature selection and hyper-parameter tuning. In: Unger, H., Sodsee, S., Meesad, P. (eds) *Recent Advances in Information and Communication Technology 2018. IC2IT 2018. Advances in Intelligent Systems and Computing*, vol 769. Springer, Cham, 33–42. https://doi.org/10.1007/978-3-319-93692-5_4.
25. Patil, A., Kokate, S., Patil, P., Panpatil, V., Sapkal, R. 2020. Crop prediction using machine learning algorithms. Int. J. Adv. Eng. Technol., 1(1):1–8.
26. Van Klompenburg, T., Kassahun, A., Catal, C. 2020. Crop yield prediction using machine learning: A systematic literature review. Comput. Electron. Agri. 177:105709.
27. Ranjani J., Kalaiselvi, V.K.G., Sheela, A., Deepika Sree, D., Janki, G. 2021. Crop Yield Prediction Using Machine Learning Algorithm. In: 2021 4th International Conference on Computing and Communications Technologies (ICCCT), Chennai, India, 611–616. doi: 10.1109/ICCCT53315.2021.9711853.
28. Ansarifar, J., Wang, L., Archontoulis, S.V. 2021. An interaction regression model for crop yield prediction. Sci. Rep. 11:17754. https://doi.org/10.1038/s41598-021-97221-7.
29. Nyéki, A., Neményi, M. 2022. Crop yield prediction in precision agriculture. Agronomy. 12:2460. https://doi.org/10.3390/agronomy12102460.
30. Zelenák, A., Szabó, A., Nagy, J., Nyéki, A. 2022. Using the CERES-maize model to simulate crop yield in a long-term field experiment in Hungary. Agronomy. 12:785. https://doi.org/10.3390/agronomy12040785.
31. Ahmad, U., Nasirahmadi, A., Hensel, O., Marino, S. 2022. Technology and data fusion methods to enhance site-specific crop monitoring. Agronomy. 12:555. https://doi.org/10.3390/agronomy12030555.
32. Chowdary, V.T., Robinson, Joel M., Ebenezer, V., Edwin, B., Thanka, R., Jeyaraj, A. 2022. A Novel Approach for Effective Crop Production using Machine Learning. In: 2022 International Conference on Electronics and Renewable Systems (ICEARS), Tuticorin, India, 1143–1147. doi: 10.1109/ICEARS53579.2022.9751798.

33. Ebenezer, V., Daniel, Arsha Anna, Lillipushpam, P. Getzi Jeba, Edwin, E. Bijolin, Thanka, M. Roshni, Devassy, Rosebel. 2023. Classification and Segmentation of Leaf Images based on Deep Learning for Peanut Plant Disease Detection. In: 2023 4th International Conference on Electronics and Sustainable Communication Systems (ICESC), Coimbatore, India, 1414–1420.
34. Edwin, Bijolin, Veemaraj, Ebenezer, Parthiban, Pradeepa, Devarajan, Joseph Pushparaj, Mariadhas, Vargheese, Arumuganainar, Ahila, Reddy, Maheshwar. 2022. "Smart agriculture monitoring system for outdoor and hydroponic environments." Indones. J. Electr. Eng. Comput. Sci. 25(3):1679–1687.
35. Thanka, M. Roshni, Edwin,E. Bijolin, Duela,J. Shiny, Ebenezer, V. 2020. "Automated diagnosis of skin disease multiclass image classification using deep convolution neural network." J. Green Eng. 10(10):7646–7660.
36. Nagasubramanian, G., Sakthivel, R.K., Patan, R., Sankayya, M., Daneshmand, M., Gandomi, A.H. 2021. Ensemble classification and IoT-based pattern recognition for crop disease monitoring system. IEEE Internet Things J. 8(16):12847–12854.
37. Thanka, M.R., Edwin, E.B., Ebenezer, V., Sagayam, K.M., Reddy, B.J., Günerhan, H., Emadifar, H. 2023. A hybrid approach for melanoma classification using ensemble machine learning techniques with deep transfer learning. Comput. Methods Programs Biomed Update. 3:100–103.
38. Joel, M.R., Ebenezer, V., Jeyaraj, K.A., Navaneethakrishnan, M., Arunadevi, R., Jenifa, D.R. 2023. Decoding and Analysing Consumer Feedback for Companies and Goods using Machine Learning. In: 2023 Third International Conference on Artificial Intelligence and Smart Energy (ICAIS), Coimbatore, India, 1085–1091.

Chapter 11

AI in Optimisation of Agricultural Output in Precision Agriculture

J. Jeyalakshmi, Vijay K., Eugene Berna I., and Sowmia K.R.

11.1 Introduction

About 38% of the Earth's surface, or about five billion hectares, is used for farming. About a third of it is farmed, while the rest is made up of grasslands and meadows where cattle may graze. Permanent crops including fruit trees, oil palm plantations, and cocoa plantations account for around 10% of the total area of agriculture. Another 21% have irrigation systems, making them better able to manage their land for agricultural purposes.

Increased demand for food is a direct result of the more than doubling of the world's population between 1961 and 2016. Land is a finite resource and is being used to capacity. From around 0.45 hectares per capita in 1961 to roughly 0.21 hectares per capita in 2016 (Figure 11.1), the global farmland area per capita has steadily dropped.

The term "cropland" refers to any area that is utilised for growing crops, whether they are annuals or perennials, and can include spaces that are left fallow from time to time or are used as temporary grazing. Livestock grazing is the primary usage of permanent meadows and pastures. Rangelands consist of a wide variety of land cover types, including both cultivated and uncultivated pastures, as well as grassland and wooded savannahs.

DOI: 10.1201/9781003504900-11

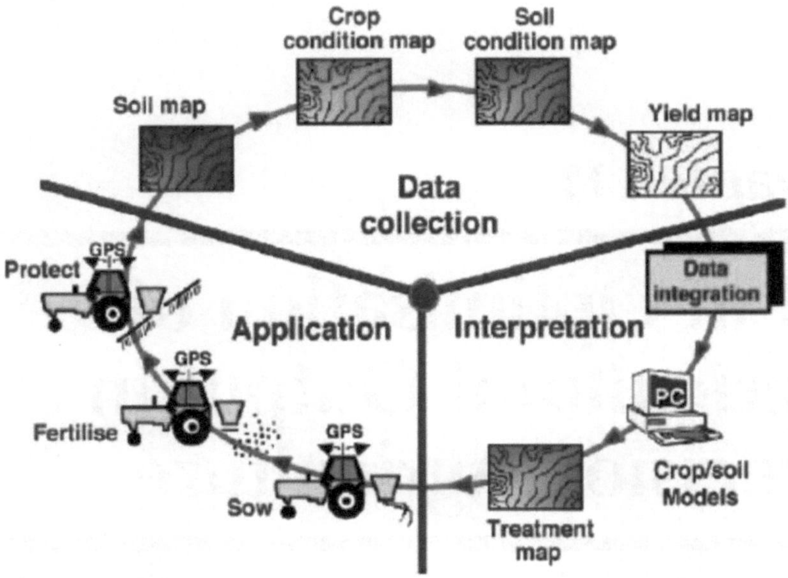

Figure 11.1 Components of Precision Agriculture.[2]

The loss of above- and belowground biomass and carbon due to land conversion from natural ecosystems to agriculture has been the leading driver of greenhouse gas emissions. The conversion of natural habitats into farmland is still one of the leading causes of habitat destruction and species extinction today. Effective land use and land management plans and strategies are necessary to maximise agricultural productivity without negatively impacting the environment through the destruction of ecosystems and the overexploitation of natural resources like soils and water.

Despite the fact that enough food is produced to feed everyone, about a billion people are malnourished or go hungry every day due to food waste, climate change, and other causes. With a projected global population of over 9.7 billion by 2050, the agriculture sector will be under increasing pressure to improve food production while reducing its ecological footprint.

The good news is that artificial intelligence (AI) could be used in farming to improve food production and distribution worldwide. AI could help farmers by evaluating data from several sources to improve decision-making, increase productivity, and reduce adverse effects on the environment. Using AI in agriculture, for instance, could reduce water use by 50% and pesticide use by 60%, according to the World Economic Forum.

India, home to some of the world's most innovative Agtech companies, has a $65 billion window of opportunity to enhance 15 agriculture databases, including information on soil health, crop yields, weather, remote sensing, storage, land records, agricultural markets, and pest imagery. To ensure a more stable and food-secure future, they demonstrate how AI is already being used in agriculture for anything from predicting crop yields to improving soil health. There are numerous opportunities for applying AI to improve crop yields.

AI plays an important part in contemporary agriculture by analysing market demand and assisting farmers in making decisions. Descartes Labs, situated in New Mexico, uses AI to aid farmers in gauging consumer interest. They provide advice on when to plant and what to grow by using machine learning algorithms to analyse satellite images and weather reports. By anticipating the market need for certain crops, this predictive analytics method helps farmers minimise crop failures and maximise earnings. While in India, a new company called Intello Labs is using AI and computer vision (CV) to improve harvest analysis, disease detection, and agricultural management. AI-driven plant growth data helps design disease-resistant and weather-adaptive crop types. The use of AI by businesses like Seed-X to collect genetic data and speed up the development of plant hybrids has the potential to completely transform the conventional hybridisation process in agriculture.

As the Dutch agritech startup AgroCares exemplifies, AI systems are radically altering agricultural practices by precisely conducting chemical soil studies and calculating missing nutrients. Nutrient Scanner data helps farmers optimise fertiliser and water usage, leading to higher crop yields with less environmental impact. In addition, AgroCares provides individualised suggestions for sustaining healthy soil over the long term. The use of AI allows for improved crop monitoring, disease forecasting, weed detection, and pest control. Precision agriculture firms like Taranis use CV and machine learning to analyse photographs of crops in order to accurately identify and categorise diseases and pests, hence reducing the need for potentially harmful broad-spectrum insecticides. AI also outperforms humans in estimating crop growth and maturity, allowing farmers to save money and maximise yields with AI-powered technology like sensors and image recognition tools.[1]

The accurate monitoring of soil conditions is made possible by integrating sensors and AI systems into farming. These sensors send data about conditions like humidity, temperature, pH, and nutritional content to AI systems. The AI evaluates the situation and gives recommendations to farmers about how to best care for their crops. For instance, it can detect when soil is dry or wet, allowing for more efficient irrigation to be planned. AI can also identify nutrient shortages and recommend the best types and amounts of fertiliser to use for maximum crop growth and harvests. Furthermore, AI helps with early pest and disease identification, alerting farmers to problems and, if required, automating the administration of pesticides. Robots powered by AI make weeding more efficient, cutting down on the need for pesticides. Microsoft's FarmVibes is only one example of how chatbots may be a useful interface for farmers by simplifying contact with consumers and distributors and providing real-time insights and suggestions based on data. Many farmers in Africa's Sahel region will reap the rewards of this bot.

11.2 Precision Agriculture and Artificial Intelligence

This strategy goes by a few different names: precision farming, precision agriculture, and precision livestock farming. The term "precision ag" can be defined in a variety of ways, but one of the simplest ways is to consider everything that facilitates more accurate and regulated crop and livestock production. Automation in the form of GPS-guided soil sampling, robotics, drones, autonomous vehicles, variable rate technologies, telematics, and software all play significant roles in this type of farm management. This strategy goes by a few different names: precision farming, precision agriculture, and precision livestock farming. One of the simplest ways to think about what is meant by the term "precision ag" is to think about everything that allows for more precise and regulated crop and livestock production. Important components of this type of farm management include global positioning system (GPS) guidance and control systems, sensors, robotics, drones, autonomous vehicles, variable rate technologies, GPS-based soil samples, automated hardware, telematics, and software.

AI's use in farming is expected to skyrocket in the coming years. AI has the potential to significantly alter the agriculture sector through, among other things, improved productivity, reduced waste, and uniform practices. From a projected $2.35 billion in 2020, the market for AI in agriculture is expected to balloon to a whopping $10.83 billion by 2025. The compound annual growth rate (CAGR) during this period is 35.6%.

One of the greatest advantages of AI in farming is that it allows for the collection and analysis of enormous datasets. Improved decision-making and increased agricultural yields are two key benefits that will help address the problem of global food insecurity.

AI could aid farmers in keeping tabs on things like crop health, weather patterns, and more. This will allow them to catch diseases early and take precautions before they can ruin a harvest. Using AI, farmers may streamline operations and sow seeds at the optimal time of year.

Wasteful spending on materials can be lowered with the use of AI as well. AI can help farmers be more eco-friendly by reducing the number of resources consumed in crop production, such as water and fertiliser. The rising threat of soil and water contamination can be mitigated with the help of this optimisation.

Despite the many obvious advantages of using AI in agriculture, the fact remains that the vast majority of farmers, especially smallholder farmers, simply do not have the means to adopt these innovations. It might be challenging for smallholder farmers to efficiently use AI systems due to a lack of technical knowledge. Many people also lack the financial means to invest in the necessary hardware and software for farming that relies on AI.[3]

Due to their prevalence, smallholder farmers' needs and limitations must be taken into account whenever AI is used in the agriculture sector. Initiatives that assist smallholder farmers get the education and resources they need to adopt AI-based agricultural methods can help close the gap. This will ensure that farmers of all stripes have access to the cutting-edge tools necessary to safeguard the future of our food supply.[4]

11.3 Crop Monitoring, Harvesting, and Management

Precision farming uses AI and CV to adjust planting, irrigation, and fertilisation based on real-time data, with several potential benefits including reduced water use, increased crop yields, and reduced pesticide use. Precision farming has been shown to enhance harvests by as much as 30%, according to one research.

Moreover, with the help of drones and satellite photos, as well as AI and CV, farmers may have access to more precise data on crop health and growth. Researchers at the University of California, Davis found that using drones equipped with cameras to keep an eye out for indicators of disease and pest infestation increased agricultural yields while decreasing the amount of pesticides needed.

Additionally, weather and soil moisture level predictions may be made with the use of AI and CV, allowing farmers to better prepare for planting and irrigation. One Indian startup, for instance, is optimising irrigation and boosting agricultural yields by 30% thanks to AI's ability to anticipate weather and soil moisture levels.

AI and CV technologies may be used to boost agricultural yields in a number of ways. Two of the most essential are monitoring crop development and spotting issues like disease and insect infestation. Photographs of crops can be inspected for signs of pests and diseases using CV. This includes things like leaf discolouration and wilting.[5]

One company, Blue River Technology, has developed a system based on CV and machine learning to monitor crops in real time and detect issues like pests, illnesses, and nutrient shortages. The system uses images captured by a camera placed on a tractor and evaluated by an AI system to determine crop yields.

If the forecasts indicate a potential crisis, such as a disease outbreak or bug infestation, it is essential to investigate the situation. In order to identify crop problems like insect infestation or disease, this programme uses a preexisting machine learning model. A camera takes pictures of the crops, which are then sent to a preprocessor and integrated into the model. Next, the model examines the forecasts to spot potential harvesting problems.

The use of AI and CV could help farmers identify problems before they cause significant crop loss. Increased agricultural yields can be achieved by optimising the use of inputs like fertilisers and insecticides.

Precision farming, which makes use of AI and computer vision technology, has the potential to greatly boost crop yields. Optimising planting, irrigation, and fertilisation through the use of real-time data on factors such as weather, soil moisture, and crop health is what precision farming is all about. Increased crop yields and more environmentally friendly farming methods could result from farmers using AI and CV to analyse this data.

In order to detect pests, diseases, and nutrient deficits in crops from above, Taranis has created a system that employs computer vision and machine learning. Increased crop yields with decreased water use are the outcome of this data being used to optimise the system's fertilisation and irrigation practices.

Figure 11.2 Crop Monitoring and Precision Agriculture (Sourced at: dtn.com).

This method calculates the ideal watering schedule for crops by employing a machine learning model that has been pre-trained using a dataset of images of those crops. The model examines the images and draws judgements based on its findings. This has the potential to help farmers increase crop yields while decreasing water consumption.[6]

Increased crop yields with less water use and pesticide use are the results of using AI and CV in precision farming. Figure 11.2 shows the Crop Monitoring and Precision Agriculture in detail.

While advances in AI and CV might dramatically alter agricultural practices and boost crop yields, they are not without their share of obstacles.

- The high price of these technologies is a major barrier to entry, especially for smaller farms.
- AI-based solutions may be less successful in areas where data on crop growth and meteorological conditions is scarce.
- The use of unmanned aerial vehicles (UAVs) and satellite imagery to monitor crop health and growth raises privacy concerns because this information could be used to track farmers' movements, among other invasions of privacy.

11.4 Resource Management

Precision farming is also being revolutionised by advances in machine learning and AI. In order to help farmers, machine learning algorithms analyse massive volumes of data. Predicting agricultural illnesses, optimising irrigation schedules, and suggesting fertilisation strategies are all within the capabilities of AI-powered systems.

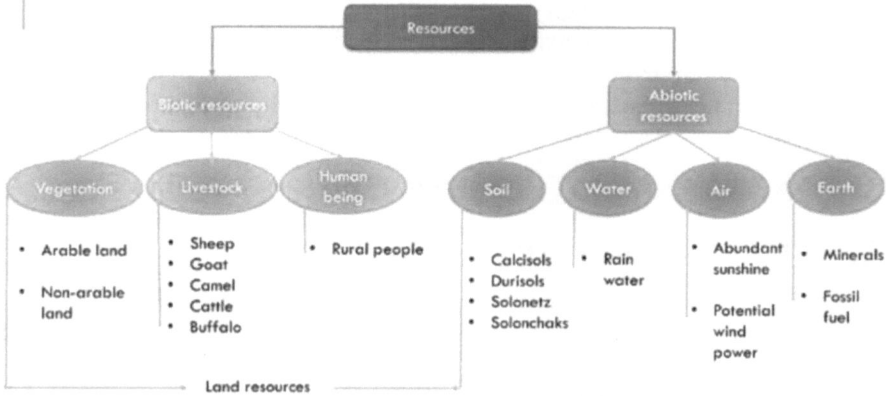

Figure 11.3 Resource Management in Agriculture.

(Sourced at: https://www.slideshare.net/RajeevTiwari89/natural-resources-management-in-dryland-agriculture-and-importance-of-water-management-in-crop-productionpdf)

AI can help farmers maximise their yields and minimise waste. Figure 11.3 shows the details of Resource Management in Agriculture.

Precision farming has a wide range of useful outcomes. Firstly, it improves efficiency in the use of available resources. Precision irrigation, fertilisation, and pest management allow farmers to waste fewer resources and have a smaller effect on the natural world. Saving money in this way helps farmers be more environmentally responsible.[7]

Secondly, the yields and quality of crops are enhanced through precision farming. Farmers may improve growing conditions by monitoring crop health and taking corrective action when necessary. As a result, productivity rises, food quality improves, and profits rise.

Precision farming also aids in saving money and increasing productivity. Input costs can be lowered if farmers maximise their use of available resources. They are able to control costs more effectively by putting resources to work just when and where they are needed. Farming operations are more streamlined and efficient as a result of automation and data-driven decision-making.

Crucially, precision farming aids in preserving the natural world. Soil erosion is lowered and fewer chemicals are needed. Farmers can help maintain agriculture and protect the environment by making better use of their resources.

More progress can be made in the field of precision farming in the future. When combined with other new technologies, agri-tech solutions may do much more. Examples include blockchain, robots, and drones. Increased precision and efficiency in resource management will have a profound impact on agricultural practices.

In short, precision farming technology has ushered in a new era of opportunity for farmers interested in the maximisation of available resources and increased output. Farmers may overcome the difficulties of modern farming and increase both sustainability and harvests by making use of data-driven insights and cutting-edge technologies.

11.5 Predictive Analytics

Statistical science is the basis for predictive analytics' seemingly magical applications in agriculture. At its core, these systems utilise high-quality data to assist farmers in estimating the likelihood of future events. [8]

Positive progress can be made with the use of predictive analytics. Predictive analytics in agriculture is no longer simply a buzzword, thanks to the availability of actionable insights that enable farmers to make more informed decisions in light of hard data.

AI and predictive analytics allow farmers to quickly and effectively analyse and respond to the massive volumes of data they collect. Predictive modelling helps farmers with a variety of issues, including preparing for unpredictable precipitation, applying the right amount of fertiliser, and knowing when to plant seeds and harvest crops. Accurately estimating the quantity, type, and timing of agricultural inputs like fertiliser and pesticide applications is simply one benefit of a comprehensive system of predictive data analytics in farming.

Improvements in agricultural yields, return on investment, sustainability, and quality of output are all possible outcomes of this data-driven decision-making. Data silos must be broken down in order for all data to be aggregated. IT and software problems are often kept in separate "silos" of information that aren't connected to one another, meaning that farmers can't reap the full benefits of 21st-century agricultural technology. In order to get the full benefits of your data, you need to break down silos and adopt a more proactive attitude.

If you want to make good use of predictive analytics tools, you'll need to standardise and link data from different silos on a single platform using a standard set of data types. Predictive analytics' positive effects on farming

From planting through harvesting, predictive analytics has numerous potential applications in agriculture. Predictive analytics and modelling can achieve the following goals:

- Farmers may choose the best crop for their fields by analysing soil, looking at weather records, and considering other factors.
- Predict crop stress periods and the ideal quantity of irrigation required at different phases of crop growth with the use of analytics.
- By comparing field management maps created using GPS technology and productivity maps, field operations can be improved.

- Using data like soil qualities and current weather, predictive analytics can anticipate pest and disease outbreaks and help farmers better protect their crops.
- Using predictive analytics to establish management zones, support optimal crop development, track season progress, and take corrective measures as needed may increase productivity and yields.
- Predictive analytics may help pinpoint which sectors and subfields have consistently low returns on investment and recommend whether or not they can be phased out of use.
- Reduce risks in the supply chain Extreme weather, drought, and shifts in insect behaviour owing to weather all have an effect on the agriculture sector. Farmers can better prepare for these problems and make informed decisions when they have access to relevant data.
- Detrimental environmental consequences can be mitigated with the use of predictive analytics, which can reveal the contexts in which environmental pollution risks are greatest, link individual activities to their cumulative impact, and suggest alternative courses of action.

11.6 Soil Management

As proven by sample data from numerous fields, the spatial variability of phosphorus, potassium, and other soil nutrients is complex, displaying surprisingly varied variability patterns between the fields. Different factors contribute to global and regional variations. Soil types, topographical variations, previous crops, and the proximity to feeding lots are typical causes of large-scale (many acres) variation. Agricultural practices such as tilling, fertilisation, and manure application also contribute to smaller-scale alteration, on the order of a few feet or even inches. Though P, K, and pH might vary widely from one field to the next, regional differences often mirror the distribution of soil types or other landscape factors. Differences in square footage measured in hundreds versus thousands are often negligible.[9]

The quest for a single, ideal method of soil sample that could be used anywhere has largely failed. Three fields' worth of soil P assessments are depicted on the maps, with each assessment based on one of three possible sample methodologies. Similar findings have been made for soil K and pH, indicating that there is no discernible trend. Large-scale grid sampling could be useful for describing nutrient availability. However, it turned out that sampling by soil type was just as useful and far cheaper. The data from most fields suggests that sampling of large cells of three to four acres in size does not represent soil nutrient levels appropriately in many fields because the variation within those areas is as large as the variation over the entire field and cell borders usually do not follow soil mapping units or landscape. It is standard practice to overestimate nutrient content and pH when collecting samples from large areas. Taking more soil cores for each composite sample will not assist.

It's possible that increasing the number of sample points or decreasing the size of the cells might not be financially viable options. The most efficient method is probably targeted grid sampling that takes into account landscape features or other field information. This strategy can be modified for use with other contexts and datasets in the real world. Useful digital resources for designing a workable sample approach include soil maps, soil test data, yield maps, and aerial photographs of both bare soil and crop canopy.

It's not a new idea to consider that economically viable soil sample techniques could not provide the level of information desired by agronomists when describing nutrient variance in soil. It has always been necessary to trade off precision for cost-effectiveness when it comes to recommending soil sample procedures. Soil testing for P and K has been effective in making fertiliser recommendations despite deficits. Soil sampling on a grid is no exception to this rule.[10]

Depending on a number of circumstances, grid sampling combined with variable-rate fertilisation may or may not increase soil fertility management and crop production profits. The ratio of detected nutrients to crop requirements, patterns of fluctuation among nutrients, recommendations for fertiliser application, expected effects from fertiliser application, and any costs connected with fertiliser application are all examples. No substantial yield benefit of variable-rate P fertilisation (or of uniform fertilisation) was identified in four tests with maize and soybeans in fields that tested optimal or high on average. Over half of Iowa's fields have P and K levels that are optimum or above, so this is to be expected. Excessive variability in a field with predominantly optimal or high finds is likely to be inconsequential because of the limited fraction of low-testing sites and the high probability of yield responses in soils testing ideal or above. The greatest potential benefit of extensive grid sampling and variable-rate fertilisation is cost savings on maintenance fertilisation, as many Iowa fields may only contain a few tiny sections of responsive soil. Farmers whose fields routinely test above optimal might benefit greatly from these methods, as they allow them to hold off on applying maintenance fertiliser until levels have returned to an acceptable range. The ecological and long-term repercussions of these triumphs will be positive as well. Due to insufficient grid sampling, farmers often tend to their fields more frequently than necessary.[11]

Data obtained at different depths lends credence to farmers' hypotheses that nitrogen changes are the primary source of production variation within a field. Maps of soil-test values, soil types, and yields were analysed statistically, and visual inspection of the maps revealed that fertility measures only partially explained the yield disparities in each area. Given the complexity of the factors at play, this result is to be expected. These yields' corresponding field metrics showed substantial variation; however, this was not always attributable to nutritional variances. Although soil-test or soil-type maps may provide an explanation for low yields, it has been demonstrated that apparent correlations are often deceptive and would lead to false findings without the use of tests and data management tools that are typically unavailable to producers. Without comparing fertilised and unfertilised treatments, it is impossible to determine which nutrients are restricting yields. This viewpoint motivated scholars to develop methods for assessing competing management styles.

11.7 Pest and Disease Management

Safeguarding food supplies, landscaping, buildings, and people's health all depend on efficient pest management. While effective, certain management measures may have unintended consequences for the ecosystem as a whole or non-target species.

Central to precision agriculture are tools that let farmers monitor their crops and make informed decisions based on that data. These tools include Geographic Information Systems (GIS), GPS, Variable Rate Technology (VRT), and Remote Sensing (RS).[12]

If the farmer is able to carry out their operation flawlessly, they will be able to micromanage the soil's nutrients, the emergence of weeds, pests, and illnesses, and eventually create the best situation to reach their intended harvests. The term "pest" doesn't seem to fit. Any organism with the potential to inflict economic or environmental harm is considered a pest. Pests in agriculture can be anything from insects and mites to rodents and birds to nematodes and weeds, all of which can cause decreased crop yields, crop damage, and the spread of disease. Urban and suburban environments are not immune to the destructive potential of pests, which may also transmit illness and provide an overall nuisance. Managing and controlling pests that pose a threat to agricultural crops, livestock, human health, or the natural environment is the goal of pest control. Pests and diseases account for 20–40% of annual crop losses globally, as reported by the FAO (Food and Agriculture Organisation of the United Nations). As a result, yearly losses in the global food chain and among farmers are predicted to total billions of dollars.[13]

Chemical treatments, biological control, cultural practices, and physical obstacles are all viable options. Pesticides and other chemical agents are used in chemical treatments. Biological pest management employs natural predators and parasites to reduce unwanted insect and animal populations. To lessen the chances of pest infestations, farmers and landscapers may employ cultural practices such as modifying their methods. Screens, netting, and fences can be used as physical barriers to keep pests out of a certain area.

In humid regions, where many pests and illnesses thrive, fungicides are frequently used to eliminate these threats to crop yields and, in the worst-case scenario, save the growing season from being ruined entirely. Varying pests progress through their life cycles at varying rates, use different strategies to counteract the impacts of their natural and manmade environments, and respond to pesticides in different ways.[14]

Weeds, insects, illnesses, and animals are just some of the many agricultural pests that can be mitigated by the application of pest management techniques.

Environmental sustainability is "the application of measures to prevent, avoid, monitor, and suppress activities that have a negative impact on the environment."

The success of any pest management plan depends on its ability to assist farmers achieve their goals while minimising losses due to pests. Insects, fungi, bacteria, and weeds are just a few examples of the organisms that might have a negative impact on crop yields. Tools and methods are used for precise monitoring, pest avoidance, prevention, and eradication in agricultural settings.

The problems can be solved by employing various tools and methods to reduce crop damage from pests. You can classify these tools and methods into two broad categories.

Methods and equipment for counting the insects. GIS, GPS, and radio telemetry (RS) are all used by the precision farming community. Tools and techniques for preventing the spread of invasive pests. GIS, GPS, radio telemetry, and virtual reality technology are only some of the tools used in the framework of precision farming for this group. There are two methods of targeted pest management. In agricultural systems that make use of precision technology, there are two ways in which pests can be managed with pinpoint accuracy. One form of agricultural pest control is known as "area-wide pest management," and it's used to head off infestations across a big swath of land.

Adopting this strategy entails employing IPM techniques and any technology that allows for remote sensing of pest incidence and development. Figure 11.4 shows pest control. Predicting the severity of an outbreak is possible with this method thanks

Figure 11.4 Pest Control (Sourced at: Wikifarmer).

to the use of satellite and UAV imagery with NDVI and hyperspectral indices, GIS software for estimating the extent of the afflicted fields, and agricultural IoTs that can remotely measure environmental parameters and the presence of pests. Successful long-term solutions, the prevention of significant outbreaks, and more sustainable management practices are what drive most producers to adopt IPM practises.[15]

- Site-specific Pest Management in agriculture

  In this method, precise pest control is carried out in a much more limited site-specific management zone, sometimes encompassing only a single agricultural plot. In addition to the aforementioned monitoring tools, the VRT is heavily incorporated into this strategy, giving farmers the ability to take targeted action.

  This type of site-specific control employs more conventional methods, such as the application of targeted insecticides, to eliminate pests and their associated damage and development. In these settings, the VRT allows farmers to lessen their financial outlay on pesticides and to lessen the environmental damage that results from their use.

11.8 Supply Chain Optimisation

Supply-chain operations are intrinsically complicated across sectors due to the number of interdependent functions, material and information transfers, and possibility for opposing goals. Fragmented incoming and outgoing networks significantly complicate the agriculture supply chain. Farmers deliver their produce to intermediate storage silos, where it is then transported to processing factories before finally reaching end consumers. There are a number of options at each decision point. Multiple choices must be made by participants in agriculture supply networks in order to maximise value.[15, 16]

It is important to us that all users, regardless of ability, have full access to this site. We are available to assist you in gathering content-related data. The difficulty of optimisation analysis becomes mired as the number of options increases for each choice. Due to supply-chain fragmentation, there are potentially thousands of outcomes because of the sheer number of alternative flows at each stage. One firm, for instance, handled more than 300 different kinds of grain across 300 silos, 7,000 additional storage locations, and 200,000 different ways to move the grains.

The complexity of the problem is heightened by the unpredictability of each potential conclusion. There are a number of external and internal elements that contribute to this unpredictability, including weather, input costs, farmer expertise, and price fluctuations caused by a worldwide supply/demand mismatch. Our analysis of crop-yield data from the past decade for sugar reveals over 150 agronomic scenarios (defined as a series of probable yields per parcel), all of which have the potential to dramatically alter future harvest allocation and planning.

11.8.1 The Usefulness of Digital Twins

The agriculture supply chain may be optimised with the help of recent developments in digital and analytics technology. More information than ever before is being collected by the agriculture sector, which includes information on agronomy, meteorology, logistics, and the volatility of market prices. The storage capacity and computational power have increased, while the cost of storing data has fallen. Both prescriptive optimisation methods and predictive data science have seen substantial development in recent years.[17]

An intriguing use of digital and analytics technologies is the creation of a digital twin of the physical supply chain (from farmers to ultimate consumers) for the sake of running digital simulations and optimisations. The complete production and distribution process, from raw materials to finished products, can be captured in a digital twin. The model's parameters can be tweaked to maximise revenue, output, efficiency, and stock levels.

Strong predictive ability is what makes the digital twin so valuable. It employs AI-based methods to perform multivariate function optimisation under user-defined limitations, checking for optimal solutions for planning and scheduling factors like lot size. In the case of a change in plans, optimisers used in planning and scheduling can be immediately re-executed. The timetable may be adjusted on the fly to account for things like urgent orders or shifts in demand [15].

The digital twin is also highly scalable, allowing for speedy iteration from prototype to production with minimal delay between iterations and early insights. The agile methodology paves the path for speedy digitisation by beginning with a minimum viable product that can be delivered in two to three months and which will satisfy 80% of customers' expectations. The rapid evolution of solutions is further ensured by short development cycles.

11.8.2 The Digital Twin: How to Create One

As the importance of digital and analytics becomes more apparent, they have become a top focus for leaders. For instance, in the industrial sector, over 50% of businesses are testing out digital programmes. While this is a positive trend overall, many businesses have difficulty taking their digital and analytics pilots to a larger scale where their full potential may be realised. Only around one-third of businesses make it from the pilot stage to the fully scaled solution, according to McKinsey data and experience.

Creating the solution's core engine, designing a user interface to aid in decision-making, and integrating the solution into the IT landscape via bidirectional interfaces with existing systems, like enterprise resource planning systems in factories, are all potential technical obstacles that businesses may face when implementing a digital twin. A three-pronged approach is the most effective way to extract value. The benefits of digital twins will be difficult to fully realise even for the most technologically capable enterprises. Two further conditions must be met for success.

Expertise in the relevant field and supply chain is required to determine the digital twin's intended purpose. In order to make educated judgements on incremental accuracy, the pace of development, and execution, it is necessary to view the problem from beginning to finish and to have access to sufficient physical and temporal data.

Efforts made to manage change from the perspective of all parties involved in agriculture, from farmers to industry bigwigs. Farmers should know that there will be shifts in the way their crops are collected and that this may result in higher prices being paid for their harvest. By optimising its supply chain with a digital twin, one agricultural player has recently impacted as many as two-thirds of farmers' crop collection dates and yielded rises of 3–5% in harvest compensation. Teams throughout the agricultural supply chain need to modify their outlook as well. Encourage the use of digital twins for planning and exception management so they may avoid having to respond to last-minute adjustments and unexpected breakdowns.

Despite the fact that technology and analytics present robust toolkits for unlocking new value in today's agriculture enterprises, the industry is extremely competitive. There are many potential hazards on the road to adoption, but the payoff for making it there is undeniable. By unlocking latent value in the agriculture supply chain, digital and analytics technology may provide established market leaders with a significant competitive edge (Figure 11.5).

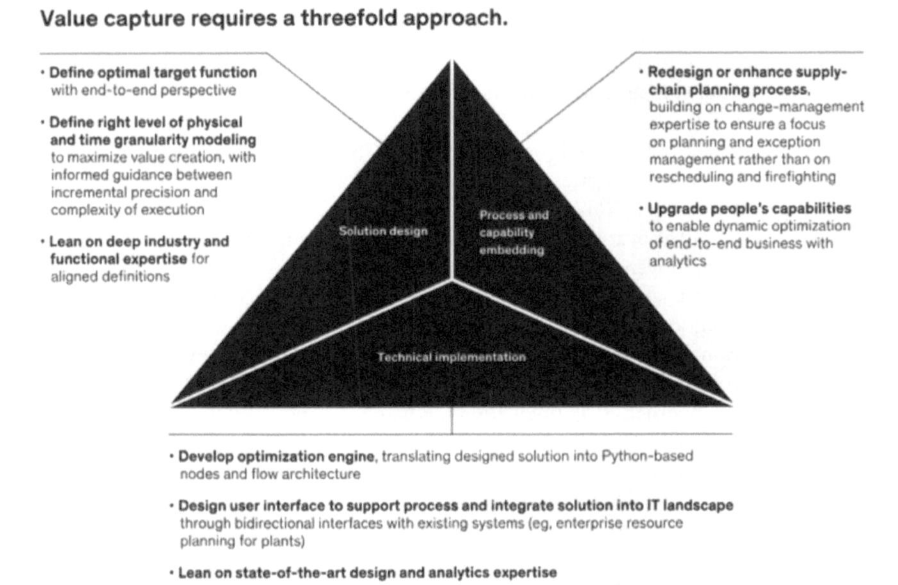

Figure 11.5 Proposed Digital Twin for Agriculture Supply Chain Optimisation by McKinsey & Company

(Sourced at: https://www.mckinsey.com/industries/agriculture/our-insights/agriculture-supply-chain-optimization-and-value-creation)

11.9 Conclusion

Using optimisation, farmers may improve the efficiency, sustainability, and output of their operations. It may be used to improve irrigation efficiency, fertiliser application, insect control, and labour productivity. Farmers can aid in feeding a rising population while safeguarding the environment if they use optimisation strategies. Optimising the usage of irrigation water in California might cut water consumption by farmers by as much as 20%. Fertiliser usage in India is being managed through optimisation, which has the potential to cut fertiliser consumption by as much as 30% without lowering crop yields. The use of optimisation in China to control pesticide application has the potential to cut pesticide application by as much as half without compromising agricultural yields. Crop rotations are being optimised in Brazil using optimisation, which has the potential to boost yields by as much as 10%. Automation of jobs enabled by optimisation in Australia has the potential to cut labour expenses by 20%. More and more novel applications of optimisation to enhance agricultural efficacy, sustainability, and production are likely as optimisation technology advances.

References

1. Tanha Talaviya, Dhara Shah, Nivedita Patel, Hiteshri Yagnik and Manan Shah. Implementation of artificial intelligence in agriculture for optimisation of irrigation and application of pesticides and herbicides. Artificial Intelligence in Agriculture, 2020, 4, 58–73.
2. Cristinel Fertu, Liviu Dobrota and Silvius Stanciu. Precision Agriculture in Romania: Facts and Statistics. Conference: 34th International Business Information Management Association Conference: Vision 2025: Education Excellence and Management of Innovations through Sustainable Economic Competitive Advantage, Madrid, Spain, November 13–14, 2019.
3. J. S. H. Al-bayati and B. B. Üstündağ. Artificial Intelligence in Smart Agriculture: Modified Evolutionary Optimization Approach for Plant Disease Identification. 2020 4th International Symposium on Multidisciplinary Studies and Innovative Technologies (ISMSIT), Istanbul, Turkey, 2020, pp. 1–6. doi: 10.1109/ISMSIT50672.2020.9255323.
4. P. Shankar, N. Werner, S. Selinger and O. Janssen. Artificial Intelligence Driven Crop Protection Optimization for Sustainable Agriculture. 2020 IEEE/ITU International Conference on Artificial Intelligence for Good (AI4G), Geneva, Switzerland, 2020, pp. 1–6. doi: 10.1109/AI4G50087.2020.9311082.
5. R. I. Mukhamediev et al. Coverage path planning optimization of heterogeneous UAVs group for precision agriculture. IEEE Access, 2023, 11, 5789–5803. doi: 10.1109/ACCESS.2023.3235207.
6. S. K. Lenka and A. G. Mohapatra. Gradient Descent with Momentum based Neural Network Pattern Classification for the Prediction of Soil Moisture Content in Precision Agriculture. 2015 IEEE International Symposium on Nanoelectronic and Information Systems, Indore, India, 2015, pp. 63–66. doi: 10.1109/iNIS.2015.56.

7. G. Nagasubramanian, R. K. Sakthivel, R. Patan, M. Sankayya, M. Daneshmand and A. H. Gandomi. Ensemble classification and IoT-based pattern recognition for crop disease monitoring system. IEEE Internet of Things Journal, 2021, 8(16), 12847–12854.
8. K. S. Moorthy, M. P. Neeraia, P. Pavitra, R. Arivumalar, K. Veeranjaneyulu and S. M. Murali. A Hybrid Data Acquisition Model for Precision Agriculture using IoT, Engineering Nanomaterials and Artificial Intelligence. 2023 7th International Conference on Trends in Electronics and Informatics (ICOEI), Tirunelveli, India, 2023, pp. 479–485. doi: 10.1109/ICOEI56765.2023.10125978.
9. S.-L. Muraru, P. Condruz and I. Calciu. Development of software for optimizing the fertilization of agricultural crop. 2020 12th International Conference on Electronics, Computers and Artificial Intelligence (ECAI), Bucharest, Romania, 2020, pp. 1–6. doi: 10.1109/ECAI50035.2020.9223174.
10. M. Gulzar, G. Abbas and M. Waqas. Climate Smart Agriculture: A Survey and Taxonomy. 2020 International Conference on Emerging Trends in Smart Technologies (ICETST), Karachi, Pakistan, 2020, pp. 1–6. doi: 10.1109/ICETST49965.2020.9080695.
11. J. B. Obien, V. Calinao, M. G. Bautista, E. Dadios, J. A. Jose and R. Concepcion. AEaaS: Artificial Intelligence Edge-of-Things as a Service for Intelligent Remote Farm Security and Intrusion Detection Pre-alarm System. 2022 IEEE 14th International Conference on Humanoid, Nanotechnology, Information Technology, Communication and Control, Environment, and Management (HNICEM), Boracay Island, Philippines, 2022, pp. 1–6. doi: 10.1109/HNICEM57413.2022.10109572.
12. V. Yadav, S. Pangaonkar, R. Gunjan and P. Rokade. Plant Pathologist - A Machine Learning Diagnostician for the Plant Disease. 2023 4th International Conference for Emerging Technology (INCET), Belgaum, India, 2023, pp. 1–5. doi: 10.1109/INCET57972.2023.10169990.
13. H. Alshahrani et al. Chaotic Jaya optimization algorithm with computer vision-based soil type classification for smart farming. IEEE Access, 2023, 11, 65849–65857. doi: 10.1109/ACCESS.2023.3288814.
14. A. Khalifeh, A. Al-Qammaz, K. A. Darabkh, L. Abualigah, A. M. Khasawneh and Z. Zinonos. An AI Based Irrigation and Weather Forecasting System utilizing LoRaWAN and Cloud Computing Technologies. 2021 IEEE Conference of Russian Young Researchers in Electrical and Electronic Engineering (ElConRus), St. Petersburg, Moscow, Russia, 2021, pp. 443–448. doi: 10.1109/ElConRus51938.2021.9396431.
15. S. P. Srinivasan, D. S. Shanthi and A. V. Anand. Inventory Transparency for Agricultural Produce through IOT. IOP Conference Series: Materials Science and Engineering, 2017, June. Vol. 211, No. 1, p. 012009. IOP Publishing.
16. N. Promsuk. Improving of the Interference Classification Techniques under the Smart Farming Environment using iSVM. 2022 19th International Joint Conference on Computer Science and Software Engineering (JCSSE), Bangkok, Thailand, 2022, pp. 1–5, doi: 10.1109/JCSSE54890.2022.9836242.
17. G. Saravana Kumar, K. K. Ramachandran, S. Sharma, R. Ramesh, K. Qureshi and K. Ganesh. AI-Assisted Resource Allocation for Improved Business Efficiency and Profitability. 2023 3rd International Conference on Advance Computing and Innovative Technologies in Engineering (ICACITE), Greater Noida, India, 2023, pp. 54–58, doi: 10.1109/ICACITE57410.2023.10182679.

Chapter 12

Data Analytics Methods for Precision Agriculture

Vijay K., Eashaan Manohar, Saiganesh B., and Krithiga P.

12.1 Introduction

In an era where technological advancements are reshaping industries at an unprecedented pace, agriculture stands on the brink of a remarkable transformation. The age-old practices of tilling soil and tending to crops have now evolved into a high-tech endeavor, thanks to the infusion of data analytics methods into the realm of precision agriculture [1, 2].

This chapter, "Data Analytics Methods for Precision Agriculture," serves as a gateway to understanding the fusion of data-driven insights with the art of cultivation. With the current pressure on farmers due to lowered cultivation area and other factors like climate change etc., it is essential to revamp the age-old practices. Precision agriculture emerges as the answer, offering an intelligent and targeted approach to farming that hinges on the judicious use of data.

Precision agriculture, often referred to as precision farming, is an advanced approach to agricultural management that utilizes modern technology, data analytics, and spatial information to optimize the efficiency and effectiveness of agricultural practices. Precision agriculture's major objective is to maximize agricultural productivity, resource efficiency, and ecological sustainability by adapting farming practices to individual plots of land. By adopting precision farming, the farmers can get to know about various aspects of farming and will be able to make many informed decisions as shown in Figure 12.1.

In the following chapters, we delve into the intricacies of machine learning algorithms, statistical models, and predictive analytics that empower farmers to make

DOI: 10.1201/9781003504900-12

Figure 12.1 Precision Agriculture.

informed decisions. We will unravel the mysteries behind yield prediction, disease detection, soil analysis, and resource allocation - all with the aid of data-driven insights.

For precision agriculture to be incorporated into farming, the huge data must be properly collected and analyzed so as to gain the maximum benefit from analysis.

It will be necessary to raise both the quantity and quality of food harvested from agriculture fields, and the application of machine learning to IoT data analytics holds the promise of bringing about these improvements. These revolutionary developments are upending conventional agricultural practices while also presenting some of the best new opportunities. The study culminates the potential and power of computing techniques such as IoT, WSN, Data Analytics, and Machine Learning in agriculture [3].

12.2 Existing Practices in Precision Agriculture

With the advent of modern technology, the scope as well as practicality of precision agriculture is being ramped up and nowadays farming aided by technology is commonplace. The current practices prevailing in precision agriculture are given in subsequent text.

12.2.1 Main Satellite Constellations, Correction Systems, and Georeferencing

Using a network of geostationary satellites in constant radio contact with one another, the GNSS (Global Navigation Satellite System) provides the foundation for precision agriculture in the field. Major nations have a network developed by them, for

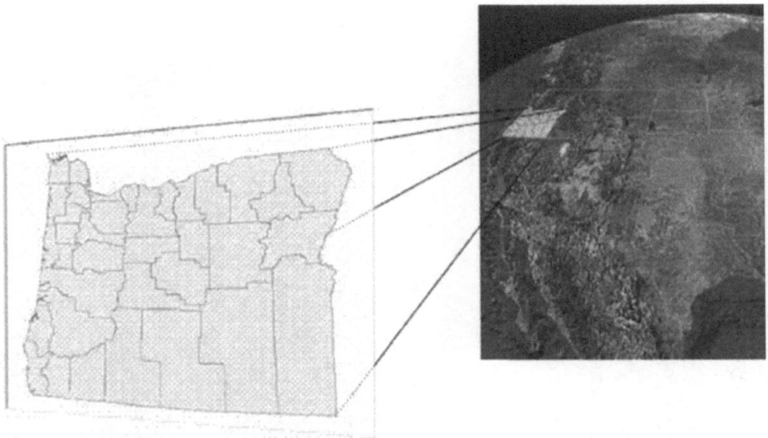

Figure 12.2 Georeferencing.

example, the GPS of the United States (31 satellites), Galileo of the European Union (24 satellites), GLONASS of the Russian Federation (24 satellites), and BeiDou of China (35 satellites) [4].

When used alone, GNSS's position determination accuracy depends on the number of active satellites in the sky and is subject to the accumulation of an error of ±30 cm as given in Figure 12.2. They are assisted by a series of minor constellations functional for signal correction called DGPS (Differential Global Positioning System). The European Union has the EGNOS network of geostationary satellites in orbit that offers a completely free correction service that reduces the error to ±10 cm.

12.2.2 GPS Guidance Systems

Assisted steering solutions certainly stand out among the major innovations introduced in agriculture. Numerous companies produce these systems, but they all use the same principle: a receiver antenna placed on the farm machine connected to the steering system, accompanied by angle sensors mounted on the wheels, and a special terminal installed in the vehicle as shown in Figure 12.3.

12.2.3 Proximal and Remote Sensing

A key feature of precision agriculture is definitely the continuous and timely monitoring of environmental and crop conditions, made possible by the multitude of sensors available on the market. Depending on the specifications of the sensors and the method of detecting environmental and crop parameters, two types of monitoring can be distinguished:

Proximal sensing or proximity monitoring: in which data is collected by sensors placed in close proximity to the crop. Common sensors in the field include weather stations, fixed cameras and video cameras, soil moisture

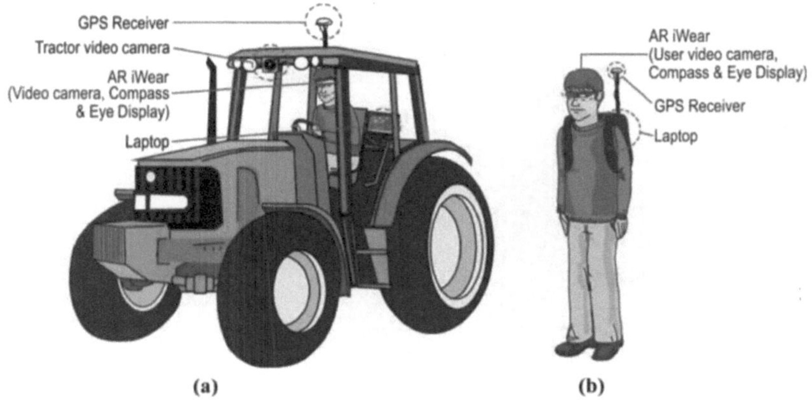

Figure 12.3 GPS Guided System.

and temperature sensors, leaf wetness sensors, and insect traps [5]. In addition, different types of sensors can be installed on board tractors: infrared sensors capable of detecting the degree of the vigor of the crop and performing the most appropriate fertilization for its needs, electrical conductivity sensors for characterizing soils and their organic matter and nutrient content, or soil compaction sensors for working at variable depths as shown in Figure 12.4.

Remote sensing or remote monitoring: made possible by drones, helicopters, airplanes, or satellites equipped with special cameras (RGB, multispectral, etc.) capable of capturing images within the visible spectrum and not visible to the human eye. Remote sensing makes it possible to operate in large and difficult-to-access areas in short periods and with high resolution and accuracy of the data. Unfortunately, using assets such as airplanes and helicopters requires specialized personnel and sophisticated, expensive technologies. It has high operational costs, so they are used for research or military purposes. In contrast, satellites provide quality imagery and are

Figure 12.4 Proximal Sensing.

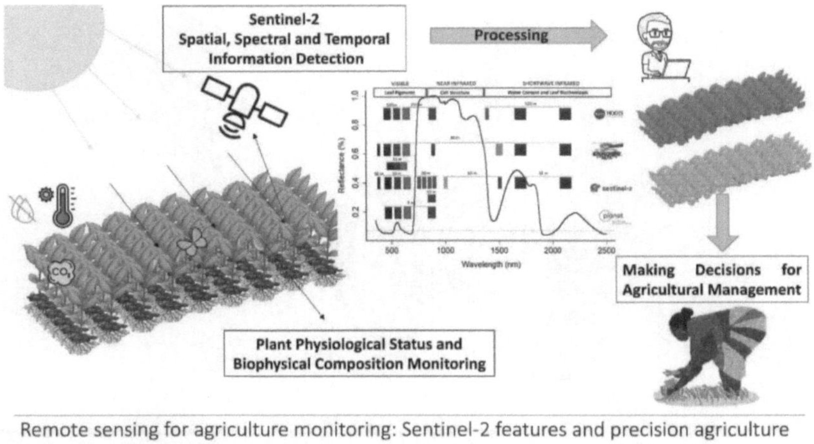

Figure 12.5 Remote sensing using Sentinel-2.

often available free of charge, as in the case of the European Copernicus environmental monitoring satellite network, which has several satellites in orbit, including two dedicated to agriculture, the Sentinel-2 as depicted in Figure 12.5. The disadvantages of satellite sensing are operational in that it provides images every 3–5 days and is climatic in nature due to interference such as clouds and precipitation.

12.3 Role of Data Analytics in Precision Agriculture

Modern digital tools and data science have enabled the collection and analysis of massive agricultural databases, which have improved the understanding of farming chores and decisions among farmers, agronomists, and other experts. In this paper, we provide a comprehensive overview of data mining's role in modern farming. While keeping the focus on yield and monitoring, we introduce the process of crop yield management and its components. We explore a wide range of prior research on the subject of data analytics after classifying the main types of data mining approaches for crop output monitoring. The effects of big data on farming are then discussed in broad terms [6].

In order to come up with digital solutions for diverse challenges in different industries, data analytics is described as the process of collecting and processing massive amounts of data to acquire and generate relevant information and insights. An overview of data analytics is shown in Figure 12.6.

The use of data analytics is essential in precision agriculture because it helps farmers draw conclusions from their data and take action accordingly. It is crucial in various aspects of precision agriculture like informed decision-making, optimized resource allocation, customized problem detection, yield prediction, etc. [7].

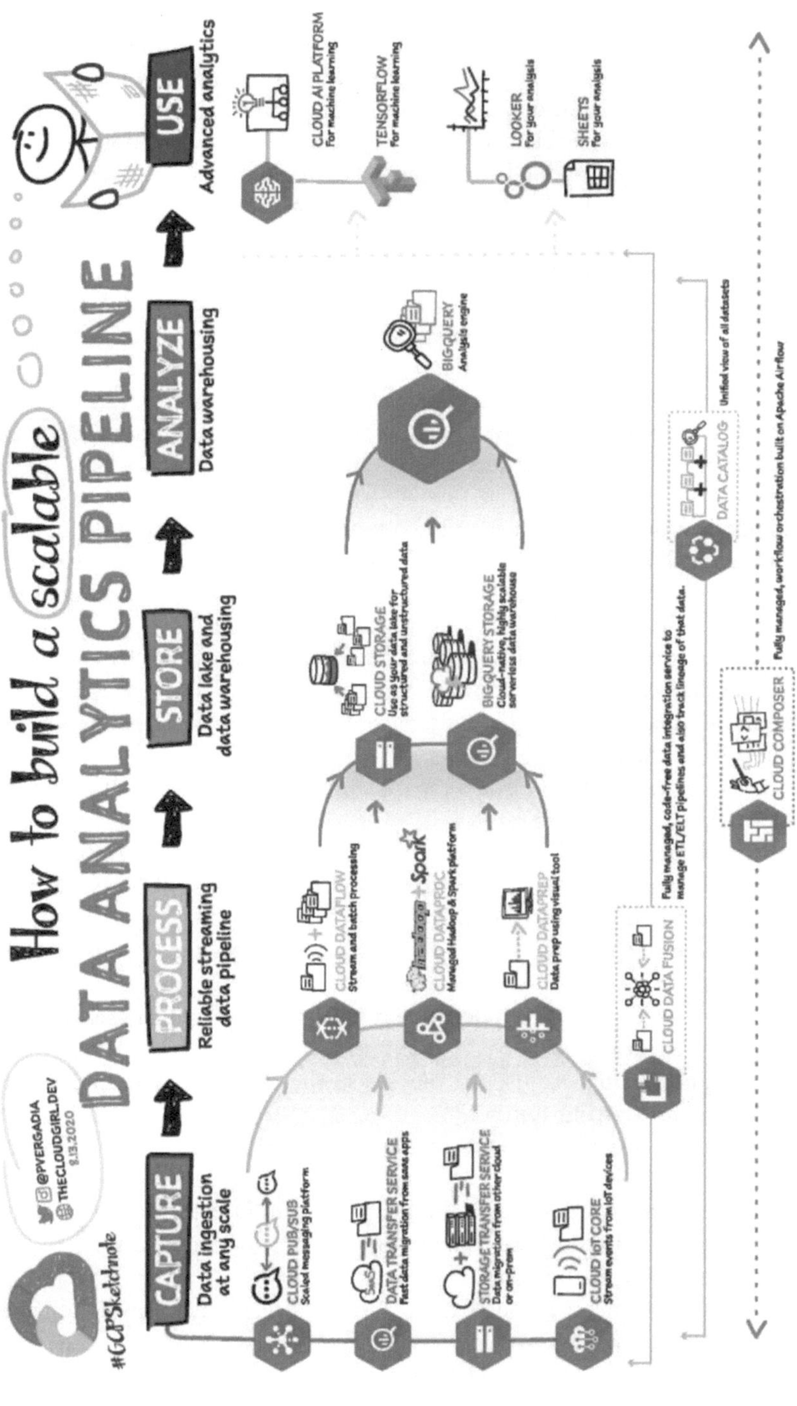

Figure 12.6 Data Analytics.

Data analytics plays a pivotal role in precision agriculture by harnessing the power of information to transform traditional farming practices into highly efficient, targeted, and sustainable operations. Through sophisticated analysis of diverse data sources as satellite imagery, sensor readings, historical records, and weather forecasts, data analytics provides invaluable insights into crop health, soil conditions, resource utilization, and pest dynamics. These insights drive informed decision-making, enabling farmers to precisely allocate resources like water, fertilizers, and pesticides where they are needed most. By customizing interventions to specific field conditions, data analytics minimizes waste, reduces input costs, and promotes environmental sustainability. Moreover, data-driven predictions of crop yields, early detection of issues, and continuous improvement strategies empower farmers to optimize productivity, mitigate risks, and innovate their approaches. When all is said and done, data analytics is the backbone of precision agriculture, radically altering how farmers manage their fields and opening the door to a more prosperous and resilient agricultural future. Data analytics entails a number of steps, including:

1. Data Collection: The first step is to collect information, which can be done by means such as sensors, drones, satellites, weather stations, and archives.
2. Data Preprocessing: Data collected must be cleaned, transformed, and organized to guarantee its accuracy and consistency.
3. Data Integration: Obtain a more complete dataset for analysis by combining information from many sources.
4. Exploratory Data Analysis: Gain first insights by seeing patterns, trends, and outliers in the data.
5. Spatial Analysis: Analyze the data's spatial linkages with the use of Geographic Information Systems (GIS).
6. Feature Extraction: Pick the right features (variables) for the analysis.
7. Data Modeling: Create prediction models through the use of statistical analysis or automated learning systems.
8. Model Training: Models are trained with past data and results that are already known.
9. Validation and Testing: Use independent validation and test datasets to assess the model's effectiveness.
10. Model Optimization: Adjust the models so that they are more precise and have greater forecasting ability.
11. Prediction and Inference: Make accurate forecasts and inferences from the taught models.
12. Decision Support Systems: Incorporate models into DSS for timely guidance.
13. Prescriptive Analytics: Make concrete suggestions for the next steps in light of model projections and actual field data.
14. Iterative Improvement: Evaluate and adjust models regularly in light of new information and outcomes.
15. Visualization: Make charts and graphs to better understand your data and analysis results.

Figure 12.7 Data Analytics in Precision Agriculture.

16. Reporting: Report back to stakeholders and policymakers with a summary of key findings and insights.
17. Implementation: Take action on the ground with suggested tactics, such as variable rate application.
18. Monitoring and Feedback: Always keep an eye on how things are going and ask for opinions.
19. Adaptation and Optimization: Optimize outcomes by adjusting strategy based on data.
20. Data Security and Privacy: Maintain strict adherence to privacy legislation and data security precautions at all times.

Data analytics in precision agriculture involves a comprehensive cycle of collecting, analyzing, modeling, and implementing insights to optimize farming practices and improve overall efficiency and sustainability as shown in Figure 12.7.

12.4 Use Cases and Working Scenarios

Certainly, here are some use cases and working scenarios that demonstrate how data analytics is applied in precision agriculture:

1. Yield Prediction and Optimization
   - Use Case: It is possible for farmers to forecast future harvests by looking at past weather, soil, and crop growth patterns. Their judgements about planting density, watering, and fertilizing can benefit from this data.
   - Working Scenario: A farmer can anticipate future production changes over a field by combining previous yield data with current weather forecasts and soil moisture levels. Because of this, we can better allocate resources and schedule harvests.

2. Variable Rate Application
   - Use Case: Prescription maps for variable rate application of inputs like fertilizers and insecticides are made possible by data analytics, greatly assisting farmers. Each area of the field has unique nutrient requirements, and these maps account for that.
   - Working Scenario: A farmer can create maps showing which areas need more or less of a given nutrient by analyzing soil nutrient data from soil samples and historical yield maps. These schematics instruct machines on how to properly apply controls.
3. Disease and Pest Management
   - Use Case: By analyzing data from distant sensors, drones, and satellites, data analytics aids in the early detection of disease and pest outbreaks. Farmers can therefore take preventative measures in specific areas.
   - Working Scenario: By analyzing temperature, humidity, and vegetation index data collected through sensors and imagery, farmers can identify areas at risk of pest infestations. They can then deploy localized interventions such as targeted spraying or removing affected plants.
4. Precision Irrigation
   - Use Case: Real-time data from soil moisture sensors, weather forecasts, and crop growth stages are analyzed using data analytics to optimize irrigation practices. This guarantees economical water consumption.
   - Working Scenario: Soil moisture sensors collect data that is analyzed alongside historical rainfall patterns and forecasted weather conditions. An algorithm then calculates the optimal irrigation schedule, reducing water wastage and ensuring adequate hydration for crops.
5. Nutrient Management
   - Use Case: Accurate nutrition management is aided by data analytics, which examines soil nutrient levels, crop growth data, and past yields to suggest dosing recommendations.
   - Working Scenario: The nutrient content of the soil is determined by collecting and analyzing samples. This is merged with details about the crops' development and past harvests. Data analytics models then propose what and how much fertilizer should be applied to various plots of land.
6. Automated Harvest Planning
   - Use Case: The ideal time to harvest different parts of a field is determined by analyzing data on crop growth rates, maturity trends, and weather forecasts.
   - Working Scenario: Predicting the best time to harvest various sections of a field requires information on crop development phases, weather forecasts, and past production statistics. This promotes maximum quality and production when harvesting crops.

7. Climate Adaptation
   - Use Case: By analyzing weather records and coming up with plans to lessen the impact of climate change, data analytics aids farmers in their efforts to adjust to the new normal.
   - Working Scenario: Farmers can prepare for changes in growth seasons, rainfall patterns, and temperature ranges by analyzing long-term weather trends and local climate data. The data helps decide when to plant, how much water to use, and what kinds of crops to grow.

These examples illustrate how data analytics is being put to use in the field of precision agriculture to better illustrate how it is helping to improve farming practices in terms of sustainability, productivity, and resource efficiency.

12.5 Data Analytics Tools and Technologies

Geographic Information Systems (GIS), Remote Sensing Platforms (satellites, drones), Sensor, Networks (soil moisture, temperature), Machine Learning Algorithms (Random Forest, SVM), Artificial Neural Networks (ANN), Data Visualization Software (Tableau, Power BI), Cloud, Computing Platforms (AWS, Azure), Big Data Frameworks (Hadoop, Spark), Predictive, Modeling Software (R, Python), Time-Series Analysis Tools (ARIMA, Prophet), Geostatistical, Analysis Software (ArcGIS), Crop Modeling Software (DSSAT, APSIM), Precision Agriculture, Software (Farm Management Systems), IoT Platforms for Agriculture, Spectral Analysis Tools (for vegetation health) are some of the few applications and technologies used in precision farming and agriculture [1].

The use of information and communication technologies (ICT) in precision agriculture has resulted in novel approaches to enhancing agricultural productivity, efficacy, and regulation while maintaining environmental stewardship, making sustainable agricultural development a crucial response to rapid population growth. The huge data analytical capabilities of big data (machine learning, deep learning, etc.) are crucial in precision agriculture, where they help agricultural practitioners understand good farming practices and make precise decisions [8].

Data analytics tools and technologies are pivotal in revolutionizing Precision Agriculture, enabling farmers to harness the power of data for optimized resource management, increased productivity, and sustainable practices. Geographic Information Systems (GIS) facilitate accurate field mapping and soil analysis, aiding in targeted planting and efficient use of fertilizers. Real-life examples showcase the impact of data analytics in agriculture. In Iowa, farmers employ GIS to create variable-rate planting maps, adjusting seed density based on soil characteristics. By analyzing historical yield data and satellite imagery, they optimize planting decisions and achieve up to 5% higher yields. Remote sensing platforms like satellites and drones provide valuable insights from above. In Brazil, farmers use drones equipped with multispectral cameras to monitor crop health. The data captured helps detect early signs of disease or nutrient deficiencies, allowing for timely interventions and preventing potential yield loss.

Sensor networks offer real-time field data. For instance, in Australia, soil moisture sensors connected to IoT platforms enable farmers to monitor soil moisture levels remotely. The data informs precise irrigation decisions, reducing water wastage and promoting efficient water use. Machine learning algorithms are transforming decision-making. In the Netherlands, a dairy farm utilizes machine learning to predict when cows are in heat. By analyzing sensor data related to feeding behavior and physical activity, the farm achieves more successful insemination rates, boosting milk production and profitability.

Predictive modeling tools further enhance precision agriculture as depicted in Figure 12.8. In the United States, a vineyard employs predictive models to anticipate disease outbreaks [9]. By analyzing weather data and historical disease patterns, the vineyard can implement targeted spraying, reducing pesticide usage while safeguarding yields. Time-series analysis is another valuable approach. In Canada, data from weather stations and soil moisture sensors are used to predict optimal planting times. This ensures crops are sown when conditions are favorable, leading to improved germination rates and overall yield.

The use of cutting-edge imaging technology is essential. In order to keep tabs on the well-being of the trees in California's orchards, hyperspectral imaging is used.

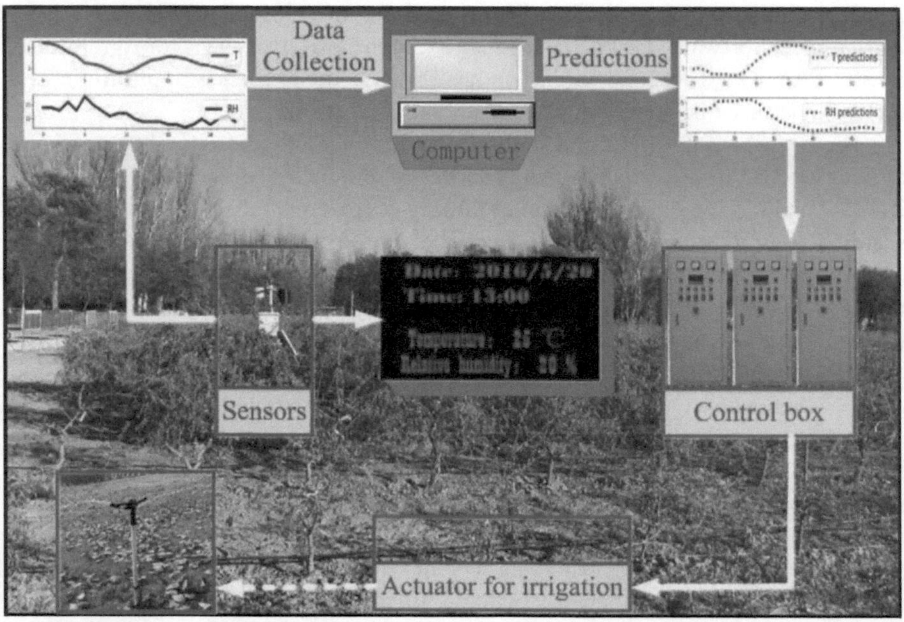

Figure 12.8 Predictive Analysis for Precision Agriculture.

The technique can pick up on minute shifts in leaf reflectance, which can be an early warning sign of stress or disease. Farmers can stop problems from spreading by treating only the damaged trees if they can identify them early. Supply chain transparency and traceability are further benefits of blockchain technology. Walmart, for one, collaborated with IBM on a blockchain-based system for tracking the provenance of fresh produce. Public health concerns are mitigated when the source of contamination is quickly determined in the case of a food recall.

Data analytics tools and technologies are driving a transformative shift in precision agriculture. These tools empower farmers with real-time insights and predictive capabilities, optimizing planting, irrigation, and resource allocation. By embracing GIS, remote sensing, machine learning, and predictive modeling, farmers globally are achieving higher yields, reduced input wastage, and more environmentally sustainable practices [10]. The integration of advanced imaging and blockchain further cements the role of data analytics in ensuring the quality, safety, and transparency of agricultural products. As the agriculture industry continues to evolve, data analytics remains at the forefront of innovation, shaping the future of farming for the better.

12.6 Data Analysis in Precision Agriculture

Data analysis in precision agriculture involves a series of steps that transform raw data into actionable insights for optimized farming practices Sensors, satellites, and weather stations are just a few of the places where information is gathered to kick off the process. This data is then preprocessed to ensure accuracy and consistency by cleaning and organizing it. The integrated data is subjected to exploratory analysis to identify initial patterns and trends. Spatial analysis helps understand the spatial relationships within the data, while feature extraction selects relevant variables for analysis. Using machine learning algorithms, predictive models are built and trained using historical data. The models are validated and tested using separate datasets. Further optimization is done to improve model accuracy. Once validated, the models are used to make predictions and draw meaningful inferences. Decision support systems are integrated, providing real-time advice based on model outputs [11, 12]. Prescriptive analytics recommends specific actions based on predictions and field conditions. The process is iterative, with continuous improvement through monitoring and adaptation based on results. Data security and privacy are maintained throughout the analysis. The insights gained from the analysis are visualized and reported to stakeholders, leading to informed decisions and improved agricultural practices [13–15].

Data analysis in precision agriculture involves the utilization of diverse and multidimensional datasets to enhance farming practices. The kinds of data employed span various categories, including spatial, temporal, and spectral data.

Spatial data encompasses geographic information collected through technologies like Geographic Information Systems (GIS), capturing field boundaries, topography, and soil types. Weather patterns, plant development stages, and past harvest totals are all examples of time-series data. Vegetation health, moisture levels, and nutrient content may all be evaluated with the help of spectral data, which is collected from various regions of the electromagnetic spectrum and typically captured by remote sensing platforms like satellites and drones.

The characteristics of such data are marked by their high volume, velocity, variety, and veracity—commonly referred to as the "4 Vs" of big data. The volume of data generated from sensors, imagery, and machinery can be immense, requiring robust storage and processing capabilities. The velocity at which data is collected, especially in real-time scenarios, demands efficient handling to ensure timely decision-making [16]. The variety of data sources, including numerical, geospatial, and image data, necessitates versatile analysis techniques and tools. Lastly, the veracity of the data pertains to the accuracy and reliability of the information collected, emphasizing the need for data validation and quality assurance as shown in Figure 12.9.

These characteristics highlight the complexity of data analysis in precision agriculture, underscoring the importance of advanced technologies and techniques to extract meaningful insights from these diverse and dynamic datasets.

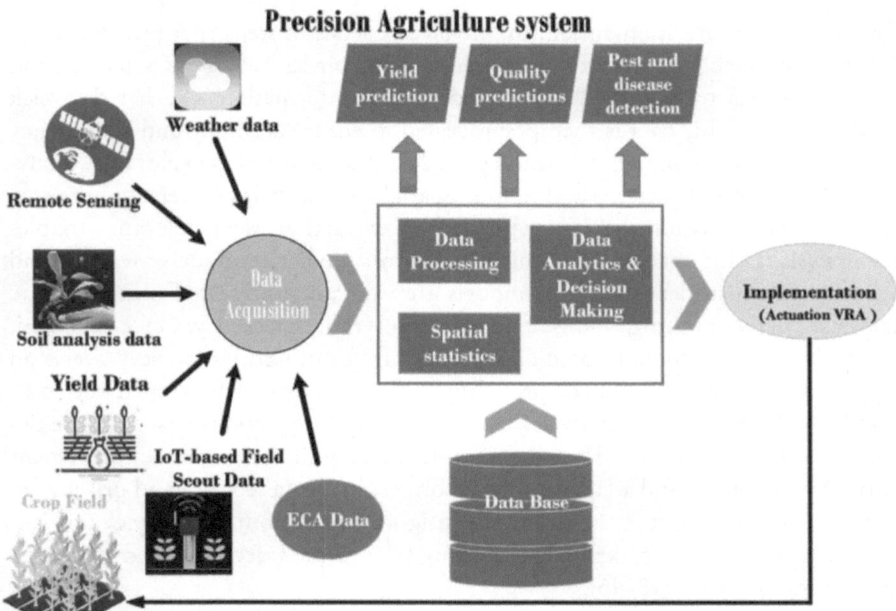

Figure 12.9 Data Analysis in Precision Agriculture System.

12.7 Case Study of Data Analysis in Precision Agriculture – India

In 2004–2005, the districts of Dharmapuri and Krishnagiri in Tamil Nadu were the pilot sites for the Precision Farming Project. Initially applied to 250 acres, the program was expanded to 500 acres in 2005–2006 and an additional 250 acres in 2006–2007. The overall budget for this project was 7.2 crores, and it was carried out over a period of three years by the Tamil Nadu Agricultural University as the nodal agency. Farmers were granted a total of Rs. 130,000 (of which Rs. 75,000 went towards drip irrigation and Rs. 40,000 towards other costs associated with crop production) as part of this initiative [17]. In three years, farmers successfully harvested five crops, two of which were planted under the close supervision of university researchers. Farmers were initially hesitant to participate in the project due to their growing dissatisfaction with the prolonged drought that had plagued their region since 2002. Farmers initially resisted the scheme because the fertilizer subsidy was only 90% at the beginning, but after seeing the success of the first 100 farmers and the high market rate for the produce obtained from this scheme, farmers began registering in large numbers for the second year, when the subsidy was reduced to 80%. Because each farmer is only allowed to receive coverage for a single unit of cultivable land (one hectare), this initiative covered a total of 247 units in 2004–05, 494 units in 2005–06, and 247 units in 2006–07. After farmers in these two areas found success with precision farming techniques, the practice spread to other districts as well. The results were bolstered by the farmer-to-farmer approach, and the rest of the state's districts demanded that the program be put into action. After its first success, the National Agricultural Development Project (NADP) expanded it to cover 40,000 hectares across the state. The project was established in 2007–2008 through a collaboration between the university and the agriculture and horticulture departments. Training has been provided for all farmers in order to empower them technically, economically, and socially in the states of Kerala, Karnataka, Andhra Pradesh, Odisha, and Maharashtra.

12.8 Data Analytics Techniques Used in Precision Farming

Various data analytics techniques are employed in precision farming to extract insights from agricultural data. These techniques enable farmers to make informed decisions that optimize resource allocation, improve yields, and promote sustainable practices [18]. Some of the key data analytics techniques used in precision farming include:

1. Regression Analysis: Relationships between variables can be better comprehended with the aid of regression models. It can be used in precision farming to foretell crop yields in response to changes in environmental variables, soil composition, and nutrient availability Crop Pest Classification [19, 20].

2. Classification Algorithms: Data classification methods sort information into many groups. These algorithms can be applied to sensor data for crop classification, disease detection, and soil type categorization in precision agriculture.
3. Clustering Analysis: When you use clustering, you can organize your data by type. Fields with comparable soil parameters or crop growth patterns can be clustered using this method for precision farming.
4. Time-Series Analysis: This method applies statistical analysis to data collected over time in order to reveal trends, cycles, and patterns. Time-series analysis is used in precision agriculture to make informed decisions about when to sow and harvest based on past weather and crop performance.
5. Geospatial Analysis: Geospatial methods dissect geographical information for the purpose of discovering hidden structures and connections. Field mapping, soil analysis, and irrigation system design are just some of the many uses for this handy tool.
6. Machine Learning: Predictive modelling makes use of machine learning algorithms including decision trees, random forests, support vector machines, and neural networks. They look at past information to speculate on things like crop development, pest infestations, and yield swings.
7. Image Analysis: Drone or satellite photos undergo analysis using image processing and computer vision methods. This aids in disease detection, crop monitoring, and vegetation cover analysis.
8. Remote Sensing: Monitoring crops, soil moisture, and other environmental variables is possible through remote sensing thanks to the use of satellites and drones to collect data. Analysis of data gathered by remote sensing devices allows researchers to learn more about the state of crops and the stresses they're facing.
9. Spatial Interpolation: This method extrapolates data from sampled areas to estimate data at un-sampled areas. To make continuous maps of soil characteristics or other factors for use in precision farming.
10. Big Data Analytics: To effectively process and analyze enormous datasets, precision farmers are turning to big data analytics tools like Hadoop and Spark [21].
11. Real-time data from sensors deployed in fields is collected by Internet of Things (IoT) devices and then analyzed to track soil moisture, temperature, and other characteristics [22].
12. Ensemble Methods: The predicted accuracy of an ensemble method is increased by using many models. This has the potential to improve yield projections and disease detection in precision farming.
13. Optimization Techniques: Finding the optimal answers to difficult issues is the job of optimization algorithms. Irrigation schedules, crop rotation strategies, and resource distribution can all benefit from these methods' application in precision farming.

There is a specific place for each of these data analytics methods in precision farming, all of which help farmers make better decisions and advance agricultural practices.

12.9 Advantages, Features and Limitations of Various Data Analytics Techniques

Regression Analysis:

- Advantage: Allows for the prediction and optimization of crop production by illuminating underlying links between factors.
- Feature: provides a numerical measure of the direction and magnitude of interdependence.
- Major Drawback: linearizes data that may not be applicable to more complicated agricultural systems.

Classification Algorithms:

- Advantage: Allows for the automatic categorization of data, which is useful for identifying crop types and detecting diseases.
- Feature: creates models from training data for assigning new data to existing categories.
- Major Drawback: The training data's quality and diversity have a significant impact on the accuracy of the model.

Clustering Analysis:

- Advantage: Groups similar data points, helping in targeted interventions and resource allocation.
- Feature: Identifies patterns based on the similarity of data points in a multidimensional space.
- Major Drawback: Requires determining the number of clusters in advance, which can be challenging.

Time-Series Analysis:

- Advantage: Analyzes trends and patterns over time, aiding in prediction and decision-making.
- Feature: Captures seasonality, trends, and irregularities in time-dependent data.
- Major Drawback: Assumes that future patterns will resemble past patterns, which might not hold in evolving agricultural conditions.

Geospatial Analysis:

- Advantage: Provides spatial insights for precise field management and resource allocation.
- Feature: Utilizes geographic data to analyze relationships and patterns within a field.
- Major Drawback: Complex analyses might require specialized software and expertise.

Machine Learning:

- Advantage: Captures complex relationships in data for accurate predictions and insights.
- Feature: Includes various algorithms like decision trees, random forests, and neural networks.
- Major Drawback: Requires substantial amounts of quality training data to develop effective models.

Image Analysis:

- Advantage: Monitors crop health and detects diseases through visual assessment.
- Feature: Utilizes image processing and computer vision techniques to analyze images.
- Major Drawback: analysis reliability depends on the availability of stable, high-quality pictures.

Remote Sensing:

- Advantage: Allows for remote crop and environmental monitoring in real time.
- Feature: Using satellite and drone imagery, crop growth and development may be monitored in real time.
- Major Drawback: Data quality may suffer from low spatial and temporal resolutions.

Spatial Interpolation:

- Advantage: Produces continuous variable maps from sparse inputs.
- Feature: Uses nearby data to make predictions for areas outside the sampling frame.
- Major Drawback: Distribution of sample points and underlying variability affect accuracy.

Big Data Analytics:

- Advantage: Capable of quickly processing and analyzing massive amounts of data
- Feature: makes use of large data technologies like Hadoop and Spark.
- Major Drawback: Implementation needs specialized infrastructure and personnel.

IoT Data Analytics:

- Advantage: Offers sensor-based, real-time field monitoring.
- Feature: gathers information from IoT gadgets and processes it for analysis.
- Major Drawback: Sensor quality and connectivity are critical to reliability and precision.

Ensemble Methods:

- Advantage: Improves model accuracy by combining multiple models' predictions.
- Feature: Combines strengths of different algorithms to address individual weaknesses.
- Major Drawback: Complexity can make interpretation and implementation challenging.

Optimization Techniques:

- Advantage: Finds optimal solutions for complex agricultural problems.
- Feature: Uses algorithms to identify best options for resource allocation and management.
- Major Drawback: Requires clear formulation of problem constraints and objectives.

Each of these data analytics techniques offers unique advantages, features, and drawbacks in the context of precision agriculture, emphasizing the need to select the most appropriate technique for specific agricultural challenges and goals [23].

12.10 Challenges and Limitations

Data analysis in precision agriculture, while promising significant benefits, comes with its share of challenges and limitations:

1. Data Quality and Variability: Agricultural data comes from various sources, and its quality can vary significantly. Inaccurate or inconsistent data can lead to flawed analysis and incorrect decisions.
2. Data Integration: Integrating data from multiple sources, such as sensors, weather forecasts, and machinery, can be complex. Different data formats and platforms might not seamlessly interact.

3. Privacy and Security: Handling sensitive farm data raises concerns about privacy and security. Farmers and stakeholders must ensure that data is protected against unauthorized access and cyber threats.
4. Infrastructure Requirements: Processing and storing large volumes of data require robust computational and storage infrastructure, which might be costly for small-scale farmers.
5. Technical Expertise: Effective data analysis requires expertise in fields like data science, GIS, and machine learning. Many farmers might lack the skills or resources to hire specialists.
6. Connectivity Issues: In remote or rural areas, poor internet connectivity can hinder real-time data collection and analysis, limiting the effectiveness of precision agriculture.
7. Scalability: Adapting data analysis solutions to larger farms or diverse landscapes can be challenging. Scaling up precision agriculture practices might require additional resources.
8. Cost: Implementing data analytics tools and technologies can be expensive, involving initial investments in hardware, software, training, and ongoing maintenance.
9. Interpretation Complexity: Extracting actionable insights from complex data can be challenging. Farmers must understand analysis results and translate them into practical on-ground actions.
10. Data Overload: The abundance of data can overwhelm farmers, leading to analysis paralysis or the inability to effectively utilize the insights.
11. Changing Conditions: Agricultural environments are influenced by dynamic factors like weather patterns and climate change. Models built on historical data might not fully account for these evolving conditions.
12. Adoption Barriers: Convincing traditional farmers to adopt data-driven approaches can be a challenge due to resistance to change, lack of awareness, or reluctance to embrace technology.
13. Data Latency: Some data sources might not provide real-time updates, leading to delays in decision-making when timely actions are crucial.
14. Model Accuracy: Predictive models, while powerful, might not always accurately capture the complexity of biological systems and natural variability.
15. Environmental Concerns: Reliance on data analysis might lead to unintended environmental consequences if not balanced with sustainable practices and holistic agricultural management [24].

Addressing these challenges requires a combination of technology advancement, education and training, policy frameworks for data sharing and privacy, and collaboration between stakeholders to ensure that data analysis in precision agriculture is effective, equitable, and sustainable [25].

12.11 Conclusion

The chapter explored the various aspects of precision agriculture by highlighting the brief as well as detailed practices that were being carried out in the agricultural sector. The chapter also shed light on what data analytics meant and how it was being used in precision agriculture. Various aspects of the data collected from the initial phase to the last analysis phase were mentioned. Precision Agriculture is the future of the agricultural sector and data analysis is an important aspect of it. Careful selection and usage of techniques must be applied so as to get the desired results. Privacy and protection of data are important along with the variety and availability of data. The comprehensive exploration of precision agriculture in this chapter has illuminated the multifaceted landscape of modern farming practices. From a concise overview to in-depth examinations, the chapter vividly portrayed the spectrum of activities encompassing this transformative approach to agriculture. Of particular significance was the intricate weave of data analytics into the fabric of precision agriculture. This chapter expounded on the very essence of data analytics, elucidating its pivotal role in driving informed decision-making within the agricultural domain.

References

1. K. R. Sowmia, S. Prithi, K. Vijay, I. Eugene Berna and B. Bhuvaneswaran, "Crop Monitoring System with Water Moisture Level using Arduino," 2023 5th International Conference on Smart Systems and Inventive Technology (ICSSIT), Tirunelveli, India, 2023, pp. 495–498, doi: 10.1109/ICSSIT55814.2023.10060913.
2. P. Samuel, A. K. Reshmy, S. Rajesh, M. Kanipriya and R. A. Karthika. (2023). AI-Based Big Data Algorithms and Machine Learning Techniques for Managing Data in E-Governance. In K. Saini, A. Mummoorthy, R. Chandrika, & N. Gowri Ganesh (Eds.), AI, IoT, and Blockchain Breakthroughs in E-Governance (pp. 19–35). IGI Global. https://doi.org/10.4018/978-1-6684-7697-0.ch002
3. Ravesa Akhter, Shabir Ahmad Sofi, "Precision Agriculture Using IoT Data Analytics and Machine Learning," *Journal of King Saud University - Computer and Information Sciences*, vol. 34, no. 8, pp. 5602–5618, 2022.
4. M. Pragadeesh, J. Adlin Layola, K. Vijay, and S. Gnanavel. "Smart and accurate agriculture using IoT." In AIP Conference Proceedings, vol. 3075, no. 1. AIP Publishing, 2024
5. https://gisresources.com/wp-content/uploads/2013/08/Georeferencing-concept2
6. N. Chergui and M. Kechadi, "Data Analytics for Crop Management: A Big Data View," *Journal of Big Data*, vol. 9, p. 123, 2022. https://doi.org/10.1186/s40537-022-00668-2
7. P. Vijay, S. Shanthi D., P. Revathy, S. Ohja, Sriram N. and Nandhiannal, "Grainchain: Agricultural Supply Chain Traceability and Management technique for Farmers Sustainability Using Blockchain Hyper Ledger", Int J Intell Syst Appl Eng, vol. 10, no. 3, pp. 141–146, Oct. 2022.

8. S. A. Bhat and N.-F. Huang, "Big Data and AI Revolution in Precision Agriculture: Survey and Challenges," *IEEE Access*, vol. 9, pp. 110209–110222, 2021. doi: 10.1109/ACCESS.2021.3102227
9. G. Nagasubramanian, R. K. Sakthivel, R. Patan, M. Sankayya, M. Daneshmand and A. H. Gandomi, "Ensemble Classification and IoT-Based Pattern Recognition for Crop Disease Monitoring System," *IEEE Internet of Things Journal*, vol. 8, no. 16, pp. 12847–12854, 2021.
10. https://www.cubictelecom.com/blog/big-data-modern-precision-farming-agtech/
11. P. Vijay, S. Shanthi D, Revathy. P, S. Ohja, Sriram. N and Nandhiannal, "Grainchain – Agricultural Supply Chain Traceability and Management Technique for Farmers Sustainability Using Blockchain Hyper Ledger," *Int J Intell Syst Appl Eng*, vol. 10, no. 3, pp. 141–146, Oct. 2022.
12. https://www.researchgate.net/publication/355161535_A_Case_Study_of_Farmers_Practicing_Precision_Agriculture_in_Dharmapuri_District_of_Tamil_Nadu
13. A. Rokade, M. Singh, P. K. Malik, R. Singh and T. Alsuwian, "Intelligent Data Analytics Framework for Precision Farming Using IoT and Regressor Machine Learning Algorithms," *Appl Sci*, vol. 12, no. 19, p. 9992, 2022. https://doi.org/10.3390/app12199992
14. B. Vandana and S. S. Kumar, "A Novel Approach using Big Data Analytics to Improve the Crop Yield in Precision Agriculture," 2018 3rd IEEE International Conference on Recent Trends in Electronics, Information & Communication Technology (RTEICT), Bangalore, India, 2018, pp. 824–827, doi: 10.1109/RTEICT42901.2018.9012549.
15. S. Singh, P. Alam, P. Kumar and S. Kaur, "Internet of Things for Precision Agriculture Applications," 2019 Fifth International Conference on Image Information Processing (ICIIP), Shimla, India, 2019, pp. 420–424, doi: 10.1109/ICIIP47207.2019.8985688.
16. A. K. Mungarwal and S. K. Mehta, "Why Farmers Today Need to Take up Precision Farming, *Down to Earth*, 2019.
17. S. Liaghat and S. K. Balasundram, "A Review: The Role of Remote Sensing in Precision Agriculture," *American Journal of Agriculture, Biological Science*, vol. 5, pp. 50–55, 2010.
18. http://blog.journals.cambridge.org/2015/07/15/a-review-of-non-invasive-techniques-to-measure-livestock-body-composition/
19. S. P. Srinivasan and D. S. Shanthi, "A Seed Yield Estimation Modelling Using Classification and Regression Trees (CART) in the Biofuel Supply Chain," *J Biomed Imaging Bioeng*, vol. 1, no. 1, pp. 8–12, 2017.
20. S. Pournima, C. Priyatharsini, G. Kirubasri and J. Manikandan, "Hybrid BILSTM Network for Improving Crop Pest Classification," 2023 International Conference on Computer Communication and Informatics (ICCCI), Coimbatore, India, 2023, pp. 1–5.
21. http://www.un.org/en/development/desa/news/population/2015-report.html (Population Estimate)
22. S. Alex David, S. Ravikumar, and A. Rizwana Parveen, "Raspberry Pi in Computer Science and Engineering Education." Lecture Notes in Electrical Engineering 492: 11–16, 2018.
23. Kadambini Katke, "Precision agriculture adoption: Challenges of Indian agriculture," International Journal of Research and Analytical Reviews, vol. 6, no. 1, pp. 863–869, 2019.
24. https://www.linkedin.com/pulse/precision-farming-india-future-agriculture-sharad-kumar-singh/
25. http://www.rkmp.co.in/content/precision-agriculture-in-india-opportunities-and-challenges

Chapter 13

Data Analytics Methods: A Hands-On Approach for Precision Agriculture

Bulbul Ahmed, Pankaj Das, Rahul Banerjee, Sahadeva Singh, and Bharti

13.1 Introduction

Methods of data analytics are a collection of procedures and techniques used to analyze and understand data in order to draw meaningful insights and make decisions based on the information acquired from the analysis. These methods are essential across various sectors, including the commercial world, the medical industry, the financial sector, and the scientific community, to name a few.

The following is a selection of the several methods that are often used in data analytics:

1. **Descriptive statistics:** The first and most fundamental stage of data analytics is known as descriptive analytics. This stage entails the analysis and interpretation of historical data in order to produce a clear and detailed account of events, patterns, and trends that have occurred in the past. It focuses on providing a response to the question "what happened?" by providing a summary of the data using measures of central tendency (e.g. mean, median, and mode) and measures of dispersion (e.g. range, variance, standard deviation, skewness, and kurtosis). This procedure also involves cleansing the data in order to verify that the data are accurate and comprehensive.

DOI: 10.1201/9781003504900-13

The most commonly used central tendency is mean. Mean, also known as the average, is computed by summing all the values inside a dataset and afterwards dividing by the total count of values. $\mu = \frac{\sum x_i}{N}$, where $\sum x_i$ is the sum of all the observations and N is a number of observations in the dataset. The method exhibits sensitivity to extreme values, often known as outliers, and is frequently used for datasets that adhere to a normal distribution. Median is obtained by arranging the values in either ascending or descending order. In the case when the number of values is even, the median may be determined by calculating the arithmetic mean of the two central values. The median is less susceptible to the influence of outliers and is used in situations when the dataset exhibits extreme values or does not conform to a normal distribution. The mode is defined as the value that has the highest frequency of occurrence within a given dataset. A dataset may have a unimodal distribution, characterized by a single mode, or a multimodal distribution, characterized by many modes. The concept of mode is particularly valuable for analyzing categorical or discrete data, since it provides a measure of central tendency. However, it should be noted that in continuous datasets, the mode may not be well-defined or may not exist.

Common measures of dispersion include a range of statistical indicators used to quantify the extent to which data points deviate from the central tendency. The range is a fundamental measure of dispersion that quantifies the variation within a dataset by calculating the difference between the highest and lowest values. The method offers a preliminary approximation of the dispersion of data, while it is susceptible to the influence of extreme values. The concept of variance involves measuring the average squared deviation between individual data points and the mean value. The aforementioned method offers a more accurate assessment of the dispersion of data, since it considers all values within the dataset. The formula for *Variance* (s^2): $s^2 = \frac{\Sigma(x_i - \bar{x})^2}{N}$ where x_i represents each data point, $\bar{x}$ is the sample mean, and N is the number of data points. The standard deviation is calculated as the square root of the variance and is often favored due to its compatibility with the original data units. The measure in question denotes the average distance between data points and the mean and is a commonly used metric for quantifying dispersion $s = \sqrt{\frac{\Sigma(x_i - \bar{x})^2}{N}}$. The coefficient of variation (CV) is a statistical metric that provides a relative measure of dispersion. It is often used to compare the variability shown by various datasets. The calculation involves the division of the standard deviation by the mean, which is then expressed as a percentage $CV = \frac{s}{\bar{x}} \times 100$. Measures of dispersion play a crucial role in the field of statistics and data analysis as they provide significant insights into the variability, consistency, and spread of data. The comprehension of these measurements enables analysts and researchers to make well-informed conclusions, evaluate the dependability of data, and detect patterns and anomalies

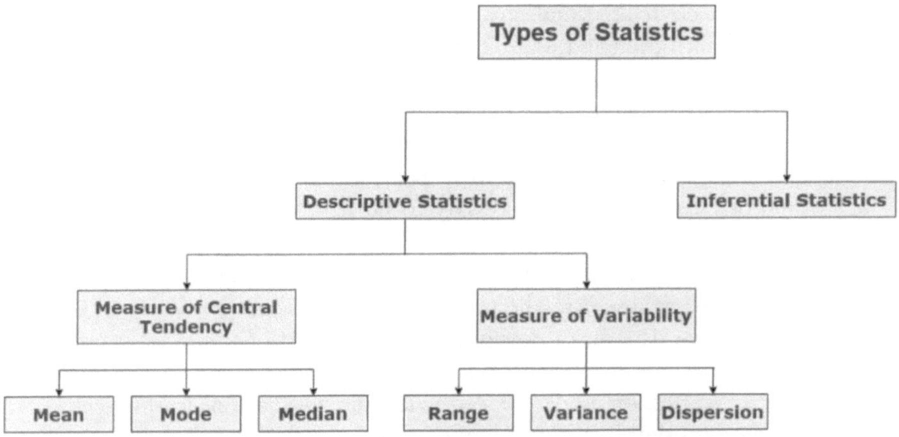

Figure 13.1 Statistics in details.

within datasets. The use of various measures of dispersion is contingent upon the characteristics of the data and the particular objectives of the investigation. The details of basic statistics are shown in Figure 13.1.

2. **Graphical representation:** Graphical representation, sometimes referred to as data visualization or graphical display of data, is a technique used to graphically portray data by using various graphical components such as charts, graphs, and maps. The basic objective of graphical representation is to enhance the comprehensibility, accessibility, and interpretability of intricate data. Bar charts are a kind of graphical representation that visually presents data by using rectangular bars of different lengths. Line charts are a visual representation of data points that are linked by lines. Pie charts are a kind of graphical representation that visually divides data into segments like wedges of a circle. These segments depict the relative proportions of different categories in relation to the whole dataset. Scatter plots are graphical representations that depict individual data points on a two-dimensional coordinate system, facilitating the identification of potential correlations between variables. Histograms are graphical representations that depict the distribution of continuous data. This is achieved by categorizing values into intervals or bins and visually presenting the frequencies or counts associated with each bin. Box plots, also known as box-and-whisker plots, are graphical representations that provide insights into the distribution of data. They effectively summarize key statistical measures such as the median, quartiles, and any possible outliers. Heatmaps are a visual representation technique that uses color intensity to depict data values. They are often utilized to visualize data arranged in a matrix or grid format.

The use of color in data visualization serves many purposes, including the emphasis on certain data points or categories, the creation of visual distinction, and the communication of significance. The incorporation of design principles, namely simplicity, clarity, and consistency, plays a pivotal role in the development of visually captivating and useful graphs and charts. The inclusion of appropriate labeling, such as axis labels, titles, legends, and data source information, is crucial in facilitating the audience's comprehension of the visualization's context and significance. The process of converting data from one format or structure to another is sometimes referred to as data transformation. Prior to generating a graph or chart, it may be necessary to do data transformation or aggregation. One potential need for effective visualization is the need to condense data into distinct categories, specific time intervals, or geographical areas in order to derive relevant insights. The use of logarithmic scaling is advantageous in situations when data encompasses a broad spectrum of values, as it enhances the informativeness of the depiction. The next step is to analyze data using techniques like Regression model, clustering techniques, machine learning, etc.

3. **Regression model:** Regression models play a pivotal role in the realm of data analysis and statistical modeling. Regression analysis is a statistical technique that entails the use of a mathematical model to assess the association between a dependent variable and one or more independent variables within a given dataset. The dependent variable, often known as the response variable, refers to the variable that is being sought to be predicted or explained. The variable in question is often represented by the symbol "Y" and serves as the dependent variable or target variable in the study. Independent variables, also known as predictors, are the factors used in order to forecast or elucidate the dependent variable. These variables are sometimes referred to as explanatory or predictive variables and are often represented by the symbol "X".

   Different types of regression models, such as simple and multiple regression, are used to represent the number of independent variables used. Simple linear regression involves the utilization of a single independent variable, whereas multiple linear regression involves the incorporation of two or more independent variables. The objective is to construct a model that examines the combined impact of many predictors on the dependent variable. The regression equation or model is mathematically represented as:

$$Y = \beta_0 + \beta_1 X + \varepsilon$$

   where β_0 denotes the intercept, β_1 signifies the coefficient associated with the independent variable X, and ε represents the error term or residuals, which encapsulate the discrepancy between the calculated and actual values. Residual analysis is a crucial step after model fitting, since it involves the examination of residuals, which are the discrepancies between observed and anticipated values. The process of residual analysis serves to validate the assumptions of a model and detect any possible outliers or influencing data points. The goodness of fit of a regression model is evaluated using several measures. The R-squared (R^2) is often used as a statistic to assess the extent to which the model accounts for the variability seen in the dependent variable. A higher coefficient of determination (R^2) indicates a more optimal alignment between the observed data and the fitted regression model. A range of regression models exists, including linear regression, logistic regression (for binary outcomes), polynomial regression (for nonlinear relationships), ridge regression, lasso regression (for regularization), and further variations. The selection of a model is contingent upon the characteristics of the data and the inquiry being pursued.

4. **Clustering techniques:** Clustering is a machine learning methodology used to group data points with similar characteristics or patterns. The objective of clustering is to divide a dataset into separate entities, known as clusters, in which data points within the same cluster exhibit more similarity to one another compared to those in other clusters. The approach under consideration is classified as unsupervised learning, indicating that it does not need the inclusion of labelled data during the training process. Its use spans across many domains such as data analysis, image processing, and natural language processing. Different clustering techniques used as:
   a. Hierarchical Clustering: It constructs a hierarchical structure of clusters, often visualized as a dendrogram. The process may be categorized as either agglomerative, which follows a bottom-up approach, or divisive, which follows a top-down approach.
   b. Partitioning Clustering: These algorithms divide data into non-overlapping clusters. K-Means and K-Medoids are popular partitioning techniques.
   c. Density-Based Clustering: These methods identify clusters as areas of higher data density, such as DBSCAN (Density-Based Spatial Clustering of Applications with Noise).
   d. Model-Based Clustering: These algorithms assume that data points are generated from a probabilistic model and attempt to fit such models to the data. Gaussian Mixture Models (GMM) is an example.
   e. Fuzzy Clustering: In contrast to traditional approaches to cluster analysis, fuzzy clustering recognizes the potential for individual data points to belong to several clusters to varying degrees.
   f. Initialization and Optimization: A number of clustering techniques need an initial estimation of cluster centroids or parameters. The optimization procedure involves a repeated refinement of these parameters in order to identify the optimal clustering solution. The convergence criteria and initialization approaches exhibit variability among different algorithms.
5. **Machine learning**: Machine learning is a specialized domain within the realm of artificial intelligence (AI) that centers on the advancement of algorithms and statistical models, which empower computers to enhance their proficiency in a given activity by means of experience or data. The technology in question serves as a foundational component for several artificial intelligence (AI) applications and has garnered extensive utilization across diverse sectors. A flow of model building using different machine learning has been shown in Figure 13.2.

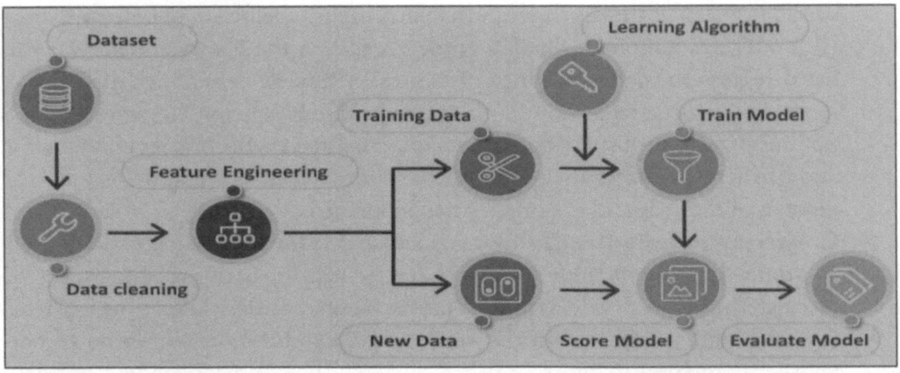

Figure 13.2 Process of machine learning model building.

13.2 Types of Machine Learning

a. **Supervised learning:** It involves the use of labeled instances, whereby the algorithm is provided with input-output pairs consisting of features and matching target values. The objective is for the algorithm to learn the relationship or mapping between the supplied inputs and outputs. Prominent algorithms are often used in several domains including linear regression, decision trees, and neural networks.
b. **Semi-supervised learning**: It is an amalgamation of supervised and unsupervised learning paradigms, whereby the model is trained using a small quantity of labeled data with a substantial quantity of unlabeled data. This particular approach proves to be particularly advantageous in situations when the process of assigning labels to data is cost-prohibitive or requires a significant amount of time.
c. **Unsupervised learning:** It refers to the process of identifying patterns or structures within data without the need for intentional supervision. Clustering and dimensionality reduction are often encountered tasks within the realm of unsupervised learning. K-means clustering and principal component analysis (PCA) serve as illustrative instances.
d. **Reinforcement learning:** It is the process through which an agent acquires the ability to make choices via iterative interactions with its environment. The system is provided with feedback in the form of prizes or penalties, which are contingent upon its behavior. In this context, algorithms such as Q-learning and deep reinforcement learning, namely Deep Q-Networks, are used.
e. **Self-supervised learning**: It refers to an unsupervised learning paradigm whereby the algorithm autonomously generates labels for the data it processes. Pretraining neural networks often involve their use in various activities.

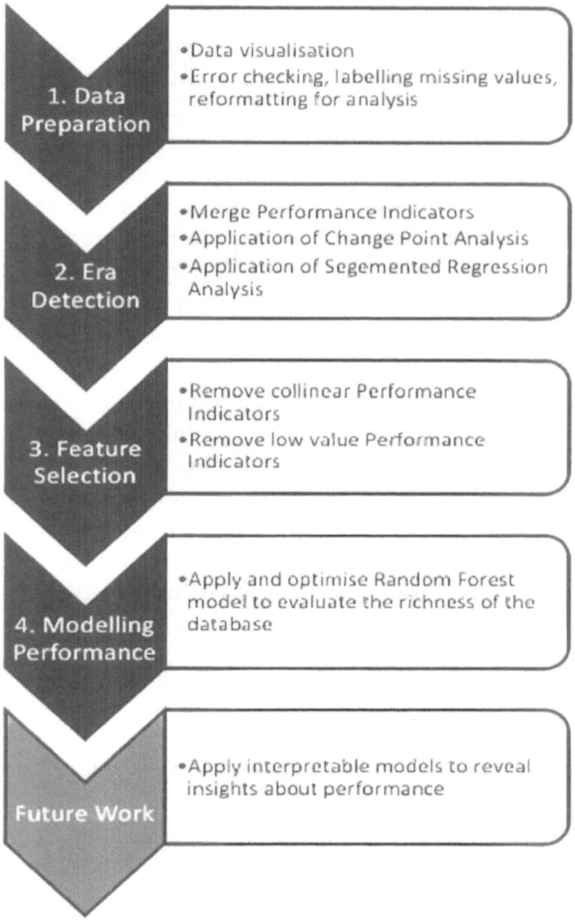

Figure 13.3 Workflow of the data analysis process.

Machine learning approaches need proper model development, feature engineering and optimization of trained models, hyperparameter tuning, etc. (Ahmed et al., 2023). The steps involved in basic data analysis are shown in Figure 13.3.

i. **Model selection:** The selection of an appropriate machine learning algorithm or model architecture for a given job is of utmost importance. The determination of this option is contingent upon the characteristics of the data, the intricacy of the issue, and the intended outcome.
ii. **Feature engineering:** The process of feature engineering encompasses the deliberate selection, transformation, or generation of pertinent characteristics derived from the original data in order to enhance the efficacy

of a machine learning model. The accuracy of the model may be greatly influenced by the use of effective feature engineering techniques.

iii. **Optimization:** During the training phase, the model is subjected to the training data and modifies its internal parameters in order to minimize a pre-established error or loss function. Optimization methods, such as the widely used gradient descent method, are often utilized for this objective.

iv. **Hyperparameter tuning:** The optimization of hyperparameters, which are parameters that are not subject to training, is crucial in order to get the highest level of performance for a model. Methods such as grid search or random search are used in order to identify the optimal hyperparameters.

6. **Inferential statistics**: Inferential statistics is a subfield within the discipline of statistics that focuses on the process of drawing logical conclusions, generating accurate predictions, and forming meaningful inferences about a larger population based on a representative sample of data. Sampling is an important component of statistical analysis, serving as a means to draw inferences about a broader population based on data and observations obtained from a smaller portion, known as a sample. Inferential statistics facilitates the ability of researchers and analysts to make well-informed judgments, evaluate hypotheses, and generate forecasts across many domains, including but not limited to science, business, social sciences, and healthcare shown in Figure 13.4.

Key components of inferential analysis are hypothesis, confidence interval, significance level, and p-value.

a. **Hypothesis testing:** Hypothesis testing is widely recognized as a fundamental application of inferential statistics. In the realm of scientific inquiry, scholars construct a null hypothesis (H0) to encapsulate the prevailing state or absence of an effect, alongside an alternative hypothesis (Ha) to encapsulate the impact

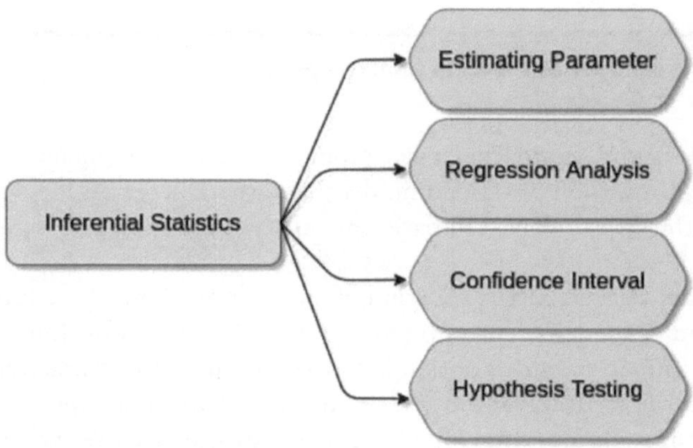

Figure 13.4 Basis of inferential statistics.

they want to investigate. Subsequently, data is gathered from the sample and subjected to statistical tests in order to ascertain if there exists sufficient evidence to reject the null hypothesis in favor of the alternative hypothesis.

b. **Confidence interval:** Confidence intervals are statistical tools that provide a range of values within which a population parameter is expected to lie, with a specified degree of confidence. An example of this concept is a 95% confidence interval for the population mean, which establishes a range of values wherever we possess a 95% level of confidence that the actual population mean is situated.
c. **Significance level (Alpha):** In the context of hypothesis testing, the significance level (often represented as α) refers to the likelihood associated with committing a Type I error, which involves erroneously rejecting a null hypothesis that is actually true. Frequently used significance values in statistical analysis are 0.05 and 0.01.
d. **P-value:** P-values quantify the level of statistical evidence against the null hypothesis. A tiny p-value, usually below the predetermined significance level, provides compelling evidence against the null hypothesis, taking into consideration factors such as sample size, effect size, and the selected significance level.

Application of data analysis in Precision Agriculture: Data analysis plays a crucial part in the field of Precision Agriculture, which is sometimes referred to as Precision Farming or Smart Farming. Precision Agriculture encompasses the use of sophisticated technology and data-centric insights to enhance and optimize several facets of agricultural operations.

Few applications of data analysis in Precision Agriculture

a. **Crop monitoring and management:** It can be achieved through remote sensing and NDVI. Remote sensing techniques including the analysis of satellite and drone pictures play a crucial role in the monitoring of crop health, identification of pest infestations, and evaluation of irrigation requirements. Also, the application of remote sensing (NDVI) is used to assess the vitality of plants and detect regions under physiological strain. The use of data analysis enables farmers to make well-informed choices pertaining to the application of fertilizers and irrigation practices.
b. **Soil analysis:** The analysis of soil data, which may include pH levels, nutrient content, and moisture levels, provides farmers with the ability to make accurate choices on the timing of irrigation and the application of soil amendments.
c. **Precision planting:** It helps to identify the ideal spacing and depth for planting seeds, which ensures equal crop emergence and maximizes yields. This is accomplished via the use of a seed planting grid, Variable Rate Planting algorithms in order to alter the planting density throughout the field and optimize seed distribution dependent on the characteristics of the soil.

There are many parameters that can be optimized such as Irrigation, Pest and Disease Management, Weather Forecasting and Risk Management, Market and Financial Analysis, Sustainability, and Environmental Impact.

13.3 Summary

Data analysis is a fundamental process with broad applications across various domains, including business, healthcare, finance, and science. It involves systematically inspecting, transforming, and interpreting data to reveal valuable insights, patterns, and trends. Key data analysis techniques encompass descriptive statistics for summarizing data and graphical representations for visualization, like histograms and line charts.

In addition, data analysis employs advanced methods such as regression models for prediction and classification, clustering techniques like K-means for grouping similar data points, and machine learning for pattern recognition and prediction. Inferential statistics, including hypothesis testing and confidence intervals, play a pivotal role in supporting decision-making based on sample data. It can be used in precision agricultural practice for crop monitoring, soil analysis, precision planting, and various management practices.

References

Ahmed, B., Haque, A., Iquebal, M. A., Jaiswal, S., Angadi, U. B., Kumar, D., & Rai, A. (2023). DeepAProt: Deep Learning Based Abiotic Stress Protein Sequence Classification and Identification Tool in Cereals. Frontiers in Plant Science, 12(627): 1–12. https://doi.org/10.3389/fpls.2022.1008756.

Ahmed, B., Rai, A., Jaiswal, S., & Iquebal, M. A. (2021). Comparative Analysis of Machine Learning and Deep Learning-Based Classification for Abiotic Stress Proteins. The Indian Journal of Agricultural Sciences, 91(6): 861–866. https://doi.org/10.56093/ijas.v91i6.114287.

Bano, S., & Khan, N. (2018). A Survey of Data Clustering Methods. International Journal of Advanced Science and Technology, 113: 133–142. http://dx.doi.org/10.14257/ijast.2018.113.14.

Bharti, C., Ahmed, B., & Maurya, A. (2021). A Review on Conservation Agriculture (CA) and Sustainable Food Production. Journal of Extension System, 37(1): 21–27. https://doi.org/10.48165/JES.2021.37103.

Bindra, K., & Mishra, A. "A detailed study of clustering algorithms," 2017 6th International Conference on Reliability, Infocom Technologies and Optimization (Trends and Future Directions) (ICRITO), Noida, India, 2017, pp. 371–376. doi: 10.1109/ICRITO.2017.8342454.

Bongiovanni, R., & Lowenberg-DeBoer, J. (2004). Precision Agriculture and Sustainability. Precision Agriculture, 5: 359–387. https://doi.org/10.1023/B:PRAG.0000040806.39604.aa.

Celik, O. (2018). A Research on Machine Learning Methods and Its Applications. Journal of Educational Technology and Online Learning, 1(3): 25–40. https://doi.org/10.31681/jetol.457046

Bharti, Das, P., Banerjee, R., Ahmad, T., Devi, S., & Verma, G. (2023). Artificial Neural Network Based Apple Yield Prediction Using Morphological Characters. Horticulturae, 9(4): 436. https://doi.org/10.3390/horticulturae9040436.

Das, P., Jha, G. K., Lama, A., & Parsad, R. (2023). Crop Yield Prediction Using Hybrid Machine Learning Approach: A Case Study of Lentil (Lens culinaris Medik.). Agriculture, 13: 596. https://doi.org/10.3390/agriculture13030596.

Garg, Ram, & Goyal, Ruchi. (2019). Inferential Statistics As a Measure of Judging the Short-Term Solvency An Empirical Study of Three Steel Companies in India (February 5, 2019). International Journal of Advanced Studies of Scientific Research, 4(1). Available at SSRN: https://ssrn.com/abstract=3329388.

Guetterman, T. C. (2019). Basics of Statistics for Primary Care Research. Family Medicine and Community Health, 7(2): e000067. doi: 10.1136/fmch-2018-000067.

Meletiou-Mavrotheris, M., & Stylianou, D.A. (2003). Graphical Representation of Data: The Effect of the Use of Dynamical Statistics Technological Tool. In *CBLIS Conference Proceedings 2003 Volume I: New Technologies and Their Applications in Education.* 296–306. Nicosia, Cyprus.

Muscatello, D. J., Searles, A., Macdonald, R. et al. (2006). Communicating Population Health Statistics through Graphs: A Randomised Controlled Trial of Graph Design Interventions. BMC Medicine, 4: 33. https://doi.org/10.1186/1741-7015-4-33.

Nagasubramanian, G., Sakthivel, R. K., Patan, R., Sankayya, M., Daneshmand, M., & Gandomi, A. H. (2021). Ensemble Classification and IoT-Based Pattern Recognition for Crop Disease Monitoring System. IEEE Internet of Things Journal, 8(16): 12847–12854.

Vetter, T. R. (2017). Descriptive Statistics: Reporting the Answers to the 5 Basic Questions of Who, What, Why, When, Where, and a Sixth, So What? Anesthesia and Analgesia. 125(5): 1797–1802. doi: 10.1213/ANE.0000000000002471.

Abdul Hakkim, V. M., Abhilash Joseph, E., Ajay Gokul, A. J., & Mufeedha, K. (2016). Precision Farming: The Future of Indian Agriculture. Journal of Applied Biology and Biotechnology, 4(6): 68–72. https://doi.org/10.7324/JABB.2016.40609.

Witte, S. R., & Witte, J. S. (2016). Statistics. 11th ed. pp. 1–456.Wiley.

Chapter 14

Smart Sensing Technologies and Data Analytics for Precision Agriculture

Deva Priya Isravel, Salaja Silas, Titus Issac, Jaspher Willsie Kathrine, and Elijah Blessing Rajsingh

14.1 Introduction

The agricultural industry needs to be modernized now more than ever as agriculture is under increased stress due to climate change to produce enough wholesome food to meet the population's rising demand [1]. Resource depletion and climate change are expected to get worse as the demand for goods rises globally [2]. The urban land growth and increased water demand from other industries have an impact on agriculture. The agricultural risk is high where the resource pressures are most intense. The current methods of production and consumption are not adapted to the effects of climate change and are pushing agriculture in a direction that will have various negative effects on human growth. The rise in income in emerging and developing nations is also increasing demand for a wider variety of food and non-food products [3].

The success of agriculture and future food production depends on how effectively natural resources are maintained and managed. The agricultural society demands improved agricultural output, preservation of the land, reduction of chemical use in crop production, management of irrigation, and efficient use of water resources.

 DOI: 10.1201/9781003504900-14

The increased effectiveness of decision-making, obtaining agricultural records, and spreading modern farming techniques are essential to increasing production quality and quantity and minimizing costs [4]. The low crop yield is caused by a variety of factors that can be mitigated using the latest advanced technologies [5, 6].

Recent advancements in agricultural technology have led to a new management strategy called precision agriculture, which makes use of a wide range of cutting-edge technologies and equipment to track a number of variables and gather data on crop growth. Precision farming creates new instruments for assessing multi-functional factors and allows the application of agricultural inputs in accordance with crop needs, soil conditions, and weather patterns. Precision farming employs precisely measured inputs to boost average yields. It is the science of increasing crop yields via the use of sophisticated sensors and analysis equipment.

Extensive research has been carried out to improve the management practices of agriculture by employing precision agriculture. It enables farmers to improve their agricultural practices and increase average yields compared to conventional methods by using fertilizers and pesticides in precise amounts, at the appropriate locations, and at the right times. Several studies show that production is improved with efficient use of chemical, energy, water, and the environment.

The rest of the chapter is structured as follows. Section 2 discusses the various applications of precision agriculture. The key components of precision agriculture are presented in Section 3. Section 4 highlights the remote sensing techniques. Section 5 examines the different data analytics techniques available for precision agriculture. The conclusion is presented in Section 6.

14.2 Applications of Precision Agriculture

This section elaborates on the various applications of precision agriculture. Modern agricultural techniques have the potential to transform the way that agriculture is done today by automating crop cultivation, water irrigation, disease management, and forecasting based on real-time monitoring of farms. A paradigm shift from manual agricultural practices to automated precision-oriented agricultural practices using recent technologies, such as the Internet of Things (IoT), AI-based agricultural analytics, and sensor and remote sensing-enabled crop monitoring and control, has been made possible by precision agriculture solutions. The major applications are shown in Figure 14.1.

14.2.1 Precision Irrigation

Precision irrigation is an innovative agriculture method that enables the distribution of water and nutrients to the plant at the appropriate time and location and in small, controlled dosages to provide it with the best growing conditions [4]. To handle dry climates, proper irrigation is essential to maintaining agricultural

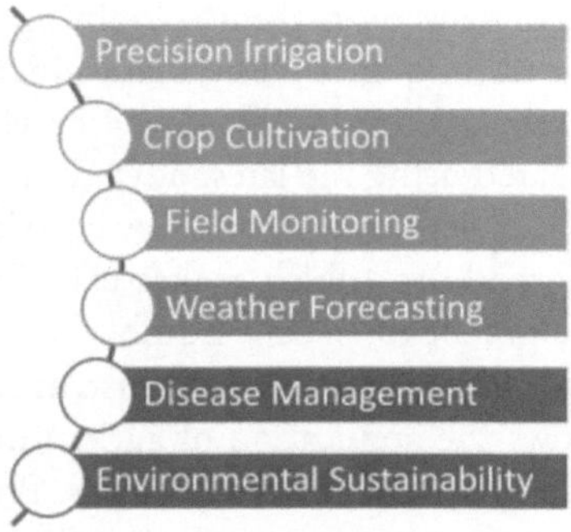

Figure 14.1 Major applications of precision agriculture.

productivity. An effective irrigation system ensures adaptability and durability. Inefficient irrigation system use can jeopardize crop output, abuse water resources, and negatively impact growers' income. Currently, most farmers rely on past performance or scant meteorological knowledge. Therefore, smart tools are required to support decisions on when and how much water to irrigate. With the integration of contemporary technology, such as computational intelligence and agro-hydroinformatics, a sustainable precision irrigation system can be created [7]. With these advanced technologies, a cost-effective irrigation system can be designed to support the effective handling of sensed data about soil, plants, and weather.

14.2.2 Crop Cultivation

Several studies show crop cultivation improvements using precision agriculture. Cultivation technologies were implemented for wheat crop management [8]. From the study, it was found that remote sensing enabled optimal production using distance-based vegetation indices. To estimate crucial agricultural characteristics like crop vigor, water stress, pest and disease severity, and soil conditions, a large number of sensors are currently being used in agriculture. Remote sensing technology was used in olive cultivation to analyze the data and make optimized decisions on farm practices [9]. LiDAR technology has emerged as one of the most cutting-edge fields, especially in laser scanning and object identification systems [8, 10]. As LiDAR can target structures or zones of interest at millimeter resolution, it can be used to identify abnormalities like surface deterioration and vegetation development. They are useful for applications like crop scouting, item detection and categorization, tree volume estimation, crop monitoring, crop maintenance, and navigation.

14.2.3 Field Monitoring

Knowledge of the entire farm cycle is required for making decisions. Sometimes the farm information cannot be obtained from one source but is rather scattered at different places. Insurmountable expenses, sensor capacity, platform stability and dependability, and an inconsistent process for analyzing enormous volumes of data are some of the challenges. Also, getting real-time data from the farm field is crucial for automating the agricultural system. Technologies such as IoT [11], data analytics [12], computer vision [13], drones [14], unmanned aerial vehicles (UAVs) [15], and Variable Rate Technology (VAR) [16] have gained tremendous attention as they involve the farmer more actively in planning and designing the field, taking pictures, and interpreting and analyzing the data. UAVs are the best equipment for field monitoring and inspecting larger farms because of their accessibility, high efficiency, and cost-effectiveness, and they pose less of a threat than traditional aerial platforms [17]. They can provide close-up inspections that are difficult or impossible to get using conventional inspection methods because they can access difficult-to-reach regions.

14.2.4 Weather Forecasting

Weather forecasting is a scientific estimation of the upcoming weather conditions at a certain moment as described by meteorological variables. Knowing the local weather conditions in advance can help safeguard crops and agricultural property and even save lives in emergency situations. Farmers can more effectively plan their daily activities when they have access to precise field-level meteorological data that enables them to assess the workability of the fields [18]. Farmers are motivated to tailor weather predictions to their own requirements for boosting earnings, reducing losses, and raising net farm revenue. Weather information assists farmers in making important decisions, such as when to sow, spray, irrigate, and harvest. The amount of solar, wind, and anaerobic energy used on a farm is often decided by farmers. Therefore, weather knowledge will help farmers build solar panels and wind turbines to gather solar and wind energy. Farmers can utilize energy more effectively by collecting and storing it. Farmers can choose how to use energy from renewable sources that will benefit their operations based on weather predictions.

For daily operations, short-term projections are often the most pertinent, and their accuracy has a big impact on whether or not their crops are successful [19]. Reliable and accurate crop evapotranspiration observation and estimation at various scales are required for modeling water supply management. The sustainable production of rice is in jeopardy due to the growing global shortage of freshwater resources for agriculture. Due to the sensitivity of rice plants to water stress, it is crucial to get short-term forest management information [20].

For planning purposes, such as choosing which crops to sow in a particular season or when to make infrastructure investments, long-term forecasts are crucial. Despite the fact that severe weather can strike suddenly, ruining crops and even affecting the viability of stored seeds, local meteorological predictions assist farmers in anticipating these events and minimizing or even avoiding their negative impacts.

14.2.5 Disease Management

Numerous diseases and pests are threatening important agricultural crops. These diseases have the power to ruin entire harvests, harm crops, and degrade the quality of fruits and vegetables. Knowing the disease you are attempting to avoid or control is crucial [21]. When diseases are not effectively managed or develop reoccurring issues, the cause is frequently incorrectly recognized. To prevent and control plant diseases, pest control mechanisms are followed. Farmers frequently have to deal with many pests or diseases, as well as newly discovered pathogenic strains that are pesticide-resistant and attack the same crop.

However, if plant diseases are accurately diagnosed and detected early, crop losses can be reduced, and specific treatments can be designed to tackle certain pathogens. As the quality and yield of crops depend on accurately identifying disease occurrences in the early stages, this requires specialized knowledge and extensive experience in plant pathology. As a result, an automated system for disease detection in crops will play a crucial role in agriculture by developing an early detection system for disease. These need-based interventions also benefit the economy and the environment. From the literature, it is evident that several data analytics [22], machine learning, deep learning methods [23, 24], and IoT [20, 25] are very prevalent in analysis and detection algorithms to accurately detect disease. Image fusion techniques are used for plant recognition and disease detection [26, 27].

14.2.6 Environmental Sustainability

Sustainable agriculture includes a wide range of agricultural techniques, including traditional practices. Environmentally friendly strategies are encouraged to be practiced among farmers through precision agriculture methodologies [28]. For environmental sustainability, it is essential to encourage consumption of a large proportion of the food produced with the aim of eradicating or at the very least decreasing food waste and its harmful effects. Food waste creates negative effects like the release of greenhouse gases into landfills and the waste of resources used to manufacture food. Different studies show that cutting-edge technologies such as Digital Twin (DT) [29] and IoT [30] are widely used to address increasing population, land degradation, climate change, and food security. These technologies are useful for choosing environmentally sustainable solutions.

14.3 Components of Precision Agriculture

The major key components that enable the automation of agriculture processes are sensors, field robots, drones, Global Positioning System (GPS) and Geographic Information System (GIS), and communication technologies. Figure 14.2 presents

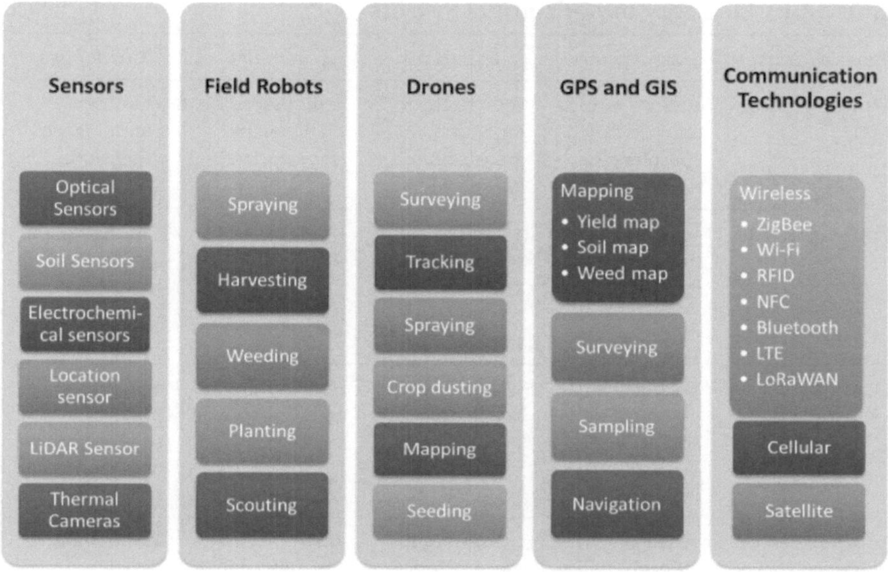

Figure 14.2 Components of precision agriculture.

the components based on their functionalities. Each of these components is detailed in the following section.

14.3.1 Sensors

A wide range of sensors are used in smart agriculture for crop monitoring, environmental monitoring, yield harvesting, weed detection, crop cultivation, soil sensing, etc. The most prevalent sensor and their uses, benefits, and challenges are detailed in Table 14.1.

14.3.2 Field Robots

Field robots automate laborious manual chores and can function completely autonomously. Smart agriculture is one of the main industries making use of this new technology. The agricultural sector deals with numerous issues. Global pricing competition necessitates ongoing pressure to cut operating expenses. At the same time, farmers are under increased pressure to raise yields due to the rapidly expanding population's growing demand for food. Field robots are an essential component of the solutions to these problems. Some of the commonly used field robots employed in precision agriculture are autonomous vehicles, robotic arms, mobile robots, and soft robots.

Table 14.1 Summary of Agricultural Sensors

Types of Sensor	*Description*	*Use Cases*	*Benefits*	*Challenges*
Optical sensor [31]	Measure various attributes of soil, plant leaves	Weed detection, moisture measurement, leaves color	Remote sensing, lightweight, miniaturized	Accuracy, connectivity, integration
Soil sensor [32]	Measure soil moisture and its profile	Irrigation scheduling, pollution control, heat stress control	Timely result, less cost	Assembling and installation
Electrochemical sensor [33]	Measure chemical data, soil nutrients	Soil and crop health, fertilizers	Eco-friendly, rapid response	Calibration, power consumption, environmental factors
Location sensor [34]	Measure altitude, longitude, and latitude	Environment monitoring, yield harvesting, equipment positioning	Real-time location data, accuracy,	Connectivity, power consumption
LiDAR sensor [35]	Measure the distance of the object	Topography, landscaping, crop growth	3D mapping capabilities, Fast scanning	Expensive, battery safety,
Thermal sensor [36]	Measure temperature of farm settings	Ambient monitoring, harvesting	Disease detection, mapping, drought monitoring	Resolution, data processing

Field robots come in a variety of shapes and perform a variety of tasks. Field robots are stepping up and assisting farmers as they face a number of mounting issues by increasing efficiency and bringing in higher crop yields. Precision agriculture is made possible by these field robots, which are utilized on small and large farms. Some of the major use cases of these field robots are weed detection, pollution monitoring, planting, seeding, harvesting, and nursery automation. Figure 14.3 represents the types of field robots and their application for different agricultural practices.

14.3.3 Drones

Drone is called UAV or unmanned aerial systems (UAS). UAV refers to a drone, whereas UAS refers to drones and any equipment they may be used with (such as flying software or cameras). Drones are the ideal instrument for a variety of

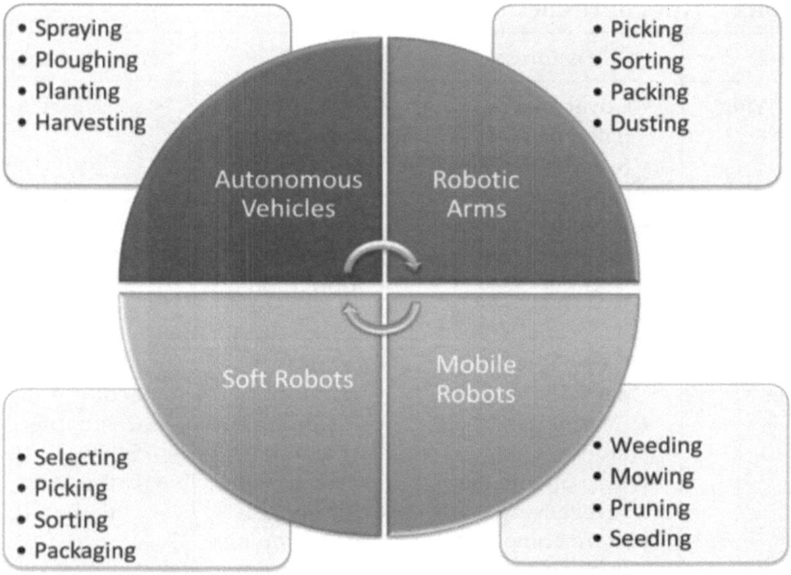

Figure 14.3 Types of field robots.

agricultural demands and present variety of opportunities for the agricultural industry. Common drone uses include pesticide application, soil sampling and fertilization, farm animal observation, real-time aerial imaging, sensor data collection, etc. In the future, airborne imaging, surveying, and transportation are all anticipated to benefit from the use of drones in agriculture. The majority of drones fall into the rotary or fixed-wing categories. A fixed-wing drone resembles an airplane identically, whereas a rotary-style drone is comparable to a helicopter that can have a single rotor or multiple rotors. A hybrid drone is another kind that has both rotary and fixed wings. Table 14.2 details the types of drones, their features, and their uses.

14.3.4 GPS and GIS

The GPS offers agricultural farmers the chance to more accurately control their land and crop production. In order to minimize unnecessary overlap and skips, GPS applications in farming include equipment guidance for sprayers, fertilizer applicators, and tillage implements. GIS are extremely useful in agriculture for a variety of reasons, one of which is their ability to quickly collect and analyze data on farmland. GIS agriculture data is gathered via field equipment on the ground as well as drones and satellites in the air [37]. GIS for agriculture are frequently used by significant agrifood stakeholders to increase short-term earnings and achieve

Table 14.2 Types of Drones

Drones	*Features*	*Uses*	*Popular Models*
Fixed-Wing Drones	• Cover a large patch of land • Faster than copters • Expensive • Large open area for takeoff and landing	• Obtain high-resolution aerial imagery • Pest and disease outbreaks • Topographic surveys	• senseFly eBee Ag • PrecisionHawk Lancaster • Parrot Disco-Pro Ag
Single-Rotor Drones	• Carry heavier payloads • Offer longer flight times • Take off and land vertically • More complex to operate and maintain • More expensive than fixed-wing drones	• Crop monitoring • Scouting • Irrigation management • Soil analysis and mapping • Plant counting	• MQ-8B Fire Scout • Camcopter S-100 • Unmanned Little Bird (ULB)
Multi-Rotor Drones	• Short flight only • Small coverage areas • Multi-dimensional insights	• Capture high-quality images • Crop mapping	• Parrot Bluegrass Fields • Sentera Omni
Hybrid Drones	• Cover large-scale farmland take off and land vertically, • Extended range and endurance	• Crop mapping • Crop scouting • Pesticide spraying	• WingtraOne • DroneVolt HeliPlane • DeltaQuad

long-term objectives of sustainable development. These two techniques are widely used to locate weeds, sample sites, detect insects and diseases, spray fertilizers, and monitor yields.

The capacity to store the in-field variations as geographically encoded data is provided by the GPS. The correct position may be determined and recorded continually. This technology takes agricultural areas and fields into account in greater detail than in the past, thus the user has access to a bigger database. A GIS application is required for storing and managing these data. Making sound agrotechnological judgments is made feasible by the GIS developed by people with computing backgrounds. With the aid of this technology, each area of the field can receive the ideal or almost ideal amount of nutrients and chemicals, as well as the correct cultivation [38]. GIS works with spatial data and uses computer-based technologies to display the gathered data as patterns and relationships. The data is obtained from different sources. Data analytics is used to transform data into information and make based on models and visualization [39].

14.3.5 Communication Technologies

The emerging technologies have a great deal of potential for smart agriculture applications. For precision agriculture applications, the different types of wireless communication technologies used are ZigBee, Wi-Fi, RFID, NFC, Bluetooth, LTE, and LoRaWAN [40, 41]. In order to achieve an improved and more environmentally sustainable agriculture, the ultimate goal is to gather, monitor, and efficiently use essential data for agricultural activities. The technologies in settings like aerial systems, cellular communication, and big data analytics with wireless have been specified. Depending on the field agriculture scenarios, communication technologies are selected and remote sensing is made possible. With the aid of machine learning and big data vision, satellite photos have proven to be a boon for precision agriculture, increasing production efficiency and assisting farmers in making the best choices at the appropriate moments. The combination of satellite images captured and big data analysis can produce astonishing outcomes and help the agricultural sector [42].

14.4 Remote Sensing Techniques

This section explores the various remote sensing techniques to access the crops health and agricultural fields [43]. Figure 14.4 depicts the remote sensing methods with the types of sensors used and their area of implementation for automating the agricultural systems. The remote sensing is carried out using various methods, including Satellite, UAV, UGV, Manned Flight, and Manned Ground vehicles. Images captured from the multispectral, hyperspectral sensor, LIDAR, and RADAR provide environmental factors that can be identified by a number of spectral signatures. Additionally, the development of land resource databases is made easier by GIS and its functional integration with remote sensing. Remote sensing and GIS technologies along with some analytical tools are used to track and analyze some indicators and determine the efficacy of restricted resources like water and land resources.

The methods of remote sensing are an effective tool for processing real-time data combined with on-site information gathered using conventional methods to work around conventional data volume restrictions. Rainfall, evapotranspiration,

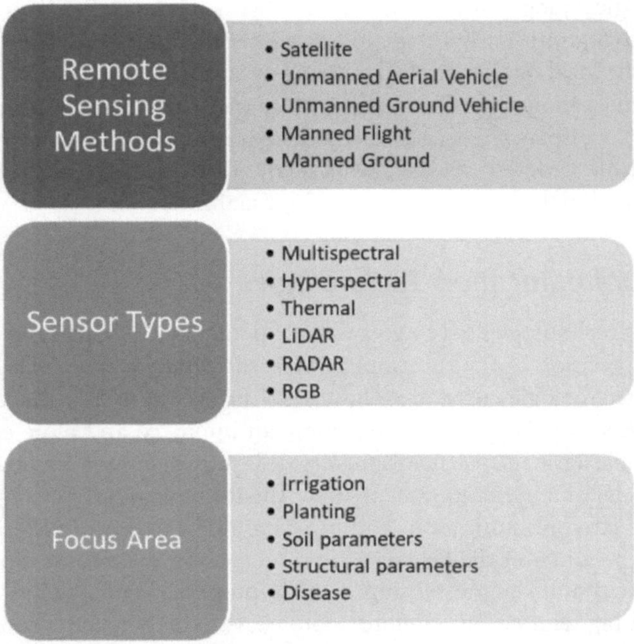

Figure 14.4 Remote sensing techniques.

changes in land use and land cover, and storms can all be monitored and assessed using remote sensing [44]. It enables the preparation of base maps, appraisal of the topography, land use maps, and classification of land degradation. Medium-high resolution remote sensing images have emerged as the primary data source for the precise extraction of agricultural information from satellite remote sensing technology.

Increasingly, early indicators of agricultural deterioration are being found using remote sensing data. Decision support also requires the use of information-sharing and communication channels, as well as other technologies for gathering, processing, distributing, and interpreting critical data [45]. In order to ensure the best use of inputs and maximize farm output, GIS is widely used to determine field variability. Thus, producers may more accurately decide what to plant and in what amounts by using remote sensing, GPS, and other forms of images and time-series data. This helps them use costly inputs like fertilizers and pesticides more effectively.

The process for creating 3D scenarios with any geometrical description and including a particular data source is covered by three-dimensional modeling applied to several remote sensing devices. As such, the assessment does not limit itself to 3D settings that are represented by mesh point clouds. The techniques used for thermal, multispectral, hyperspectral, and visible photography help in reconstructing 3D settings and improve in decision-making process [26].

Crop disease detection and mapping have long been accomplished using remote sensing technology. In addition to being used for early diagnosis and within-season treatment of some crop diseases, aerial and satellite images collected during seasons of cultivation can also be utilized for the control of reoccurring diseases in subsequent seasons [16]. If the disease is stable, site-specific fungicide applications can be made to afflicted regions using precision agriculture technology. The goal of farmers using precision agriculture techniques is to increase crop yields.

14.5 Data Analytics

The application of big data analytics to agriculture is an exciting development, and it is interesting to see how the sector develops and changes over the years. Big data analytics has the power to revolutionize the agricultural industry by giving farmers and agribusinesses the information and understanding they require to make better decisions. The technology helps farmers and agribusinesses make better decisions that boost productivity, cut waste, and enhance sustainability. Precision agriculture is a modern approach to farming that leverages technology to optimize crop production, reduce waste, and improve the sustainability of agriculture.

To increase production, agricultural data collection and analysis using the right statistical methods are essential. Often the agricultural data is gathered from secondary sources like websites, published papers, journals, and other sources, or it can be gathered through primary surveys. Typically, the data is vast, varied, complex, and rich. Because of this, analyzing the data is difficult. Not only may data kinds and formats convey heterogeneity, but it can also be gathered with varying equipment of varying caliber. Large-scale data collection and analysis might present certain difficulties, though, including those related to data security, privacy, integration, and quality. Furthermore, compared to extremely current data, historical data may be defined using various sets of properties. When data is gathered from several locations and times, it may result in discrepancies in naming standards and measurements. Additionally, the data can be offline, classified as static and historical, or online, classified as weather data collected at regular intervals such as weather data or satellite imagery, which is classified as geospatial data because it is spatiotemporal. The reality of data analytics and precision agriculture has significantly changed. The gathering and application of massive volumes of data are prioritized in order to improve agricultural decision-making.

A majority of the aforementioned precision agriculture components in the previous section generate data. And the generated raw data from the precision agriculture components could not be visualised or interpreted directly. Data analytics is the process that comes in aid performing operations including (i) refining, (ii) processing, (iii) interpreting, and (iv) visualizing the data.

Data analytics in precision agriculture is a broad field that involves the use of statistical and computational methods to extract insights from data. Data analytics in precision agriculture can help make informed production decisions, including

- crop selection, planting, growing, and harvesting;
- fertilizer selection, utilization;
- pest and weed prediction, prevention, and eradication;
- financial decision-making (cost of overall agriculture process including end product price);
- weather forecasting (predicting flood, rain, snow, hail storm, etc.);
- Asset management (land identification, soil classification, etc.).

The major types of data analytics techniques employed in precision agriculture are elaborated in this section and shown in Figure 14.5. The data analytic methods with respect to precision agriculture and its applications are presented in the following sections.

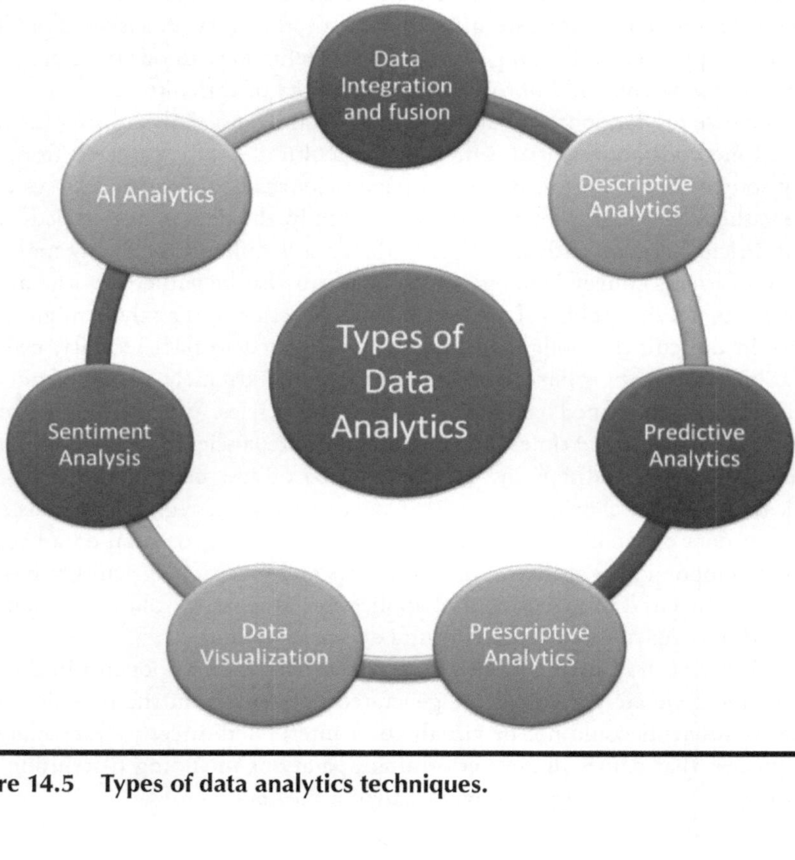

Figure 14.5 Types of data analytics techniques.

14.5.1 Data Integration and Fusion

Data fusion is a data-driven strategy that combines related outputs gathered from several instruments and takes synergies from multiple sources into consideration. It improves the performance and robustness of models used for data interpretation to generate more accurate knowledge about process conditions and performance. Applying suitable data-driven modeling techniques helps in deciphering intricate data matrices for improved process control. The use of multi-sensor data fusion for granule and production processes enables critical analysis of current developments and potential future directions [46]. It assists in handling big analytical datasets in continuous operations in an effective manner. The commonly used fusion methods in precision agriculture are image fusion and sensor fusion.

Image fusion, a subset of information fusion has since found extensive application in the fields of medicine, remote sensing, and the armed forces. With the aid of image fusion technology, significant information from two or more photos can be combined to create a comprehensive agricultural information set. The images captured by a single sensor have low resolution and insufficient information. For instance, infrared pictures record an object's thermal radiation, which is unaffected by variations in light, weather, or other external factors. However, because infrared images usually have a lesser spatial resolution, it is difficult to identify fine details. Therefore, the accuracy of detection is not good.

The use of picture fusion is performed in crop recognition and detection, planting area calculation, plant diseases and pest detection, and livestock health evaluation and classification. Utilizing image-related data from the same scene recorded by several sensors, image fusion techniques take the images as a research object and produce a fused image that may be viewed by humans or subjected to additional computer processing and analysis [27]. Pixel level, feature level, and decision level are the three processing levels that make up image fusion. In order to create a fused image that is more informative, pixel-level image fusion directly synthesizes each sensor's output signal while maintaining stringent registration.

A farm robot that uses object identification, picture fusion, and data augmentation with fused images to correctly estimate a target plant's growth statistics. The smart farm robot makes use of an end-to-end, real-time object detector based on deep learning [47]. Image fusion employing both RGB and depth pictures to differentiate the target plant from other plants with better robustness and accuracy is utilized. Despite the great density of fruits and vegetables in these farms, the data augmentation based on the fused RGB and depth information helps to accurately estimate the growth information from smart farms.

A sensor fusion is another data analytics technique that integrates information from several sources to determine the optimal application and accurate output leading to the collection of more dependable field data. In order to combine and process different types of data, sensor fusion approaches need to rely on sophisticated algorithms, such as artificial intelligence (AI) algorithms [48]. The purpose of this sensor fusion system was to carry out three distinct tasks: (i) object detection, which involved determining whether an object was present; (ii) object classification, which involved categorizing the detected object as belonging to a citrus tree or not and estimating the tree's canopy height and leaf density; and (iii) fruit detection and counting, which involved calculating how many fruits the identified tree produced.

14.5.2 Descriptive Analytics

Descriptive analytics is used to summarize and describe past events. It helps identify patterns and relationships in data and is often used in business intelligence and data mining. Application based on descriptive data analytics can help agriculturist keep track of their field activities. Descriptive analytics helps to keep track of information about planting, fertilizing, harvesting, etc. through sensors. Analytics in terms of the weather forecast, vegetation, the growth rate of plants growth, etc. aids farmers in precise decision-making about their crops. A scouting app ensures easy and simplified scouting by allowing farmers to set, manage, and monitor tasks remotely. It helps to keep track of important information about crops, land cover, land use, field boundary, harvest dynamics, yield prediction, cloud mask, soil type, etc.

Descriptive techniques are frequently used to analyze data and can include constructing tables of means and quantiles, measures of dispersion such as variance or standard deviation, and cross-tabulations or crosstabs. These techniques can be used to examine many different hypotheses, often related to observed differences across subgroups. Specialized descriptive techniques are also used to measure segregation, discrimination, and inequality of data.

14.5.3 Predictive Analytics

Predictive analytics uses statistical models and machine learning algorithms to analyze historical data and make predictions about future events [49]. It is often used in forecasting, such as predicting crop yields in precision farming. Variable rate fertilizer or seed application is a technique powered by predictive data analytics that decides the different amounts of fertilizer or seed to different parts of their fields based on the agricultural productivity map. This technique can help end users optimize their yields and reduce input costs. A widely used predictive analytics tool is regression analysis, which can determine the relationship between two variables (single linear regression) or three or more variables (multiple regressions).

A dataset that includes information from multiple sources may have multiple dimensions. However, when evaluating data for particular goals, care must be taken to choose pertinent dimensions. Hierarchy, sequential properties, and dependence relationships across the dimensions form the foundation of data analysis [50]. From the earlier works, we can understand that Spatial distribution mapping frequently uses the multivariate statistical approach and GIS interpolation techniques [51]. Soil distribution patterns and soil contamination evaluations are gradually using multivariate statistical analysis and GIS-based methodologies.

14.5.4 Prescriptive Analytics

Prescriptive analytics uses optimization and simulation techniques to determine the best course of action for a given situation. Near real-time change detection is another important tool that can help end users including farmers make timely data-driven decisions thanks to remote problem area detection and precision scouting through sensors, field robots, and drones. It is used to optimize solutions and make recommendations for crop selection, crop planting, and harvesting, based on the results of predictive models [39]. The end users can make timely data-driven decisions thanks to remote problem area detection and precision scouting through sensors, field robots, and drones. With this data analytics, end users can quickly identify areas of their fields that are experiencing problems and take corrective action before the problems become more severe. Specific techniques used in prescriptive analytics include optimization, simulation, game theory, and decision-analysis methods [12]. Prescriptive analytics aims to find the best solution given a variety of choices. It can be used to help farmers make more informed decisions about their crops by providing recommendations based on data analysis.

14.5.5 Data Visualization

Data visualization is an essential step in communicating data to farmers and other co-workers. It helps the team members to make decisions effectively. Reporting system management teams usually use predetermined template views to track performance. However, performance dashboards aren't the only applications for data visualization. For instance, a word cloud can be used by an analyst conducting text mining to identify important ideas, patterns, and undiscovered connections in unstructured data. As an alternative, use a graph structure to show how different things in a knowledge graph relate to one another. Different forms of data can be represented in a variety of ways. A wide variety of open-source tools is available for data visualization in order for users to evaluate and comprehend. In recent years, GIS has evolved as a platform for data visualization related to remote sensing for modularization, web access, data-based management, mobility, and miniaturization [52].

Also, including data visualization tools that facilitate regular large data exploration and interaction is a significant challenge for NextGen. When examining enormous datasets, the conventional workflow of loading a file, analyzing it, and computing some analytical parameters might not function well. To locate pertinent results, the analyst might have to experiment with a few different procedures and techniques. A weak coupling between analysis and visualization might cause issues with large data since the processing time of the data can be longer than the data transport time. A subfield of computer science known as "visual analytics" combines interactive visual interfaces, data management, visualization, and analytical algorithms. Visual analytics tools and techniques are used to understand information from large amounts of frequently updated and ambiguous data. Sometimes effectively communicating the assessments made in a timely manner is challenging when there is conflicting data [53]. The usage of many technologies, such as grids, cloud computing, distributed computing, and general-purpose graphics processing units, has led to the convergence of data management and analysis tools.

14.5.6 Sentiment Analysis

Sentiment analysis is one of the most extensively studied topics in natural language processing. Sentiment analysis is the process of employing text mining and natural language processing tools to categorize opinions and mine subjective texts with emotional overtones. Sentiment feature extraction, sentiment polarity classification, sentiment retrieval, generalization, and other tasks are typically included in the tasks examined in sentiment analysis. Due to its great potential, sentiment analysis is applied in the agricultural sector to analyze and understand the farmer's emotions. Sentiment trend analysis of online after-sale reviews of agricultural products may be a useful tool for consumers assessing the quality of these goods as well as for businesses looking to improve and enhance their offerings [54]. The sentiment analysis targets issues such as non-standard expressions and sparse features in agricultural product reviews. Speech rules are used as the basis for a recognition technique to determine consumers' emotional inclinations when assessing agricultural products and to extract consumer desire for certain features from online evaluations. To understand the sentiments shared on social media by the public regarding agricultural produce, protests by farmers [55], etc., can be analyzed using sentiment analysis.

14.5.7 AI Analytics

With a large amount of data, machine learning has developed into an excellent predictive analytical tool. Smart farming encompassing modern agricultural equipment, tools, and algorithms, requires AI analytics such as machine learning and deep learning. A branch of artificial intelligence known as "machine learning" enables a system to learn from input data intelligently in order to detect patterns, relationships, and make judgments without the need for explicit programming [56]. Utilizing powerful machine learning algorithms, we integrate data on pesticides, agricultural productivity, and climate to create a decision-making system. Deep learning is the most extensively utilized application in agriculture [57]. As a result, agricultural productivity can be increased, managed, and enhanced. Deep learning is widely used to handle problems involving picture categorization, feature

Table 14.3 Summary of Data Analytics Techniques for Precision Agriculture

	Data Integration and fusion	*Descriptive Analytics*	*Predictive Analytics*	*Prescriptive Analytics*	*Data Visualization*	*Sentiment Analysis*	*AI Analytics*
Precision Irrigation	✓		✓		✓		✓
Crop Cultivation	✓	✓	✓	✓	✓		✓
Field Monitoring	✓	✓		✓	✓		✓
Weather Forecasting	✓		✓		✓		✓
Disease Management	✓	✓	✓		✓	✓	✓
Environment Sustainability	✓	✓		✓	✓	✓	✓

extraction, transformation, and pattern analysis. Machine learning may be used to classify plant diseases based on different attributes. Preprocessing, such as picture enhancement, color correction, and segmentation, is required before features can be extracted successfully. After feature extraction, a variety of classifiers can be applied. Using different models, real-time measures can be predicted precisely and in a timely manner. Table 14.3 summarizes the different data analytics techniques that can be applied for different applications of precision agriculture. By using these types of analytics together, farmers can optimize their yields and reduce input costs while making more informed decisions about their crops.

14.6 Conclusion

Smart sensing and data analytics are revolutionizing the field of precision agriculture. By leveraging advanced smart sensing technologies and artificial intelligence-based big data analytics, the end-users including farmers can optimize their yields and reduce input costs. This chapter explored the various applications of precision agriculture. The key components of autonomous agriculture are detailed with its use cases. The sensing and data analytics techniques required for improving precision agriculture are addressed in detail. We discovered from the study that every technology has distinct advantages and directions for progress. Choosing the right technology for a specific purpose will yield high productivity and sustainability.

References

1. C. Prakash, L. P. Singh, A. Gupta, and S. K. Lohan, "Advancements in smart farming: A comprehensive review of IoT, wireless communication, sensors, and hardware for agricultural automation," *Sens. Actuat. A Phys.*, vol. 362, p. 114605, 2023, doi: 10.1016/j.sna.2023.114605.
2. R. Seppelt, S. Klotz, E. Peiter, and M. Volk, "Agriculture and food security under a changing climate: An underestimated challenge," *iScience*, vol. 25, no. 12, p. 105551, 2022, doi: 10.1016/j.isci.2022.105551.
3. Y. Jararweh, S. Fatima, M. Jarrah, and S. AlZu'bi, "Smart and sustainable agriculture: Fundamentals, enabling technologies, and future directions," *Comput. Electr. Eng.*, vol. 110, p. 108799, 2023, doi: 10.1016/j.compeleceng.2023.108799.
4. J. Zhang *et al.*, "Challenges and opportunities in precision irrigation decision-support systems for center pivots," *Environ. Res. Lett.*, vol. 16, no. 5, 2021, doi: 10.1088/1748-9326/abe436.
5. E. Ece, I. Eş, and F. Inci, "Microneedle technology as a new standpoint in agriculture: Treatment and sensing," *Mater. Today*, vol. 68, pp. 275–297, 2023, doi: 10.1016/j.mattod.2023.07.002.
6. D. Bhavsar, B. Limbasia, Y. Mori, M. Imtiyazali Aglodiya, and M. Shah, "A comprehensive and systematic study in smart drip and sprinkler irrigation systems," *Smart Agric. Technol.*, vol. 5, p. 100303, 2023, doi: 10.1016/j.atech.2023.100303.
7. E. A. Abioye *et al.*, "Precision irrigation management using machine learning and digital farming solutions," *AgriEngineering*, vol. 4, no. 1, pp. 70–103, 2022, doi: 10.3390/agriengineering4010006.
8. G. Rivera, R. Porras, R. Florencia, and J. P. Sánchez-Solís, "LiDAR applications in precision agriculture for cultivating crops: A review of recent advances," *Comput. Electron. Agric.*, vol. 207, p. 107737, 2023, doi: 10.1016/j.compag.2023.107737.
9. J. H. Al-Ammary, and M. E. Ghanem, "Information and communication technology in agriculture: Awareness, readiness and adoption in the Kingdom of Bahrain," *Arab Gulf J. Sci. Res.*, vol. 2, no. 1, pp. 182–197, 2023, doi: 10.1108/AGJSR-07-2022-0113.
10. T. L. Webster, and G. Dias, "An automated GIS procedure for comparing GPS and proximal LIDAR elevations," *Comput. Geosci.*, vol. 32, no. 6, pp. 713–726, 2006, doi: 10.1016/j.cageo.2005.08.009.
11. J. Xu, B. Gu, and G. Tian, "Review of agricultural IoT technology," *Artif. Intell. Agric.*, vol. 6, pp. 10–22, 2022, doi: 10.1016/j.aiia.2022.01.001.
12. N. Nyoman Kutha Krisnawijaya, B. Tekinerdogan, C. Catal, and R. van der Tol, "Data analytics platforms for agricultural systems: A systematic literature review," *Comput. Electron. Agric.*, vol. 195, p. 106813, 2022, doi: 10.1016/j.compag.2022.106813.
13. D. Tirkey, K. K. Singh, and S. Tripathi, "Performance analysis of AI-based solutions for crop disease identification, detection, and classification," *Smart Agric. Technol.*, vol. 5, p. 100238, 2023, doi: 10.1016/j.atech.2023.100238.
14. A. Hafeez *et al.*, "Implementation of drone technology for farm monitoring & pesticide spraying: A review," *Inf. Process. Agric.*, vol. 10, no. 2, pp. 192–203, 2023, doi: 10.1016/j.inpa.2022.02.002.
15. A. Rejeb, A. Abdollahi, K. Rejeb, and H. Treiblmaier, "Drones in agriculture: A review and bibliometric analysis," *Comput. Electron. Agric.*, vol. 198, p. 107017, 2022, doi: 10.1016/j.compag.2022.107017.

16. C. Yang, "Remote sensing and precision agriculture technologies for crop disease detection and management with a practical application example," *Engineering*, vol. 6, no. 5, pp. 528–532, 2020, doi: 10.1016/j.eng.2019.10.015.
17. A. Mukherjee, S. Misra, and N. S. Raghuwanshi, "A survey of unmanned aerial sensing solutions in precision agriculture," *J. Netw. Comput. Appl.*, vol. 148, p. 102461, 2019, doi: 10.1016/j.jnca.2019.102461.
18. M. Chen *et al.*, "A reinforcement learning approach to irrigation decision-making for rice using weather forecasts," *Agric. Water Manag.*, vol. 250, p. 106838, 2021, doi: 10.1016/j.agwat.2021.106838.
19. M. Javaid, A. Haleem, R. P. Singh, and R. Suman, "Enhancing smart farming through the applications of Agriculture 4.0 technologies," *Int. J. Intell. Networks*, vol. 3, pp. 150–164, 2022, doi: 10.1016/j.ijin.2022.09.004.
20. N. K. Nawandar, and V. R. Satpute, "IoT based low cost and intelligent module for smart irrigation system," *Comput. Electron. Agric.*, vol. 162, pp. 979–990, 2019, doi: 10.1016/j.compag.2019.05.027.
21. H. Waheed, N. Zafar, W. Akram, A. Manzoor, A. Gani, and S. U. Islam, "Deep learning based disease, pest pattern and nutritional deficiency detection system for 'Zingiberaceae' crop," *Agric.*, vol. 12, no. 6, 2022, doi: 10.3390/agriculture12060742.
22. R. Akhter, and S. A. Sofi, "Precision agriculture using IoT data analytics and machine learning," *J. King Saud Univ. Comput. Inf. Sci.*, vol. 34, no. 8, pp. 5602–5618, 2022, doi: 10.1016/j.jksuci.2021.05.013.
23. P. Bharman, S. Ahmad Saad, S. Khan, I. Jahan, M. Ray, and M. Biswas, "Deep learning in agriculture: A review," *Asian J. Res. Comput. Sci.*, vol. 13, no. 2, pp. 28–47, 2022, doi: 10.9734/ajrcos/2022/v13i230311.
24. A. Vij, S. Vijendra, A. Jain, S. Bajaj, A. Bassi, and A. Sharma, "IoT and machine learning approaches for automation of farm irrigation system," *Procedia Comput. Sci.*, vol. 167, pp. 1250–1257, 2020, doi: 10.1016/j.procs.2020.03.440.
25. G. Nagasubramanian, R. K. Sakthivel, R. Patan, M. Sankayya, M. Daneshmand, and A. H. Gandomi, "Ensemble classification and IoT-based pattern recognition for crop disease monitoring system," *IEEE Internet of Things Journal*, vol. 8, no. 16, pp. 12847–12854, 2021.
26. J. M. Jurado, A. López, L. Pádua, and J. J. Sousa, "Remote sensing image fusion on 3D scenarios: A review of applications for agriculture and forestry," *Int. J. Appl. Earth Obs. Geoinf.*, vol. 112, 2022, doi: 10.1016/j.jag.2022.102856.
27. D. Li, Z. Song, C. Quan, X. Xu, and C. Liu, "Recent advances in image fusion technology in agriculture," *Comput. Electron. Agric.*, vol. 191, p. 106491, 2021, doi: 10.1016/j.compag.2021.106491.
28. S. Lago-Olveira, S. R. M. El-Areed, M. T. Moreira, and S. González-García, "Improving environmental sustainability of agriculture in Egypt through a life-cycle perspective," *Sci. Total Environ.*, vol. 890, 2023, doi: 10.1016/j.scitotenv.2023.164335.
29. W. Purcell, T. Neubauer, and K. Mallinger, "Digital twins in agriculture: Challenges and opportunities for environmental sustainability," *Curr. Opin. Environ. Sustain.*, vol. 61, p. 101252, 2023, doi: 10.1016/j.cosust.2022.101252.
30. S. A. Jaishetty, and R. Patil, "IoT sensor network based approach for agricultural field monitoring and control," *IJRET*, vol. 5 no. 6, pp. 45–48, 2016. Available: http://ijret.esatjournals.org

31. R. Umapathi, B. Park, S. Sonwal, G. M. Rani, Y. Cho, and Y. S. Huh, "Advances in optical-sensing strategies for the on-site detection of pesticides in agricultural foods," *Trends Food Sci. Technol.*, vol. 119, pp. 69–89, 2022, doi: 10.1016/j.tifs.2021.11.018.
32. S. Datta, and S. Taghvaeian, "Soil water sensors for irrigation scheduling in the United States: A systematic review of literature," *Agric. Water Manag.*, vol. 278, p. 108148, 2023, doi: 10.1016/j.agwat.2023.108148.
33. A. Rabak *et al.*, "Sensor system for precision agriculture smart watering can," *Results Eng.*, vol. 19, p. 101297, 2023, doi: 10.1016/j.rineng.2023.101297.
34. M. Z. Ali, H. J. Chu, Tatas, and T. J. Burbey, "Spatio-temporal estimation of monthly groundwater levels from GPS-based land deformation," *Environ. Model. Softw.*, vol. 143, no. 1, p. 105123, 2021, doi: 10.1016/j.envsoft.2021.105123.
35. S. Debnath, M. Paul, and T. Debnath, "Applications of LiDAR in agriculture and future research directions," *J. Imaging*, vol. 9, no. 3, 2023, doi: 10.3390/jimaging9030057.
36. S. Khanal, J. Fulton, and S. Shearer, "An overview of current and potential applications of thermal remote sensing in precision agriculture," *Comput. Electron. Agric.*, vol. 139, pp. 22–32, 2017, doi: 10.1016/j.compag.2017.05.001.
37. M. I. Ramos, A. J. Gil, F. R. Feito, and A. García-Ferrer, "Using GPS and GIS tools to monitor olive tree movements," *Comput. Electron. Agric.*, vol. 57, no. 2, pp. 135–148, 2007, doi: 10.1016/j.compag.2007.03.003.
38. M. Neményi, P. Mesterházi, Z. Pecze, and Z. Stépán, "The role of GIS and GPS in precision farming," *Comput. Electron. Agric.*, vol. 40, no. 1–3, p. 45–55, 2003, doi: 10.1016/S0168-1699(03)00010-3.
39. R. Sharma, S. S. Kamble, and A. Gunasekaran, "Big GIS analytics framework for agriculture supply chains: A literature review identifying the current trends and future perspectives," *Comput. Electron. Agric.*, vol. 155, pp. 103–120, 2018, doi: 10.1016/j.compag.2018.10.001.
40. W. Tao, L. Zhao, G. Wang, and R. Liang, "Review of the internet of things communication technologies in smart agriculture and challenges," *Comput. Electron. Agric.*, vol. 189, p. 106352, 2021, doi: 10.1016/j.compag.2021.106352.
41. H. El Bilali, and M. S. Allahyari, "Transition towards sustainability in agriculture and food systems: Role of information and communication technologies," *Inf. Process. Agric.*, vol. 5, no. 4, p. 456–464, 2018, doi: 10.1016/j.inpa.2018.06.006.
42. D. Vieira, R. Orjuela, M. Spisser, and M. Basset, "Positioning and attitude determination for precision agriculture robots based on IMU and two RTK GPSs sensor fusion," *IFAC-PapersOnLine*, vol. 55, no. 32, p. 60–65, 2022, doi: 10.1016/j.ifacol.2022.11.115.
43. E. Anastasiou, A. T. Balafoutis, and S. Fountas, "Trends in remote sensing technologies in olive cultivation," *Smart Agric. Technol.*, vol. 3, p. 100103, 2023, doi: 10.1016/j.atech.2022.100103.
44. Y. Pan, X. Wang, L. Zhang, and Y. Zhong, "E2EVAP: End-to-end vectorization of smallholder agricultural parcel boundaries from high-resolution remote sensing imagery," *ISPRS J. Photogramm. Remote Sens.*, vol. 203, pp. 246–264, 2023, doi: 10.1016/j.isprsjprs.2023.08.001.
45. M. P. Wachowiak, D. F. Walters, J. M. Kovacs, R. Wachowiak-Smolíková, and A. L. James, "Visual analytics and remote sensing imagery to support community-based research for precision agriculture in emerging areas," *Comput. Electron. Agric.*, vol. 143, pp. 149–164, 2017, doi: 10.1016/j.compag.2017.09.035.

46. M. Asachi, and M. Alonso Camargo-Valero, "Multi-sensors data fusion for monitoring of powdered and granule products: Current status and future perspectives," *Adv. Powder Technol.*, vol. 34, no. 7, p. 104055, 2023, doi: 10.1016/j.apt.2023.104055.
47. S. Cho *et al.*, "Plant growth information measurement based on object detection and image fusion using a smart farm robot," *Comput. Electron. Agric.*, vol. 207, p. 107703, 2023, doi: 10.1016/j.compag.2023.107703.
48. V. Partel, L. Costa, and Y. Ampatzidis, "Smart tree crop sprayer utilizing sensor fusion and artificial intelligence," *Comput. Electron. Agric.*, vol. 191, p. 106556, 2021, doi: 10.1016/j.compag.2021.106556.
49. I. K. Nti, A. Zaman, O. Nyarko-Boateng, A. F. Adekoya, and F. Keyeremeh, "A predictive analytics model for crop suitability and productivity with tree-based ensemble learning," *Decis. Anal. J*, vol. 8, p. 100311, 2023, doi: 10.1016/j.dajour.2023.100311.
50. S. Hira, and P. S. Deshpande, "data analysis using multidimensional modeling, statistical analysis and data mining on agriculture parameters," *Procedia Comput. Sci.*, vol. 54, p. 431–439, 2015, doi: 10.1016/j.procs.2015.06.050.
51. A. Saha, B. Sen Gupta, S. Patidar, and N. Martínez-Villegas, "Optimal GIS interpolation techniques and multivariate statistical approach to study the soil-trace metal(loid) s distribution patterns in the agricultural surface soil of Matehuala, Mexico," *J. Hazard. Mater. Adv.*, vol. 9,. p. 100243, 2023, doi: 10.1016/j.hazadv.2023.100243.
52. Y. Huang, Z. Chen, T. Yu, X. Huang, and X. Gu, "Agricultural remote sensing big data: Management and applications," *J. Integr. Agric.*, vol. 17, no. 9, p. 1915–1931, 2018, doi: 10.1016/S2095-3119(17)61859-8.
53. S. J. C. Janssen *et al.*, "Towards a new generation of agricultural system data, models and knowledge products: Information and communication technology," *Agric. Syst.*, vol. 155, pp. 200–212, 2017, doi: 10.1016/j.agsy.2016.09.017.
54. Y. Cao, Z. Sun, L. Li, and W. Mo, "A study of sentiment analysis algorithms for agricultural product reviews based on improved BERT model," *Symmetry (Basel).*, vol. 14, no. 8, 2022, doi: 10.3390/sym14081604.
55. A. S. Neogi, K. A. Garg, R. K. Mishra, and Y. K. Dwivedi, "Sentiment analysis and classification of Indian farmers' protest using twitter data," *Int. J. Inf. Manag. Data Insights*, vol. 1, no. 2, p. 100019, 2021, doi: 10.1016/j.jjimei.2021.100019.
56. R. Aworka *et al.*, "Agricultural decision system based on advanced machine learning models for yield prediction: Case of East African countries," *Smart Agric. Technol.*, vol. 2, p. 100048, 2022, doi: 10.1016/j.atech.2022.100048.
57. I. Ahmed, and P. K. Yadav, "A systematic analysis of machine learning and deep learning based approaches for identifying and diagnosing plant diseases," *Sustain. Oper. Comput.*, vol. 4, p. 96–104, 2023. doi: 10.1016/j.susoc.2023.03.001.

Chapter 15

Knowledge Visualization Tools Relating to Agriculture

Prithi Samuel, A.K. Reshmy, Eugene Berna I., Ravikumar S., and Udendhran R.

15.1 Introduction

The prevalence of big data, the proliferation of machine learning applications, the pervasiveness of artificial intelligence in everyday speech, and the widespread availability of internet-connected devices in the home are all signs that the future has arrived. High-performance computing is becoming offered as a service, and the cloud is being utilized to run enterprises and organizations (Rodriguez-Mazahua et al, 2016). In contrast to other industries, agriculture is marked by a high degree of uncertainty due to climate and environmental factors, as well as a constantly shifting supply and demand for food related to crop and livestock growing and breeding times (Astill et al., 2020; Cockburn 2020; Kamble et al., 2020; Newton et al., 2020; Pylianidis et al., 2021; Saiz-Rubio and Rovira-Más, 2020) are only a small subset of the many applications of big data technologies in the agriculture sector. They found that big data analysis was still in its infancy and hadn't been widely used in farming at the time. In addition, they noted a lack of consistency and administration in agricultural information, a shortage of reliable infrastructure, and a disparity in people skills and resources.

 DOI: 10.1201/9781003504900-15

The bulk of the reviewed big data apps were either still in the prototype phase or were too narrow in scope to properly handle agricultural risk. Therefore, the target audience had not yet warmed up to them totally. Since then, advances in knowledge, algorithms, best practices, and computational power have allowed for improved large data management technologies (Oussous et al., 2018). Numerous cutting-edge big data and smart technology scenarios for the agriculture sector have been developed in various parts of the world (Lezoche et al., 2020).

15.2 Big Data Visualization

15.2.1 Big Data's Use in the Agriculture Sector

Big data in the agricultural sector is entirely dependent on utilizing information, technology, and analytics to provide farmers with meaningful information. Big data can be used to gather information about the agriculture sector, or it can help any particular area or segment increase its effectiveness. Big data uses data mining techniques to generate such crucial data. Using this process, you may distill large amounts of data into consumable forms while identifying the key patterns in the data. The big data mechanism uses a variety of contemporary systems, including machine learning statistics, artificial intelligence, and more (Rodríguez-Mazahua et al., 2016; Sjoukje et al., 2022).

15.2.1.1 Important Issues in Applying Big Data in the Agriculture Sector

- One of the main challenges with farm management information systems is the generation of qualitative data. Big data in real time can only do so much to solve this issue.
- Professional domain specialists and data scientists were necessary due to the vast amount of unstructured and heterogeneous data.
- Developing nations are having trouble coming up with solutions that farmers can afford.
- A successful firm needs to sustainably integrate data from numerous sources, but this is one of the Herculean jobs.

15.2.2 Big Data's Effects on the Agriculture Sector

Farmers can save money by using smart farming and precision agriculture techniques, but they can also seize new economic opportunities. Figure 15.1 shows the big data effects on the agriculture sector.

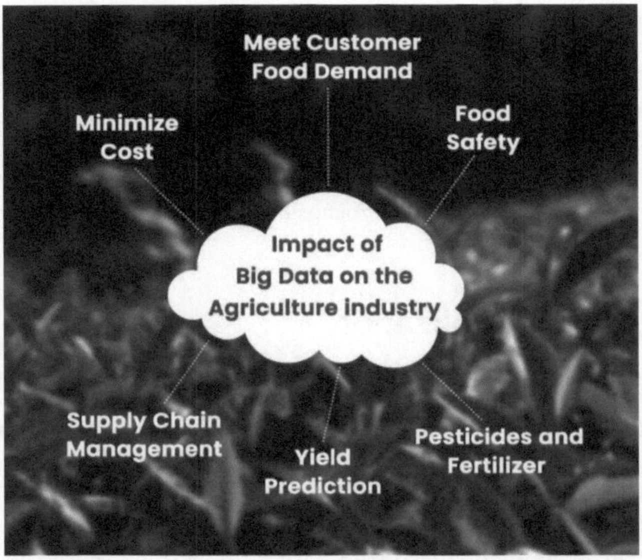

Figure 15.1 Big Data's effects on the agriculture sector.

15.2.2.1 Supply Chain Governance

The introduction of Big Data helps to improve the efficiency of the agricultural supply chain by providing better options for tracking and optimizing delivery routes for trucks. This facilitates easy tracking of food products from their point of origin to the consumer, helping to prevent food waste. Farmers get up-to-the-minute data on their products, while stores and wholesalers get reliable data on market trends they can use to improve their offerings. In conclusion, using big data in agriculture helps companies keep up with the right level of output, optimize their food distribution systems, and save transit time.

McKinsey estimates that annually, one-third of the food produced for human consumption is either wasted or lost. A sad fact considering the industry's efforts to fix the supply-demand imbalance. To combat this, food production and distribution cycles need to be accelerated. Big data can help supply chain efficiency by tracking and optimizing the paths used by delivery trucks.

The agricultural industry is being affected by several sorts of technology. While smart farming can encompass all aspects of the agricultural supply chain, precision agriculture focuses more on how crops are produced (Jerhamre et al., 2022). During the manufacturing process, automated systems monitor and display data to demonstrate performance and flag issues with critical machinery. When handling perishables like seeds, plants, and food, preserving their freshness is a top priority. Fleet management software is adjusted by farmers and distributors utilizing big data to increase delivery reliability. Advanced mapping of animal and vehicle locations is made possible by big data tracking technology, smart meters, and GPS-based analytics.

15.2.2.2 Fulfilling Customer Food Demand

The greatest method to meet the rising food demand is to utilize the acreage that already exists in an efficient manner. But the question is, how does big data help? Changing weather patterns, rainfall, fertilizer use, rainfall, and other aspects that affect agricultural productivity are all recorded by farmers using big data (Kamilaris et al., 2017). Farmers may use this information to make educated decisions that will help them get the most out of their land.

15.2.2.3 Fertilizer and Pesticides

Chemicals are used in modern farming to safeguard and nurture crops. However, if pesticides and fertilizers are not handled correctly, they can potentially have harmful effects on the soil and plants. Big Data enables farmers to choose the best pesticides, at the right time and place. Farmers can prevent the abuse of chemicals by using this kind of monitoring. Additionally, it helps farmers make more money by sparing them the price of buying needless pesticides.

The negative impacts of pesticide use on the ecosystem have caused much debate. Using big data, farmers can be guided as to when, where, and how much pesticides to use (Ravikumar et al., 2023; Sowmia et al., 2023) to mitigate this problem.

Farmers can reduce their reliance on chemicals in food production by keeping track of their use as per regulatory regulations. Profitability rises as a result of the fact that fewer pests and weeds are a boon to agricultural output.

15.2.2.4 Forecasting Yield

When discussing the technologies and methods used to collect and analyze data on the weather, chemicals, vegetation, and other elements, we talk about yield prediction. Sensor-based data collection reduces the requirement for human labor and enhances yield prediction.

15.2.2.5 Reduce Costs

When sophisticated technology is utilized in agriculture, the farm and its resources inevitably develop. This will have an immediate effect on the farmers' expenses. Big data increases productivity, which helps farmers increase key procedures while also saving money.

15.2.2.6 Food Security

The primary issues that millions of individuals face each year are diseases and illnesses related to food. Farmers can rapidly see germs and contamination signals thanks to modern farming techniques. Data on humidity, temperature, and chemical use that affect the health of developing plants can be collected to make this determination.

Big data applications have a broad range, and we've only just scratched the surface of their potential. A true game changer in agricultural techniques could be the capacity to track tangible objects, gather real-time data, and foresee possible outcomes. Let's have a peek at some practical applications of big data.

Big data's insights into food production enable consumers to feel confident about the security and safety of their food supply. Devices like sensors, drones, and smartphones are used to collect data in fields and smart agricultural facilities at precise locations. Because of this, businesses may collect and consider accurate information about aspects like humidity, temperature, chemical exposure, and more. Customers can also discover more about the product's origins, production methods, and creators with the help of data. Extra motivation for manufacturers and logistics experts to maintain quality.

15.2.3 How Big Data May Benefit Farming

Policymakers and business leaders are looking to innovations like the Internet of Things (IoT), big data, analytics, and cloud computing to help soften the blow of rising food demand and shifting climatic patterns.

The data collection phase of this process is aided by IoT devices. Sensors attached to farm equipment, fields, soil, and plants collect data in real time.

Second, analysts link the vast amounts of data they collect with other data saved in the cloud, such as weather data and pricing models, to spot patterns.

In the end, the problem can be contained with the help of these tendencies and insights. They help pinpoint existing problems including operational gaps and soil quality concerns, and they contribute to the development of prediction algorithms that can offer early warnings of impending crises.

At a CAGR (Compound Annual Growth Rate) of 16.2%, the agricultural analytics market is expected to expand from $585 million in 2018 to $1236 million by 2023.

15.3 Utilization of Precision Agriculture and Big Agriculture Data

Precision agriculture instruments have become increasingly important to farmers in recent years. Precision in crop monitoring, machine operation and inspection, field assessment, and input management are all made possible by the use of global positioning system (GPS) signals and other technology. Productivity and income might be significantly boosted as a result of this.

As the technology spreads, farmers are collecting massive amounts of data—far more than can be processed using conventional methods of analysis. Therefore, many businesses in the precision agriculture industry must meet the growing demand for their products and services, which includes anything from hardware to research and consulting. With the abundance of data on particular processes and domains come questions regarding their use, their users, and their goals.

University of Nebraska–Lincoln researchers recently looked into these barriers to adoption and farmer attitudes toward precision agriculture technologies. The university's Undergraduate Creative Activities and Research Experience program awards students with modest funding for their work on research and creative projects under the guidance of faculty mentors.

Farmers who attended the 2015 Nebraska Extension Fremont Corn Expo, the 2015 Nebraska Extension Precision Ag Data Management Workshops, the 2015 Nebraska Extension Ag Technology Conference, and the 2015 Nebraska Extension Crop Production Clinics were surveyed for this study. This is why the survey's population comprised Nebraska farmers who were involved with UNL Extension rather than being selected at random.

Overall, 135 proposals were presented across all sessions, with 126 of those being viable options based on the reported county of operation in Nebraska. The survey's preliminary findings and analysis offer a useful insight into Nebraskans' attitudes toward and use of precision technology at the moment. In a typical operation of 1,507 acres, producers who responded to the study reported having an average of 1,247 acres of row crops. Forty-seven percent of those acres were privately owned, 26% were rented for cash, 22% were leased for crop shares, and 5% were farmed on a custom basis. The three easternmost crop districts in Nebraska received the most responses (79%). These are also the regions where the vast bulk of Nebraska's farmers and agricultural output are located.

15.3.1 Adoption of Precision Agriculture Technology

Several questions were directed at the creators with regards to their usage of technology. Several precision agriculture and ag data management technologies are currently available, and their adoption rates are summarized in Figure 15.2.

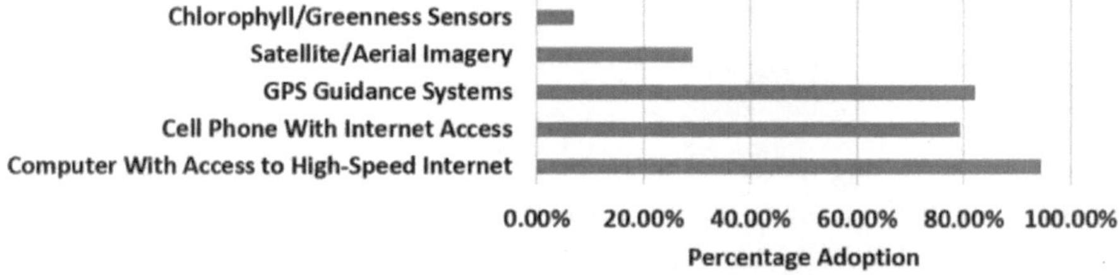

Figure 15.2 Use of precision agriculture technology.

Two examples of widely accepted technology found in the survey replies include methods for monitoring soil (98%) and access to the internet at high speeds (94%). To know if they are used specifically for managing precision agriculture techniques, more study is required. Seventy-five percent of those reporting using soil sampling also mentioned using grid sampling processes, but soil sampling might include many different approaches. Yield monitors, maps, and GPS navigation devices followed as the next most popular practices, each with acceptance rates of over 80%. In order to collect baseline information for analysis and management purposes in the future, a common first step is to use yield monitors and maps.

Field productivity is increased, operator fatigue is reduced, and applications are made more precisely thanks to guiding systems. Many respondents to the poll indicated that they often employed autosteer and automatic section controls and that their usability was much enhanced by guidance systems. In terms of planting (population or hybrid), fertilization, lime (rate and product), and irrigation, 68% of respondents have used variable rate (VR) technology. Drones are being used in more and more industries for exploration, analysis, and management decisions, but satellite photography and plant sensing are being largely ignored.

Precision agricultural technology used by farmers, and how it is used, will be the subject of future research and in-depth analysis. The use of yield monitors is suggested as a first step in compiling a production history that may then be layered with data on factors like as soil composition, climate, hybrid and variety performance, and other production methods. Having access to these data layers may make using guidance systems and variable rate applications more appealing. While yield monitors and variable rate applications may improve operator performance and reduce wasteful input usage, autosteer and section controls may do the same thing regardless of their use. More study is needed to better understand the producers' decisions and the technology that drives the spread of others.

15.3.2 Which Maps Can Be Used by Agricultural Businesses?

Older farmers are more knowledgeable about the land and can assess the soil with their unaided eyes. However, when dealing with hundreds of georeferenced indicators and thousands of square kilometers, agribusinesses should rely on various forms of precise maps to monitor key land features on a broad scale.

15.3.2.1 Soil Zone Maps

In order for farmers to be aware of detrimental irrigation and other soil issues that could harm their long-term production plan, they need soil zoning to observe what is happening at the soil surface. Providers of mapping services for agriculture can create soil zone maps by gridding out fields and taking soil samples from each grid block, which is depicted in Figure 15.3. You'll have more grid blocks for more precise soil zone maps as you gather more samples.

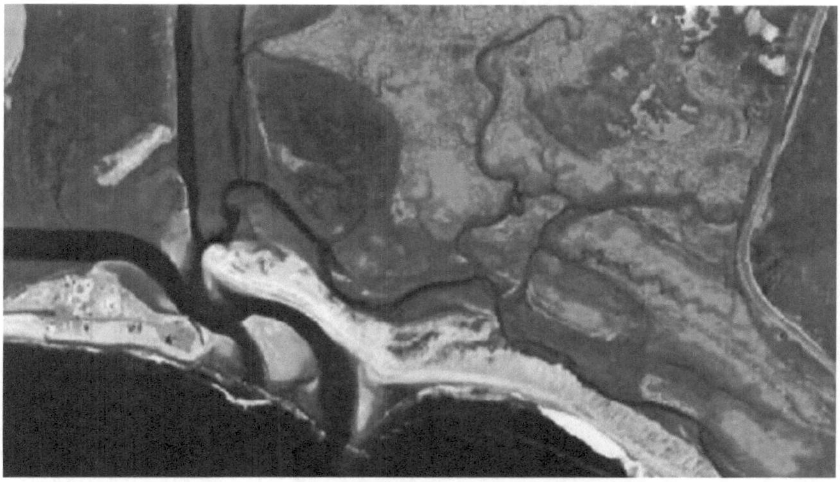

Figure 15.3 Soil zone map.

15.3.2.2 GIS and GPS Maps

GIS and GPS maps are a gold mine of reliable data for farmers in terms of resource availability, crop yield, soil moisture, and harvesting progress. As can be shown in Figure 15.4, agribusinesses can use GIS technologies and web analytics systems to foresee crop growth and future agricultural production. Google Earth Pro and other international commercial products are vital to the field of agricultural geoinformatics, as are specialized GIS software solutions.

Figure 15.4 GIS and GPS maps.

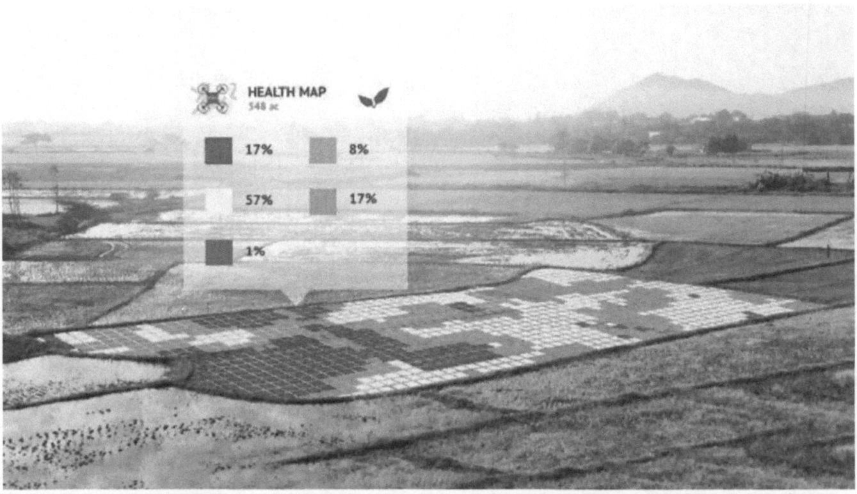

Figure 15.5 Satellite and drone imagery.

15.3.2.3 Satellite and Drone Imagery

Images from satellites and drones are perfectly synced. For instance, farmers may use satellite data to forecast weather changes in advance and prepare fields for heavy rain, saving the loss of nutrients and seeds. In order to track drainage and locate crops harmed by the weather, they can also compare satellite photos taken before and after rain, which is shown in Figure 15.5. In contrast to satellite observations, photos captured by drones at lower altitudes are not frequently obscured by clouds. In order to take exact photographs of crop damage, examine crop conditions and homogeneity, or spray herbicides with smart spraying technologies, agribusinesses may deploy fleets of drones to fly across fields.

15.3.2.4 Crop Yield Maps

Agribusinesses can evaluate vegetation and crop production using a combination of satellite photos and GIS data, which is depicted in Figure 15.6. To manage crops wisely throughout the seasons and pinpoint characteristics that limit output, they may differentiate between soils and dense canopies. These maps for precision agriculture provide a visual aid for identifying relationships between crop yields and markers of field condition.

15.3.2.5 Green Vegetation Indices (NDVI Maps)

The most precise and thorough measurements of the amount of vegetation in fields, the definition of moisture levels, and the classification of soil health are provided by

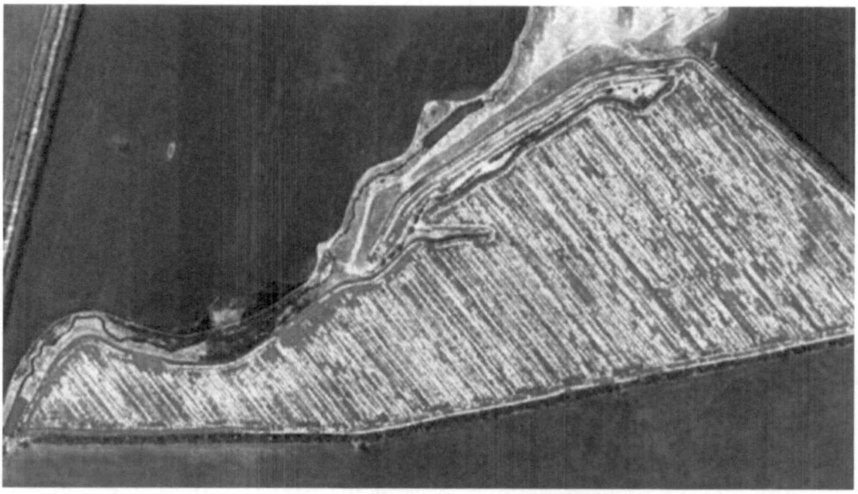

Figure 15.6 Crop yield maps.

NDVI (Normalized Difference Vegetation Index) maps, which are a powerful tool for agribusinesses, which is illustrated in Figure 15.7. The environmental, sociological, physical, and other elements that affect crop development can be displayed on these maps. For AgTech predictive software programs that choose where to take soil samples and what places to choose to expand fields, NDVI maps are essential. The most important aspect in enhancing agricultural earnings and preventing many possible issues with flooding and dry soils is these maps.

Figure 15.7 Green vegetation indices (NDVI maps).

15.4 Cartographic Visualization and Visualizations of Uncertainties of Agricultural Sensor-Based Information

A critical feature of modern precision agriculture is the cartographic presentation of agricultural sensor-based data. It entails the use of geographic information systems (GIS), remote sensing, and sensor data to generate maps and visualizations that assist farmers in making educated crop and land decisions. Here's a rundown of how this procedure works:

15.4.1 Data Collection:

The collection of data is the initial stage in cartographic representation of agricultural sensor-based information. Weather stations, soil moisture sensors, drones, and satellite imagery are just some of the possible sources of such data. These sensors monitor factors like temperature, humidity, soil quality, and plant vitality.

15.4.1.1 Data Integration:

The data integration may involve data cleansing and transformation to guarantee accuracy and consistency. All data should be georeferenced, which means it should be linked to particular geographic coordinates.

15.4.1.2 GIS Software:

GIS software is used to process and evaluate the data. You can use ArcGIS, QGIS, or any of the other well-known GIS programs. Maps, spatial analyses, and various data visualizations are all possible within these applications.

Because of its usefulness in managing, analyzing, and visualizing geographical data, GIS software has become indispensable in the agricultural sector, assisting farmers, agronomists, and agricultural researchers with matters such as crop cultivation, soil quality, weather conditions, insect management, and more. Here is a quick rundown of GIS software for farmers:

a. **ArcGIS by Esri:**

ArcGIS is among the most popular GIS programs. It has everything you need for geographical analysis, mapping, and keeping track of data.

Field mapping, crop monitoring, soil analysis, and yield prediction are just a few of the precision agriculture applications for ArcGIS. Data-driven decisions, irrigation schedules, and resource maximization are all aided by this tool.

b. **QGIS (Quantum GIS):**

QGIS is an open-source GIS program with sophisticated mapping and spatial analytic features. It is well-known for its easy-to-use interface and versatility.

QGIS is used in agriculture for activities such as land parcel mapping, soil categorization, and geospatial analysis of crop disease transmission. Because it is open-source, it is available to a wide spectrum of users.

c. **GRASS GIS:**

GRASS GIS is another open-source GIS program that is well-known for its strong geospatial analytic capabilities. It excels in raster analysis in particular.

Agricultural Applications: In agriculture, GRASS GIS is used for topography analysis, hydrological modeling, and land use planning. It can aid in determining the appropriateness of land for various crops and irrigation systems.

d. **ENVI (by Harris Geospatial):**

ENVI is image analysis software that is specialized in remote sensing. Its primary goal is to extract information from satellite and aerial pictures.

Agricultural Applications: ENVI is used in agriculture to analyze crop health, identify illness, and perform precision farming. It can monitor agricultural conditions by analyzing multispectral and hyperspectral data.

15.4.1.3 Google Earth Engine:

Google Earth Engine is a web service that provides users with a wealth of satellite imagery and geographical information. It's great for looking at the big picture.

Google Earth Engine is used for crop monitoring, land cover categorization, and change detection in agriculture. Because of its scalability, it is appropriate for regional and worldwide agricultural evaluations.

15.4.1.4 Agisoft Metashape (Formerly Photoscan):

Agisoft Meta shape is photogrammetry software that is used to create 3D models and Ortho mosaics from aerial photos.

It is used in precision agriculture to generate high-resolution maps of fields and to monitor crop health, drainage, and topography.

15.4.1.5 Farm Management Software with GIS Integration:

Farm management software solutions like Farm Logs and Granular provide GIS tools for farm planning, resource management, and production analysis.

Agricultural Applications: By adding spatial context to their data, these integrated systems assist farmers expedite their operations from planting through harvest.

GIS software in agriculture helps data-driven decision-making, improves crop management methods, optimizes resource consumption, and promotes sustainable farming. Farmers and agricultural professionals may use these technologies to increase crop yields, save money, and decrease environmental impact.

15.4.1.6 Spatial Analysis:

GIS software can perform spatial analysis on agricultural data to find patterns and trends. It can, for example, assist in identifying regions of a field that are under water stress or detect disease spread inside a crop.

15.4.1.7 Map Creation:

The process of developing and producing visual representations of geographic information, which can incorporate a variety of data kinds and topics, is known as map production. These maps are effective tools for communication, analysis, and decision-making in many domains, including geography, cartography, and urban planning. Here's a quick rundown of the essential phases in map creation:

- **Selection of Map Type:** The sort of map you wish to build is determined by your goals and the data you have. Thematic maps (e.g., population density, temperature), topographic maps (height and terrain characteristics), and choropleth maps (data shown by regions or areas) are examples of common map types.
- **Design and Symbolization:** Making judgments on how to graphically portray data is part of map design. Colors, symbols, line styles, and typefaces are all options. Map readers can simply grasp the information offered thanks to effective symbolization.
- **Scale and Projection:** To better suit the purpose and region of interest, maps are frequently constructed at specified scales (e.g., 1:10,000) and projections (e.g., Mercator, Robinson). These options have an impact on the map's accuracy and look.
- **Map Layout:** The placement of features such as the title, legend, scale bar, and north arrow is determined by the map's layout. A well-designed map improves reading and usage.
- **Cartographic Generalization:** Depending on the map's scale and purpose, generalization may be necessary. This process involves simplifying complex geographic features to make them legible at smaller scales. For example, roads might be generalized from detailed paths to simplified lines.
- **Map Production:** The map is created using cartography software or graphic design tools after the design and data are complete. These capabilities enable the exact positioning of data and design components on the map canvas.
- **Quality Control:** It is critical to evaluate the map for flaws, inaccuracies, and inconsistencies before finishing it. This stage ensures that the map appropriately shows the information intended.
- **Distribution and Publication:** Once completed and reviewed, the map can be distributed or published in a variety of ways, such as print, digital, or interactive internet maps. The appropriate distribution strategy is determined by the intended audience and purpose.

- **Update and Maintenance:** Maps should be updated on a regular basis to reflect changes in geographic data or to accommodate new information. This is particularly critical for dynamic and real-time mapping applications.
- **Thematic Maps:** These maps depict certain topics or factors, such as soil moisture levels, agricultural production forecasts, or insect infestations. Color coding and symbols are frequently employed to communicate information.
- **Heatmaps:** Heatmaps are used to depict the intensity or concentration of a certain component, such as temperature or insect numbers, over time and space. Color gradients are used to show different levels of intensity.
- **Contour Maps:** Contour maps are used to depict data that varies continuously, such as elevation or soil pH. Contour lines link locations of equivalent value, assisting farmers in understanding topography or soil properties.
- **Overlay Maps:** These maps use numerous layers of data to highlight correlations between various factors. Overlaying soil quality data with meteorological data, for example, can aid in determining appropriate planting timings.

  To summarize, map development is a multi-step process that comprises gathering, analysing, creating, and producing visual representations of geographic data. The purpose is to effectively portray spatial connections and data to enhance decision-making and communication in a variety of domains and applications.
- **Real-Time Monitoring:** Some agricultural systems include real-time sensor data monitoring and map updates. This enables farmers to make fast decisions based on changing field conditions.
- **Decision Support:** The fundamental purpose of cartographic visualization in agriculture is to assist farmers in making decisions. They may utilize these maps and visualizations to make educated decisions regarding crop irrigation, fertilization, pesticide application, and other crop-related issues.
- **Historical Analysis:** The historical data gathered and shown on maps can be utilized for long-term planning and trend analysis throughout time. It can assist farmers in making changes to their techniques based on prior results.

15.5 Conclusion

The introduction of big data analytics has had an impact on and is still having an impact on the major economic sectors. Big Data deployment in the agriculture sector is still in its early phases, but it has already proven useful in tackling the main issues. Farmers and stakeholders should collaborate to create and adopt new data-driven and data-enabled crop management approaches in order to fully benefit from big data.

Data-driven agriculture increases productivity and aids in the fight against issues related to agriculture, such as food demand and starvation. Companies supported by reputable technology providers have already proved the practical advantages of smart farming. Investment volume in this area only serves to reinforce big data's growing trend in the retail and data-driven agriculture development sectors.

The latest data on how producers in Nebraska feel about big data and precision farming issues can be found in a recent study. Preliminary data and analysis show that yield mapping and guidance control systems are just the tip of the iceberg for precision agriculture's broad implementation. Images and sensing systems for plant or field monitoring and diagnostics, variable rate application systems, and other technologies can be used to further investigate the correlation between the adoption of these technologies and other technologies.

The rise of precision mapping as a tool in the agricultural technology revolution. Increasing numbers of privately owned small farms and big agroholdings are using AgTech companies' GIS and GPS mapping technologies in an effort to increase crop productivity. The incorporation of precision technologies into agribusinesses' operations and products is seen as an easy way to increase their consumer base. Long-term financial success can be achieved with the help of precise geodata extracted from farm maps.

References

Astill, Jake, Rozita A. Dara, Evan D.G. Fraser, Bruce Roberts, and Shayan Sharif. "Smart poultry management: Smart sensors, big data, and the internet of things." Computers and Electronics in Agriculture 170, (2020): 105291.

Cockburn, Marianne. "Application and prospective discussion of machine learning for the management of dairy farms." Animals 10, no. 9 (2020): 1690.

Jerhamre, Elsa, Carl Johan Casten Carlberg, and Vera van Zoest. "Exploring the susceptibility of smart farming: Identified opportunities and challenges." Smart Agricultural Technology 2, (2022): 100026.

Kamble, Sachin S., Angappa Gunasekaran, and Shradha A. Gawankar. "Achieving sustainable performance in a data-driven agriculture supply chain: A review for research and applications." International Journal of Production Economics 219, (2020): 179–194.

Kamilaris, Andreas, Andreas Kartakoullis, and Francesc X. Prenafeta-Boldú. "A review on the practice of big data analysis in agriculture." Computers and Electronics in Agriculture 143, (2017): 23–37.

Lezoche, Mario, Jorge E. Hernandez, Maria del Mar Eva Alemany Díaz, Hervé Panetto, and Janusz Kacprzyk. "Agri-food 4.0: A survey of the supply chains and technologies for the future agriculture." Computers in Industry 117, (2020): 103187.

Newton, Joanna E., Ruth Nettle, and Jennie E. Pryce. "Farming smarter with big data: Insights from the case of Australia's national dairy herd milk recording scheme." Agricultural Systems 181, (2020): 102811.

Oussous, Ahmed, Fatima-Zahra Benjelloun, Ayoub Ait Lahcen, and Samir Belfkih. "Big data technologies: A survey." Journal of King Saud University-Computer and Information Sciences 30, no. 4 (2018): 431–448.

Pylianidis, Christos, Sjoukje Osinga, and Ioannis N. Athanasiadis. "Introducing digital twins to agriculture." Computers and Electronics in Agriculture 184, (2021): 105942.

Ravikumar, S., K. Vijay, B. Bhuvaneswaran, Babu Rajendiran, and S. Vidhya. "Water Quality Monitoring and Notification System Based on Arduino, MATLAB and Visualization Techniques." In Recent Trends in Computational Intelligence and Its Application: Proceedings of the 1st International Conference on Recent Trends in Information Technology and its Application (ICRTITA, 22), p. 431. CRC Press, 2023.

Rodríguez-Mazahua, Lisbeth, Cristian-Aarón Rodríguez-Enríquez, José Luis Sánchez-Cervantes, Jair Cervantes, Jorge Luis García-Alcaraz, and Giner Alor-Hernández. "A general perspective of big data: Applications, tools, challenges and trends." The Journal of Supercomputing 72, (2016): 3073–3113.

Saiz-Rubio, Verónica, and Francisco Rovira-Más. "From smart farming towards agriculture 5.0: A review on crop data management." Agronomy 10, no. 2 (2020): 207.

Sowmia, K. R., S. Prithi, K. Vijay, I. Eugene Berna, and B. Bhuvaneswaran, "Crop Monitoring System with Water Moisture Level using Arduino," 2023 5th International Conference on Smart Systems and Inventive Technology (ICSSIT), Tirunelveli, India, 2023, pp. 495–498, doi: 10.1109/ICSSIT55814.2023.10060913.